Springer Mineralogy

More information about this series at http://www.springer.com/series/13488

Yuriy Litvin · Oleg Safonov
Editors

Advances in Experimental and Genetic Mineralogy

Special Publication to 50th Anniversary of DS
Korzhinskii Institute of Experimental
Mineralogy of the Russian Academy
of Sciences

Springer

Editors
Yuriy Litvin
Mantle Laboratory
DS Korzhinskii Institute
of Experimental Mineralogy
Russian Academy of Sciences
Chernogolovka, Moscow Oblast, Russia

Oleg Safonov
Metamorphism Laboratory
DS Korzhinskii Institute
of Experimental Mineralogy
Russian Academy of Sciences
Chernogolovka, Moscow Oblast, Russia

ISSN 2366-1585   ISSN 2366-1593   (electronic)
Springer Mineralogy
ISBN 978-3-030-42861-7   ISBN 978-3-030-42859-4   (eBook)
https://doi.org/10.1007/978-3-030-42859-4

This Springer imprint is published by the registered company Springer Nature Switzerland AG
The registered company address is: Gewerbestrasse 11, 6330 Cham, Switzerland

Dmitriy S. Korzhinskii. First Director-Organizer of the Institute of Experimental Mineralogy

Preface

Researchers of the DS Korzhinskii Institute of Experimental Mineralogy (June 2019). *First row, from left* Vitaliy Chevychelov, Ekaterina Brichkina, Mariya Golunova, Alexey Kotelnikov, Yuriy Shapovalov, Oleg Safonov, Mikhail Zelenskiy, Anastasiya Kuzuyra, Vera Ermolaeva, Tatiana Setkova, Vladimir Balitskiy, Sergey Pivovarov. *Second row, from left* Aleksander Redkin, Yuriy Litvin, Nikolay Bezmen, Olga Shaposhnikova, Valentina Butvina, Tatiana Kovalskaya, Vera Devyatova, Anastasiya Kostyuk, Anna Spivak. *Third row, from left* Dmitriy Khanin, Konstantin Van, Valentina Korzhinskaya, Natalia Kotova, Paver Bukhtiyarov, Eduard Persikov, Kiril Shmulovich, Aleksander Simakin, Olga Karaseva. Galina Akhmedzhanova, Dmitriy Varlamov. *Back rows, from left* Galina Kashirtseva, Evgeniy Limanov, Liudmila Sipavina, Andrey Plyasunov, Natalia Suk, Nikolay Gorbachev, Evgeniy Osadchiy, Veniamin Polyakov, Pavel Gorbachev, Galina Bondarenko, Leonid Lakshtanov, Dilshad Sultanov, Tatiana Bublikova, Mikhail Voronin

This book *Advances in Experimental and Genetic Mineralogy* is devoted to 50th anniversary of the DS Korzhinskii Institute of Experimental Mineralogy of the Russian Academy of Sciences. This book has been prepared by the researchers of the institute and gives an indication of a variety of its present-day experimental and theoretical inquiries.

The Institute of Experimental Mineralogy has been organized at a period of August 1–October 1, 1969, in Chernogolovka town as a part of the Near-Moscow Scientific center of the USSR Academy of Sciences accordingly to the suggestion of academician Dmitry Sergeevich Korzhinskii who becomes the first director of the institute. D. S. Korzhinskii has made a basic contribution to working out the methods of physicochemical analyses for the metasomatic, metamorphic and magmatic processes under conditions of the Earth's crust and mantle. His pioneering investigations being directed onto the development of physicochemical methodology in studying and cognition of the basic regularities of the geologic processes date back to a beginning of the 1930s. In succeeding years, D. S. Korzhinskii and his disciples have actively developed this line of inquiry, and the results of their studies gained general recognition.

By the 1950–1960s, it dated a creation of the specialized laboratories for experimental studies of the deep-seated materials and processes under high pressure and high temperature in Moscow in the Institute of the Earth Physics (Prof. M. Volarovich), Institute of Geochemistry and Analytical Chemistry (Prof. N. Khitarov) and Institute of Geology of Ore Deposits, Petrography, Mineralogy and Geochemistry (Prof. I. Ostrovskiy) and in Novosibirsk in the Institute of Geology and Geophysics (Dr. I. Malinovskiy). At the end of 1960s, it has evolved the Laboratory of Experimental Mineralogy in the Institute of Solid State Physics (Prof. I. Ivanov) in the scientific center in Chernogolovka. Then in 1969, by the presentation of academician D. S. Korzhinskii, it has adopted a decision of reorganization of the laboratory into the Institute of Experimental Mineralogy for the fundamental experimental and theoretical studies in the fields of genetic mineralogy, physicochemical petrology and physical geochemistry. At the advent of the institute, it were taken into account the diversity and range of the magmatic, metamorphic, hydrothermal and ore-forming processes in the substance of the Earth's crust and mantle which have been going on at natural conditions over a wide range of temperature, pressure and reduction–oxidation potentials.

The disciples and colleagues of D. S. Korzhinskii become the first research workers and leaders of the scientific laboratories of the new institute. There are the well-known scientists Vilen Zharikov, Aleksey Marakushev, Leonid Perchuk, Ivan Ivanov, Georgiy Zaraiskiy, Ivan Nekrasov. Their theoretical and experimental searches were of considerable importance in the making the prospective themes of scientific researches by the institute collective. In the institute, it is worked out and brought into being the unique complex of hydrothermal, gas and solid-phase apparatus of high pressure and temperature. The research activity of the institute has focused onto the experimental studies of physicochemical properties of minerals and simplified and multicomponent mineral systems for the purpose of application of their results to the elaboration of physicochemical genetic models for the minerals, mineral assemblies and rocks, including the ore-bearing ones, at the deep-seated horizons of the Earth's crust and mantle.

The book *Advances in Experimental and Genetic Mineralogy* is originated by the scientists of the DS Korzhinskii Institute of Experimental Mineralogy (Russian Academy of Sciences) and reflects the present-day lines of experimental and theoretical studies. The book presents the fundamental experimental data and development of the experiment-based theoretical conclusions and physicochemical models for the natural hydrothermal, metasomatic, metamorphic, magmatic and ore-producing processes of the Earth's crust, upper mantle, transition zone and lower mantle. The topics discussed in the book concern the interaction of oil and aqueous fluids that is revealed by the methodology of aqueous-hydrocarbonic inclusions in synthetic quartz and applied to the natural evolution of the oil; the determination of solubility and interphase partitioning of trace and strategic elements and their components; and the experimental substantiation of physicochemical mechanisms for ultrabasic–basic evolution of the deep-mantle magmatic and diamond-forming systems. Experimental studies of physicochemical properties of supercritical water and hydrothermal fluids, viscosity of acidic ultramafic magmatic melts, peculiarities of metamorphism of basic rocks, kinetics of mineral nucleation in silicate melts and hydrothermal solutions, influence of the complex $H_2O–CO_2–HCl$ fluids onto melting relations of the mantle–crust rocks have been marked with novel results and conclusions. The book is of interest for the Earth scientists, lecturers and students specialized in experimental and genetic mineralogy, petrology and geochemistry.

Chernogolovka, Russia   Oleg Safonov
                                                                           Yuriy Litvin

Contents

Abbreviations

Ab	Albite $NaAlSi_3O_8$
Alm	Almandine $Fe_3Al_2Si_3O_{12}$
An	Anorthite $CaAl_2Si_2O_8$
Ancl	Anorthoclase $(Na, K)AlSi_3O_8$
And	Andalusite Al_2SiO_5
Brd	Bridgmanite $MgSiO_3$
CaPrv	Ca-perovskite $CaSiO_3$
Carb, Carb*	Assembly of several carbonates
Chr	Chromite $FeCr_2O_4$
Coe	Coesite SiO_2
Cpx	Clinopyroxene $(Ca, Na)(Mg, Fe)(Si, Al)_2O_6$; $[(Di \cdot Hd \cdot Jd)_{ss}]$
Crn	Corundum Al_2O_3
D	Diamond C
Di	Diopside $CaMgSi_2O_6$
Dol	Dolomite $CaMg(CO_3)_2$
En	Enstatite $MgSiO_3$
Fa	Fayalite Fe_2SiO_4
FBrd	Ferrobridgmanite $(Mg, Fe)SiO_3$
Fo	Forsterite Mg_2SiO_4
FPer	Ferropericlase $(Mg, Fe)O$
FRwd	Ferroringwoodite $(Mg, Fe)_2SiO_4$
Fs	Ferrosilite $FeSiO_3$
Gros	Grossularite $Ca_3Al_2Si_3O_{12}$
Grt	Garnet $(Mg, Fe, Ca)_3(Al,Cr)_2Si_3O_{12}$; $[(Prp \cdot Alm \cdot Gros)_{ss}]$
Hd	Hedenbergite $CaFeSi_2O_6$
Jd	Jadeite $NaAlSi_2O_6$
Kfs	K-feldspar $KAlSi_3O_8$
Ky	Kyanite Al_2SiO_5
L	Liquid, melt
Ma	Mathiasite $(K, Ba, Sr)(Zr, Fe)(Mg, Fe)_2(Ti,Cr,Fe)_{18}O_{38}$

Mcl	Microlite $NaCaTa_2O_6F$
Mgs	Magnesite $MgCO_3$
Mic	Microcline $KAlSi_3O_8$
MWus	Magnesiowustite $(Fe, Mg)O$
Ol	Olivine $(Mg, Fe)_2SiO_4$; $[(Fo\cdot Fa)_{ss}]$
Olg	Oligoclase $(Na, Ca)(Si, Al)_4O_8$
Omph	Omphacite, Jd-rich clinopyroxene
Opx	Orthopyroxene $(Mg, Fe)SiO_3$; $[(En\cdot Fs)_{ss}]$
Par	Paragonite $NaAl_3Si_3O_{10}(OH)_2$
Pchl	Pyrochlore $(Ca, Na)_2(Nb, Ta)_2O_6(O, OH, F)$
Per	Periclase MgO
Pf	Pyrophyllite $Al_2Si_4O_{10}(OH)_2$
Phl	Phlogopite $KMg_3AlSi_3O_{10}(F,OH)_2$
Pl	Plagioclase $(NaAlSi_3O_8\cdot CaAl_2Si_2O_8)_{ss}$; $(Ab\cdot An)_{ss}$
Pri	Priderite $(K, Ba)(Ti, Fe, Mn)_8O_{16}$
Prp	Pyrope $Mg_3Al_2Si_3O_{12}$
Qz	Quartz SiO_2
Rwd	Ringwoodite Mg_2SiO_4
Srp	Serpentine $(Mg, Fe, Ni, Al, Zn, Mn)_{2-3}Si_2O_5(OH)_4$
Sti	Stishovite SiO_2
Tnt	Tantalite $(Mn, Fe)(Nb, Ta)_2O_6$
Wds	Wadsleyite Mg_2SiO_4
Wol	Wollastonite $CaSiO_3$
Wus	Wustite FeO
Yim	Yimendite $K(Cr, Ti, Mg, Fe, Al)_{12}O_{19}$

Physical Symbols

A_{Kr}	The Krichevskii parameter
f_1^*	The pure water fugacity
f_2	The fugacity of a solute
$G^o(g)$	The Gibbs energy of a compound in the ideal gas state
G_2^∞	The Gibbs energy of a compound in the state of the standard aqueous solution
K_D	The vapor–liquid distribution constant
k_H	Henry's constant
m	Molality (a number of moles of a substance in 1000 g of water)
$N_w \approx 55.508$	The number of moles of H_2O in 1 kg of water
P_1^*	Pressure of saturated water vapor
T_c	The critical temperature of pure water

Greek Symbols

ρ_c The critical density of pure water

$\rho_1^*(L)$ The pure water density along the liquid side of the saturation vapor–liquid curve

ϕ_1^* The fugacity coefficient of pure water

ϕ_2^∞ The fugacity coefficient of a solute at infinite dilution in water

Part I
Crystallization and Properties of Minerals and Mineral-Forming Solutions

Chapter 1
Phase Composition and States of Water-Hydrocarbon Fluids at Elevated and High Temperatures and Pressures (Experiment with the Use of Synthetic Fluid Inclusions)

V. S. Balitsky, T. V. Setkova, L. V. Balitskaya, T. M. Bublikova, and M. A. Golunova

Abstract The article considers new approaches and methods for studying the phase composition and states of aqueous-hydrocarbon fluids in the temperature range of 240–700 °C and pressures of 5–150 MPa respectively. The essence of the approach is to conduct experiments in autoclaves on the formation of aqueous-hydrocarbon fluids through the interaction of hydrothermal solutions with bituminous and high-carbon rocks, as well as with crude oil. Simultaneous, in the same autoclaves, the quartz crystals and less other minerals with trapped aqueous-hydrocarbon inclusions are growing. These inclusions are the main objects of research that are carried out using modern micro thermometry methods, especially in combination with conventional and high-temperature FT-IR spectroscopy and microscopy with UV and natural light. This allows in situ monitoring of changes in the phase composition and states of aqueous-hydrocarbon fluids in a wide range of PT-parameters, to determine the effect of temperature and volume ratios of water and hydrocarbon (liquid and gas) phases on the occurrence of multiphase, three-phase and biphasic heterogeneous liquid and gas-liquid phases and homogeneous fluids. Studies have shown that with certain compositions of synthetic aqueous-hydrocarbon inclusions at heated and cooled is often appeared so-called imaginary homogenization in inclusions. This event associated with the periodic alignment of the refractive indices and densities of the aqueous and hydrocarbon liquid phases. The influence of the volume ratios of the aqueous and hydrocarbon phases on the cessation of cracking and metamorphic transformations of oil when the fluid goes into a homogeneous supercritical state, and the resumption of these processes after fluid heterogenization as a result of the temperature drop were established. In addition, a comparison of the effect of the PT-condition on the phase composition and the state of the hydrocarbon made it possible to estimate the maximum depths of the oil in the Earth's interior. Depending on the volume ratios of the water and oil phases this depth turned out to be equal from 14 to 22 km.

V. S. Balitsky (✉) · T. V. Setkova · L. V. Balitskaya · T. M. Bublikova · M. A. Golunova
D.S. Korzhinskii Institute of Experimental Mineralogy, Russian Academy of Sciences,
Academician Osipyan street 4, Chernogolovka, Moscow region 142432, Russia
e-mail: balvlad@iem.ac.ru

Y. Litvin and O. Safonov (eds.), *Advances in Experimental and Genetic Mineralogy*, Springer Mineralogy, https://doi.org/10.1007/978-3-030-42859-4_1

Keywords Water-hydrocarbon fluids · Oil stability · Depths of oil formation ·
Hydrothermal experiment

1.1 Introduction

It is known that oil in natural conditions is constantly associated with water. And this
is not surprising, because these compounds, although sharply different in composition, but both are liquids that are stable in a wide range of temperatures and pressures. As liquids, they are subject to the same hydrodynamic laws and this largely
predetermines their joint migration and concentration in certain areas of the Earth's
crust.

Direct evidence of the joint migration of hydrothermal solutions, oil and hydrocarbon gases is found in many areas of modern volcanic and thermal activity, ocean
floor spreading zones and other tectonically active parts of the Earth (Pikovskii et al.
1987; Simoneit 1995; Bazhenova and Lein 2002; Rokosova et al. 2001). The temperature of such fluids often reaches 400 °C and higher. At the same time, oil and gas
provinces often show signs of hydrothermal activity. They are expressed in carburiz-
ing, sulphidization, quartzing and argyllization of rocks containing oil and gas fields.
Sometimes these changes are accompanied by accumulations of uranium, mercury,
antimony, gold and other minerals (Ivankin and Nazarova 2001). At the same time,
gas, liquid and solid hydrocarbons (HC) are quite often found in magmatic and metamorphic rocks, contact-metasomatic formations, pegmatites and hydrothermal veins
(Balitsky 1965; Beskrovnyi 1967; Ozerova 1986; Florovskaya et al. 1964; Chukanov
et al. 2006). In addition to independent emissions, they are present in the fluid inclusions of various vein and ore minerals. All this indicates that hydrothermal solutions
in the Earth's interior often interact with bituminous rocks and crude oil. The nature
of such interactions at elevated and high temperatures and pressures is still not sufficiently studied. This is especially true for the phase composition and states of natural
aqueous-hydrocarbon fluids, which are practically inaccessible for direct observation
because of the great depths and associated high TR-parameters. However, many of
their characteristics can be recreated with varying degrees of confidence in experimental studies using synthetic fluid inclusions in quartz and other minerals grown
simultaneously with the interaction of hydrothermal solutions with bituminous rocks
and crude oil (Balitsky et al. 2005, 2007, 2011, 2014, 2016; Teinturier et al. 2003).

Similar researches were carried out in the last 20 years in the Minerals Synthesis and Modification Laboratory of the D.S. Korzhinskii Institute of Experimental Mineralogy and led to the creation in the laboratory one of the new scientific
directions connected with the solve of various problems of HC geochemistry. This
article was also implemented within the framework of the abovementioned scientific direction. Its purpose is to demonstrate the possibility of using synthetic fluid
inclusions in quartz to study and predict the phase composition and states of deep
water-hydrocarbon fluids under various *PT*-conditions and to estimate on this basis

the oil and gas bearing capacity of bituminous rocks and the maximum depths of oil in the Earth's interior.

1.2 Approach to the Study of the Phase Composition and States of Water-Hydrocarbon Inclusions in Quartz and Methods of Their Research

1.2.1 Basic Approach

The problems set in the article were solved on the basis of the approach developed by the authors of the article specifically to study the interaction of various bituminous rocks and crude oil with hydrothermal solutions (Balitsky et al. 2005, 2007). The essence of the approach is to conduct experiments on the formation of artificial aqueous-hydrocarbon fluids by means of interaction of hydrothermal solutions with bituminous rocks and crude oil at the same time growing quartz crystals with water-hydrocarbon inclusions. Such inclusions are essentially microsamples of mother solutions selected and hermetically preserved under *PT* conditions of experiments without disturbances in the system of dynamic equilibrium. Further inclusions are used to study in situ behavior, phase composition and states of their fluids by thermobarogeochemical methods (Ermakov 1972; Roedder 1984; Mel'nikov et al. 2008), especially microthermometry in combination with conventional and high-temperature local FT-IR spectroscopy and microscopy using UV lighting (Balitsky et al. 2016). It was assumed that if all syngenetic inclusions contained the same phases with the same volumetric ratios, they were captured from homogeneous fluids. When different phases or identical phases were observed in the inclusions, but with different volumetric ratios, this indicated that the mother fluids were in a heterogeneous state. Quartz was chosen to capture inclusions because it has high mechanical strength and chemical resistance - properties that are necessary both in the formation of fluid inclusions and subsequent thermometric studies. An important factor for the choice of quartz as a carrier of fluid inclusions is the relative ease of its growth on seed in aqueous solutions of different compositions in a wide range of temperatures (240–900 °C) and pressures (from the pressure of saturated steam to 150–200 MPa) (Balitsky et al. 2005).

1.2.2 Methods of Growing Quartz Crystals with Synthetic Water-Hydrocarbon Inclusions

All experiments on growing quartz with water-hydrocarbon inclusions were carried out by the hydrothermal method of temperature difference. Heat-resistant autoclaves of 30, 50, 100 and 280 ml volume made of stainless steel and Cr–Ni alloy were used in experiments. Autoclaves were heated in electric shaft furnaces with two heaters. Depending on the design and size of the furnaces, they were equipped with 3 to 10 autoclaves at the same time. Temperature in furnaces and autoclaves was maintained and controlled by a set of standard devices (TYP 01 T4, TYP R3 and Thermodat-25M1). Temperature accuracy was ± 2 °C. The necessary pressure was set by pouring the solution with filling coefficients, which were determined by P-V-T diagrams for solutions of the corresponding composition (Samoilovich 1969), and in the absence of such—by tabular data P-V-T—dependencies for pure water (Naumov et al. 1971).

Crystals were grown on seed bars, mainly ZY- and ZX-orientations with the length of 140–210 mm, width of 5–8 mm and thickness of 2–4 mm. The absence of an autoclave diaphragm, which usually separates the dissolution and growth zones of quartz, led to a continuous convective mixing of the solution during the entire experiment and the formation of crystals of an unusual wedge-liked shape, due to a gradual increase in the bottom-up growth rate as saturation increases (Fig. 1.1). Entering of autoclaves into the working mode was carried out with the speed of 50–70 °C/h. After reaching the set temperature the autoclave was maintained for 12–24 h in isothermal conditions or with a small (10–15 °C) inverse temperature difference. This made it possible to prepare pickling channels in the seed bars, which were infected with fluid inclusions when the direct temperature difference was restored (Balitskaya and Balitsky 2010).

An important methodological task of the conducted researches was to find the conditions under which the formation of water-hydrocarbon inclusions would occur in quartz. The inevitability of the formation of such inclusions in quartz predetermined the choice of crystallographic orientations of seed cuts, stimulating the appearance of coarse regenerative relief on the growing surfaces (Balitsky et al. 2005). Anisotropy

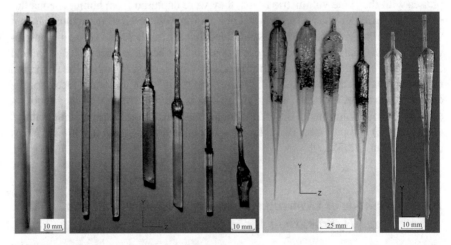

Fig. 1.1 Examples of quartz crystals with water-hydrocarbon inclusions, grown in the temperature range of 240–670 °C and pressures of 5–150 MPa in alkaline and weakly alkaline solutions in the presence of bituminous rocks and crude oil

of the growth rates of the faces composing the pyramids of growth of such relief led at first to the appearance between them of micro-cavities, which at the subsequent overgrowth turned into fluid inclusions. The most actively similar inclusions are formed at growth of quartz on seed bars of *ZY*- and *ZX*-orientations. Regeneration pyramids of quartz relief growth are composed, as a rule, of faces of positive {r} and negative {z} rhombohedrons, hexagonal prism {m}, trigonal positive {+a} and trigonal negative {−a} prisms and trigonal positive {+s} dipyramids (Fig. 1.2a, b). In addition, fluid inclusions originate from particles of altered rocks, solid bitumen and oil droplets, which are usually deposited at the boundary between seed bars and newly formed quartz (Fig. 1.3). Fluid inclusions formed in the etching channels of dislocations in seed quartz bars are also no less informative for studies.

To obtain water-hydrocarbon inclusions captured in quartz crystals, two series of experiments were conducted. In the experiments of the first series, liquid and gas hydrocarbons were supplied to the hydrothermal solution from bituminous rocks. The oil shale from the Kashpirskoye (near the town of Syzran) and Leningradskoye (near the town of Slantsy) deposits and high-bituminous rocks of the Bazhenov Formation (Western Siberia) and Domanikovyh sediments (Volga-Ural Province) were

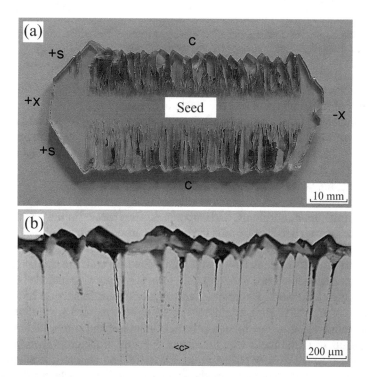

Fig. 1.2 Regeneration pyramids of quartz relief growth (**a**) and regeneration relief of rough faces of the basal pinacoid (**b**). Rough regeneration relief of rough faces, especially the basal pinacoid, favors to the generation of inclusions

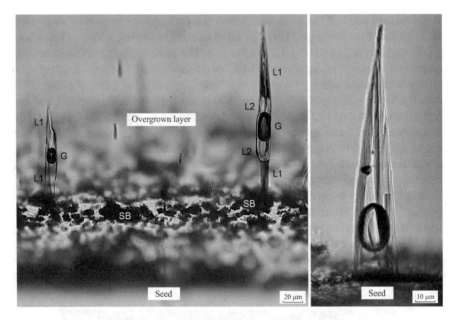

Fig. 1.3 The origin and formation of fluid water-hydrocarbon inclusions on particles of solid bitumen, oil droplets, altered bituminous rocks, micas and other minerals

used. Besides, hydrothermal treatment experiments were carried out with bituminous claystones of the Central Oil and Gas Basin of France.

The elementary composition of oil shale from the Kashpirskoye and Leningrad-skoye deposits includes, respectively (wt%): carbon—75.6 and 59.7–70; hydrogen—9.4 and 6.0–7.8; sulfur—1.4 and 6.0–14.2; nitrogen—0.5 and 1.7–2.5 and oxygen—13.1 and 22.3 (Lapidus and Strizhakova 2004). Organic matter, ash and total sulfur, respectively, are (wt%): 35.0, 46.0, 1.6 and 32.5, 57.5 and 5.8.

The Bazhenov Formation includes siliceous clay-carbonate rocks with hydro-mica, kaolinite, montmorillonite and carbonates. The share of initial organic carbon in them varies from 8 to 12 wt%, and the residual carbon decreases by 1–2 wt%. The content of chloroform bitumen rarely exceeds 1.0–1.5 wt%, although in general the soluble part of rocks reaches 10–14 wt%. Household deposits are similar in composition to the Bazhenov Formation in many respects. They also include bituminous siliceous clay-carbonate rocks enriched with organic matter. Its share is 2.5–10.0 wt% (Zaidelson et al. 1990). High bituminous rocks of both regions are attributed to oil and gas producing strata due to high content of microoil and combustible gas.

Fragments of initial rocks with the size of 5–8 mm in the cross section with the total mass of 10–12 g were placed in the lower (hotter) zone of the autoclave. In the same place on the perimeter of the inner wall of the autoclave charge quartz bars of 80–100 mm long, 4–8 mm wide and 2–4 mm thick were installed, which are necessary for feeding the solution with silica. Another ZY- and/or ZX-orientation quartz bar with a length of up to 208 mm was hung from a metal frame along the vertical axis of the autoclave. This bar was a seed for the build-up of newly formed quartz with captured water-hydrocarbon inclusions.

Pure water, aqueous solutions of sodium chloride (10 and 20 wt% NaCl), sodium bicarbonate (5 and 7 wt% $NaHCO_3$) and sodium carbonate (3 and 5 wt% Na_2CO_3) were used as initial solutions in experiments of this series. Solutions interaction with rocks was carried out at temperatures of 240/280, 300/320, 330/350, 360/380 and 370/400 and 420/450 °C at a pressure of 7, 20, 75, 90, 100 and 120 MPa, respectively. Here and below, the temperatures of the upper and lower end of the autoclave are indicated via a slant line, respectively.

In the second series of experiments, water-hydrocarbon inclusions in quartz were obtained by interaction of hydrothermal solutions directly with crude oil. Oil from the Bavlinskoye (Tatarstan) field was selected for testing. It included methane (67%), naphthenic (21%) and aromatic (12%) HC. Asphaltene content (wt%) did not exceed 7.5, sulfur 3.4 and resins 12.8. Specific weight is 0.9 g/cm^3. The share of oil in the initial mixtures changed from 0.01 to 70 vol%.

The scheme of autoclave loading was slightly different from the experiments with oil shale and bituminous rocks. At first, the charge quartz and seed bars were placed in the autoclave. Then it was filled with clean water or aqueous solution of a given composition and then—with crude oil in given proportions in relation to the flooded water phase. The ratio of aqueous solutions to oil in an autoclave was controlled after the completion of experiments on the boundaries of their separation. These boundaries were clearly fixed on the grown crystals as their growth occurred only on that part of seed bars which were in an aqueous solution). Quartz crystals did not grow on the areas of the bars that were in the oil. Moreover, the seed bars under such conditions were partially dissolved by the penetration of water vapour into the oil. In addition, the areas with an overgrown quartz layer colored yellow with different intensities allowed to determine the levels of water and oil phases arising directly during the experiments and after the autoclave cooling.

Experiments of this series were carried out at temperatures from 280/300 to 650/670 °C and filling of autoclaves from 83 to 50%. This created pressure in autoclaves from 14 to 150 MPa, respectively.

Special experiments on autoclave heat treatment of primary water-hydrocarbon inclusions with sharply different volumetric ratios of water (L1) and oil (L2) phases were also carried out, schematically shown as follows: L1 ≥ G > L2 and L2 > L1 ≥ G. Primary quartz crystals with such inclusions were grown at relatively low temperatures (240/260 and 280/300 °C) and saturated steam pressures (5 and 7 MPa respectively). Further, after detailed study of primary water-hydrocarbon inclusions, they were subjected to repeated autoclave thermobaric treatment in pure water at temperatures of 300, 320, 350 and 380 °C and pressure of 100 MPa for 12–15 days. Comparison of the results of experiments allowed to find out the influence of *PT*-parameters and volumetric ratios of water and oil phases on the processes of cracking and metamorphic transformations of oil, which were not manifested during short-term (60–90 min) heating and cooling of inclusions in a microthermometric chamber.

1.2.3 Methods of Studying Products of Experiments, Phase Composition and States of Synthetic Water-Hydrocarbon Inclusions

After completion of the experiments, the products of interaction of oil shale and other bituminous rocks and crude oil with hydrothermal solutions were studied under binocular (MBS-9) and polarization (Amplival po-d) microscopes. Diagnostics of solid phases was carried out with wide use of radiographs and microprobe analysis. Initial oil and oil-like hydrocarbons from the experiments were characterized by IR spectra recorded on the Avatar 320 FT-IR spectrometer by Nicolet Company and by chromatographic analyses using Perkin Elmer Clarus 5000 chromatograph with Sol-gel 60 cm capillary column and helium carrier gas. From grown quartz was prepared polished plates with thickness of 0.5–2.0 mm to study fluid and solid inclusions. Behavior, phase composition and states of inclusions were studied in situ at their heating and cooling (Mel'nikov et al. 2008; Prokof'ev et al. 2005) in the measuring microthermometric complex on the basis of microthermal camera THMSG-600 by Linkam (England) and microscope Amplival (Germany), equipped with an additional source of UV light, a set of long-focus lenses, video camera and control computer. The complex allows to observe in real time the behavior and phase states of fluids inclusions in the temperature range from -196 to $+600$ °C, record video films with continuous automatic recording of temperature and rate of temperature rise and fall. However, actually thermometric measurements were stopped at temperatures of 405–410 °C, because at higher temperatures fluid inclusions lost their tightness due to their microscopic cracks, up to complete destruction.

Liquid and gaseous HC in fluid inclusions were identified by fundamental bands of IR spectra in the range of 6000–2600 cm^{-1}, recorded with IR microscope Continuum and single-beam FT-IR spectrometer Nicolet, Nexus with minimum aperture size of 5 μm (resolution 4 cm^{-1}). Distribution of hydrocarbons in inclusions was controlled with the help of microspectrophotometer QDI 302 by CRAIC company on the basis of microscope LEICA DM 2500 P, as well as microscope ZEISS AXIO Imager (Germany), equipped with an additional source of UV light.

1.3 Results and Discussion

1.3.1 Phase Composition and States of Synthetic Aqueous-Hydrocarbon Inclusions in Quartz Grown at Interaction of Hydrothermal Solutions with Bituminous Rocks

The interaction of hydrothermal solutions with oil shale and other bituminous rocks was studied in 20 experiments lasting from 14 to 60 days. Fragments of initial rocks

Fig. 1.4 Quartz crystals of spontaneous nucleation covered with numerous spherical secretions of solid bitumen (asphalt)

after the experiments were usually completely or partially destroyed and turned into clay-like material. It contained hard and less often viscous bitumen in the form of shiny black spherical secretions (often hollow) and irregularly shaped clots (Fig. 1.4). Dimensions of solid bitumen emissions ranged from a hundredth of a millimetre to a few millimetres across. They were completely dissolved in chloroform and, according to diffractograms, were amorphous, which allowed them to be attributed to asphalt. In autoclaves after experiments at temperatures above 320 °C the residual gas pressure was fixed up to 0.3–0.5 MPa. A small clap, sometimes with partial escape of mother solution, accompanied autoclave opening. In the chromatograms of the samples of the sampled gas, methane and sharply subordinated amounts of propane and ethane were determined. On the surface of the residual solution, an oily film of light yellow to yellow-orange and almost black color was observed. The same film was observed on the surface of grown quartz crystals (see Fig. 1.1). Chromatograms and FT-IR film spectra practically do not differ from those of conventional crude oil.

The grown quartz crystals, as noted above, have an unusual wedge-liked shape (see Fig. 1.1), are very defective and contain numerous water-hydrocarbon inclusions. There are two main types of inclusions. The inclusions of the first type are confined to the pickling cavities of the growth dislocations (Fig. 1.5a, b). The inclusions have a drop-shaped, needle-shaped, spindle-shaped and tubular shape and are stretched in the direction of the optical axis of quartz crystals, often with a deviation from it by 5–15°. The diameter of inclusions varies widely—from the first to 20 μm, and the length—from 30 to 1500 μm. Often such inclusions completely cross the seed bar. The composition and phase volumetric ratios of such inclusions are characterized by a wide variety of inequalities, which can be represented by: L1 > G, L1 $\geq$ G > L2, L1 $\geq$ G > L2 $\geq$ L2 $\geq$ SB, L1 $\geq$ G, L2 $\geq$ L1 > G $\geq$ SB and L2 > L1 $\geq$ G $\geq$ SB > L3, where L1 is an aqueous solution, G is a gas (mainly methane, water vapour and liquid hydrocarbons), L2 and L3—liquid hydrocarbons (synthetic oil), SB—solid bitumen (Fig. 1.6). Such diversity of phases and their volumetric ratios is

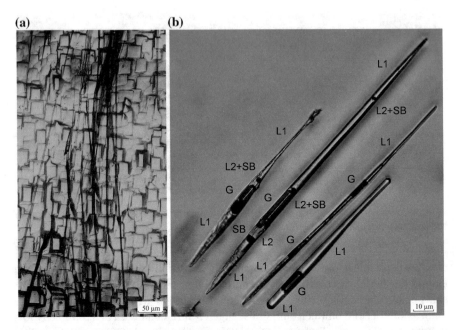

Fig. 1.5 The etching cavities of growth dislocations formed in the process of entering the autoclave into the operating mode (**a**) and water-hydrocarbon inclusions of the first type formed in the cavities (**b**). Inclusions contain aqueous solution, liquid and gas hydrocarbons and solid bitumens in various volume ratios, sometimes with complete absence of some phases

probably connected with different duration of formation of etching cavity of growth dislocations and time of their refilling.

The phases in the inclusions have distinct partition boundaries and are determined on the basis of local FT-IR spectra. Liquid hydrocarbons are also clearly identified by bright blue fluorescence in UV light. It is obvious that the formation of inclusions with such a variety of phases and their volumetric ratios could occur only in heterogeneous fluids.

The formation of the second type of inclusions occurred in the overgrown layer of newly formed quartz, as a rule, near its boundary with the seed bar. They were less frequently recorded in the later growth zones of the basal pinacoid {c} and positive trigonal prism {+a}, and very rarely in the growth sectors of the positive dipyramid {+s}, main rhombohedrons {r} and {z} and hexagonal prism {m}. The inclusions are mainly of elongated conical and tubular shape (Fig. 1.7). However, in later growth zones, irregular and complicated form inclusions begin to predominate. Their length varies from 15–20 to 100 μm, rarely reaching 1000–1500 μm in diameter from 5 to 30 μm

The composition and volume ratios of phases in inclusions differ significantly from those of the first type. As a rule, these are four-phase inclusions with distinctly distinguishable phases of aqueous solution (L1), gas (G), mainly methane, liquid hydrocarbons (L2) and solid bitumen (SB) with volumetric phase ratios $L1 \geq G \gg L2$

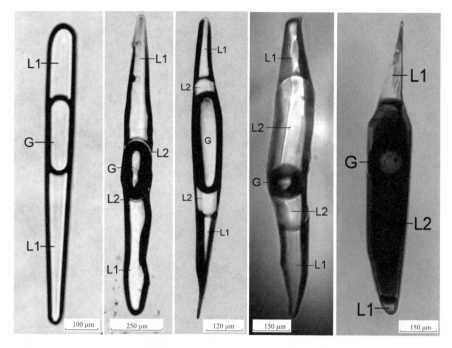

Fig. 1.6 Synthetic water-hydrocarbon inclusions of the first type formed in heterogeneous water-oil fluids at temperatures of 240–310 °C, pressure 8–30 MPa. The volume ratio of phases: from L1 > G ≫ L2 to L2 > G ≈ L1

> SB. It is noteworthy that the proportion of liquid hydrocarbons inclusions formed in sub and supercritical fluids increases, while the proportion of solid bitumens decreases. However, in syngenetic inclusions, the real ratios of these phases are also subject to changes, although not as significant as in earlier inclusions in seed rods.

1.3.2   Phase Transformations and Phenomena of Imaginary Homogenization in Water-Hydrocarbon Inclusions at Their Heating and Cooling

The behavior, phase composition and states of the most characteristic fluid inclusions of both types were studied by microthermometric method up to 400 °C at pressures in inclusions up to 90 MPa in combination with FT-IR spectroscopy and microscopy with natural and UV light.

First type inclusions. Such inclusions, as mentioned above, are concentrated in seed bars. Most often these are four-phase inclusions with volumetric phase ratios L1 > G ≥ L2 > SB, where L1/G ≈ 4, L1/L2 ≈ 12 and L1/SB ≈ 18. When heated to 100 °C, the phase composition and state of them remain practically unchanged (Fig. 1.8). The

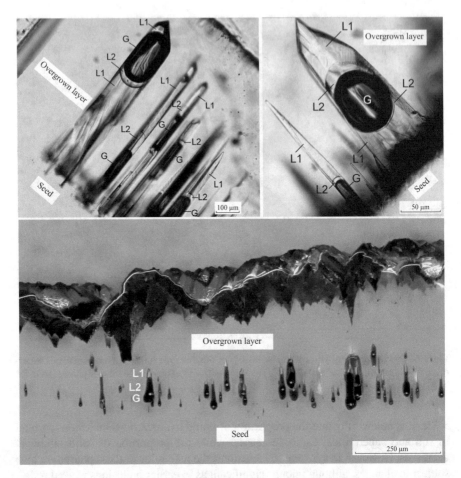

Fig. 1.7 Synthetic water-hydrocarbon inclusions of the second type formed in heterogeneous water-oil fluids at temperatures of 320–340 °C, pressure 50–80 MPa. The volume ratio of phases: from L1 > G ≫ l2 > SB to L2 ≈ SB ≫ G ≈ L1

boundaries of the divide between all phases are clear. In the interval of 110–200 °C the volume of the gas phase (G) has sharply decreased by 8%. At the temperature above 210 °C the volume of liquid hydrocarbon phase (L2) began to decrease, and the phase G, on the contrary—to increase rapidly. The complete disappearance of phase L2 occurred at 220 °C due to its dissolution in phase G. This episode was accompanied by the release of numerous, rapidly disappearing gas bubbles. The new phase, designated as (G + L2), is pale yellow and has a clear boundary with the water phase (L1). When the temperature rises to 250 °C, the boundary between phases L1 and (G + L2), gradually pale, disappears completely. Visually, this is very similar to achieving homogenization of the solution. However, with further heating up to 270 °C, this boundary appeared again in the same area where it had disappeared. In the local FT-IR spectrum of inclusions, where the water phase L1 was located, there

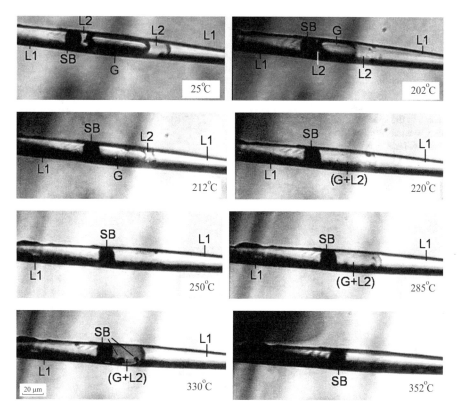

Fig. 1.8 The thermogram of water-hydrocarbon inclusions with volume ratios of phases L1 > G > L2 > SB, where L1/G ≈ 4, L1/L2 ≈ 12 and L1/SB ≈ 18. The thermogram illustrates the imaginary homogenization of hydrocarbon inclusion in quartz. Conditions of inclusions formation: 5 wt% $NaHCO_3$, 330/360 °C, ~80 MPa, the oil shale from the Kashpirskoye deposit

are only absorption bands of molecular water. But in the area where the allegedly extinct phase (G + L2) was located, the absorption bands around 2972, 2949 and 2887 cm^{-1}, typical of crude oil, are clearly visible (Fig. 1.9a–c). Further increase of temperature up to 330 °C has led to colouring of this area of inclusion in yellow colour with an orange shade. At the same time, there was a decrease in the phase volume (G + L2), which at 350 °C was about 30% of its initial volume. The inclusion exploded at 352 °C.

In other, similar inclusions in terms of composition, intensive fluorescence appeared in the areas of the disappeared phase of liquid hydrocarbons when viewed in UV light, while in normal and polarized light there were no signs of the presence of liquid hydrocarbons. The disappearance of the boundary between the water (L1) and hydrocarbon (G + L2) phases at 250 °C, in our opinion, is not due to the real homogenization of the fluid, but is a consequence of the alignment of densities and refractive indices of these liquid phases in the process of heating the inclusion.

Fig. 1.9 FT-IR absorption spectra of water-hydrocarbon inclusion (see Fig. 1.8), **a**—methane and liquid hydrocarbons dissolved in methane, **b**—liquid hydrocarbons, **c**—aqueous phase. In the spectra of all phases, an absorption band near 2347 cm^{-1} is observed, due to the presence of CO_2 in the inclusion

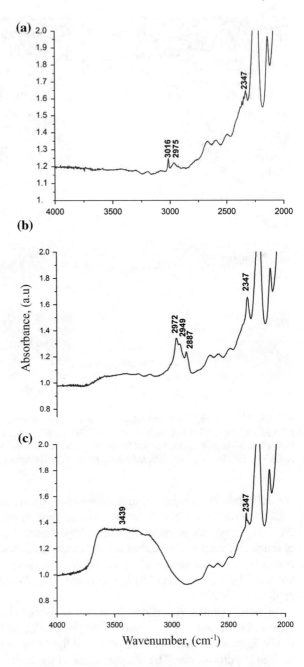

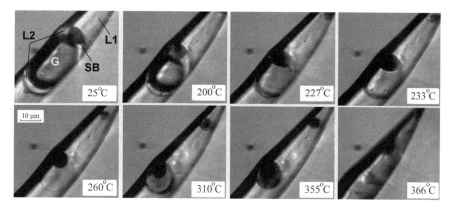

Fig. 1.10 The thermogram of water-hydrocarbon inclusion of a similar nature as shown on Fig. 1.8, but the complete disappearance of the boundary between the gas and water phases was not achieved

The behavior of the next inclusions with similar phase volumetric ratios is close to that described above (Fig. 1.10). However, there was no complete disappearance of the boundary between the gas and water phases here, although at 260 °C the boundary between them became barely distinguishable. With a further increase in temperature, the gas phase, decreasing in volume, regained its yellow color, which changed to bright orange-yellow as the temperature increased. Then, the inclusion exploded at 366 °C.

Destruction of inclusions begins as shown earlier (Naumov et al. 1966), when the pressure inside inclusions rises to 80–90 MPa. The temperature should be 320 and 360 °C at the density (ρ) of the solution $\rho = 0.8$ g/cm^3 and $\rho = 0.6$ g/cm^3 respectively. Therefore, the microthermometric study of other water-hydrocarbon inclusions did not raise the temperature of heating above 320 °C. This preserved the inclusions from destruction and made it possible to trace their behavior not only at the increase, but also at the decrease of temperature. Thus, heating of one of the four-phase inclusions with volumetric ratios of phases L1 > G $\geq$ L2 > SB led at first, at 190 °C, to disappearance of the boundary between the phases of liquid hydrocarbons (L2) and aqueous solution (L1) (Fig. 1.11). Further, at 202 °C, the gas HC phase (G) disappeared completely. This created the illusion of complete homogenization of the solution. However, a close look at the inclusion showed that the area with the missing phases L2 and G had a subtle yellowish tinge. It was assumed that in this case the G phase dissolved in the L2 phase with the appearance of a new liquid hydrocarbon phase marked as (L2 + G). And really at the further increase of temperature to 250 °C intensity of yellow colouring of a phase (L2 + G) began to increase with simultaneous occurrence of its border with phase L1. At 310 °C, the phase (L2 + G) acquired an orange shade. Its volume decreased by about 30%. Due to the risk of depressurization of the switch at a higher temperature, heating was stopped and replaced by cooling. This first led to the loss of the orange shade of the phase (L2 + G) and then to its decomposition at 187 °C into L2 and G. At 60 °C, the phase composition and volumetric ratios of the phases in inclusion are fully restored.

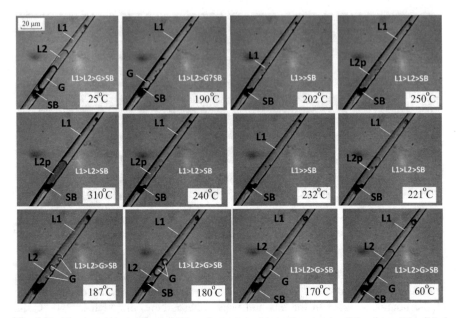

Fig. 1.11 The thermogram of a multiphase inclusion with a volume ratio of the phases L1 > L2 > G > SB shows the behavior of the phases at heated and cooled in the temperature range 25–310–60 °C until the appearance of imaginary homogenization and inclusion destruction. Conditions of inclusions formation: 5 wt% NaHCO$_3$, 330/360 °C, ~80 MPa, the oil shale from the Kashpirskoye deposit

Another indisputable proof that the phase of liquid hydrocarbons does not disappear in inclusions at certain temperatures (and corresponding pressures), but remains in the form of an invisible phase, was obtained by simultaneous observation of inclusions in the thermometric chamber in normal and UV light. In particular, in contrast to the disappearance of the hydrocarbon phase (L2 + G) and the onset of apparent homogenization observed during heating in the ordinary light, the presence of this phase in the inclusion is unequivocally proved by its bright fluorescence in UV light (Fig. 1.12). At the same site in the local FT-IR spectrum there are absorption bands typical for liquid hydrocarbons.

In general, all this indicates that at certain temperatures (and corresponding pressures) the disappearance and appearance of liquid and gas hydrocarbons inclusions can occur not only in connection with the real homogenization of the fluid, but also in the alignment of the density and refractive indices of two or more liquid phases due to their joint dissolution. This phenomenon we call as imaginary homogenization.

Other important and interesting from the point of view of generation and extraction of liquid and gas hydrocarbons from oil shale and other bituminous rocks is the experimental establishment of formation at 180–220 °C and pressures of about 10–20 MPa of homogeneous phases (G + L2) and (L2 + G). Their difference is determined by the predominance of gas, mainly methane, and liquid hydrocarbons

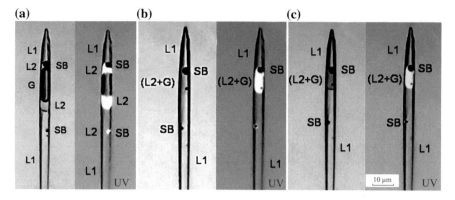

Fig. 1.12 Fragment of the thermogram of water-hydrocarbon inclusion with simultaneous use of natural and UV light at **a**—30 °C, **b**—286 °C and **c**—305 °C. The combination of illumination allows to definitely seeing the apparent disappearance of the liquid hydrocarbon phase in daylight and its appearance (with bright fluorescence) in UV rays

in the initial fluid. Heterogenization of such phases with separation of liquid and gas hydrocarbons occurs at 187–200 °C.

Second type inclusions. The behavior of the second type of inclusions in the process of heating and cooling is much easier than in the first type of inclusions. This is demonstrated in the thermograms of the most characteristic inclusions in the overgrown layer of quartz crystals (Fig. 1.13). All fluid inclusions of this type include aqueous solution, gas (mainly methane) and liquid hydrocarbons. The volumetric phase relations are subject to the inequality of L1 > G ≫ L2 ≫ SB. As well as in inclusions of the first type, the character of behavior of fluids inclusions at heating and cooling does not change. But the absolute temperatures of occurrence and disappearance of gas and liquid phases of hydrocarbons inclusions can differ by tens of times. This indicates the heterogeneous state of the fluids during the capture of inclusions.

It should be especially noted that there are cases of hidden presence of liquid hydrocarbons in fluid inclusions at room temperature. Such inclusions were observed in the quartz grown at interaction of bituminous argillites of the Bazhenov Formation and the Central Oil and Gas Basin of France with pure water and slightly alkaline water solutions at temperatures from 350/380 to 420/450 °C and pressures from 80 to 120 MPa, respectively. The presence of liquid HC in fluid inclusions is usually recognized by the characteristic yellow or yellow-orange color. However, in many inclusions in quartz grown in sub- and supercritical fluids during their interaction with bituminous argillites, liquid hydrocarbons were not detected in ordinary and polarized light. But, as the study of numerous inclusions has shown, they are clearly fixed in UV light, combined UV and polarized light due to very intense fluorescence in the area adjacent to the boundary of the water phase (L1) and gas (G) phases (Fig. 1.14). The proof of the fact that the observed intensive fluorescence is caused by the presence of liquid hydrocarbons is the appearance of fluorescent inclusions of

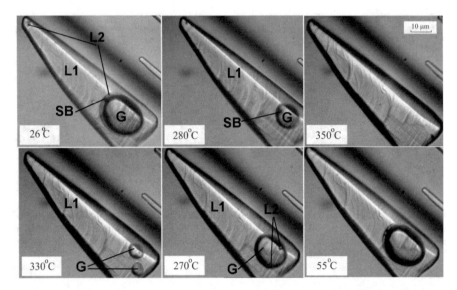

Fig. 1.13 Thermogram of multiphase inclusion with the ratio of the phases L1 > G ≫ L2 > SB formed in overgrown layer of quartz shows the behavior of the phases at heated in the temperature range of 26–350 °C and cooling in the range of 350–55 °C. Conditions of inclusions formation: 5 wt% $NaHCO_3$, 330/360 °C, ~80 MPa, oil shale of the Leningrad deposit

absorption bands in FT-IR spectra, which are usual for crude oil. At the same time, in the FT-IR spectra of the gas phase of such inclusions one can observe absorption bands of methane (near 3016 cm^{-1}) and carbon dioxide (near 2347 cm^{-1}). In the spectra of the rest of the inclusions outside the phases of liquid and gas hydrocarbons there is a diffuse band in the range of 3000–3600 cm^{-1} with a maximum at 3400 cm^{-1}, indicating the water composition of the phase. Mass viewing of such inclusions in quartz, grown during the interaction of sub- and supercritical fluids with bituminous mudguards, allows us to draw a conclusion about the broad development of this phenomenon, which was not previously noted.

In addition, the hidden presence of liquid hydrocarbons is constantly observed in inclusions formed in quartz, grown in sub and supercritical water-oil mixtures. In normal and polarized light, such inclusions usually have water and gas, mainly methane, and solid bitumen inclusions. The latter, as a rule, form myriads of the smallest (first μm) droplet-shaped inclusions in the earliest quartz growth zones (Fig. 1.15a, b). There are no liquid and gas HC phases in such inclusions. However, in the later growth zones of the same crystals, much larger (up to 50–70 μm) two-phase inclusions with water (L1) and gas (G) phases are observed in normal and polarized light. When observing in UV light it is visible that the gas (methane) phase contains condensate of liquid hydrocarbons, which have a bright fluorescent glow. In FT-IR spectra of the same inclusion sites there are absorption bands typical for crude oil.

Fig. 1.14 The presence of oil, including gas condensate in synthetic fluid inclusions in quartz grown in weakly alkaline aqueous sub- and supercritical fluids when they interact with bituminous argillites of the Bazhenov Formation and the Central Oil and Gas Basin of France at temperatures of 350–450 °C and pressure 80–120 MPa, respectively

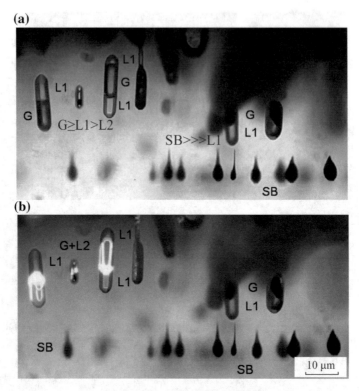

Fig. 1.15 Two types of inclusions in quartz crystal in natural light (**a**) and UV light (**b**). Intensive fluorescence of inclusions under UV light reveals the hidden presence of liquid oil hydrocarbons in fluid inclusions

Also consider the behavior of inclusions with volumetric phase ratios L2 > L1 ≥ G > L3 ≫ SB (Fig. 1.16). Such inclusions with oil-like liquid prevailing 3–5 times over all other phases are often found in quartz formed during interaction of hydrothermal solutions with oil shale. The inclusions are located in the overgrown quartz layer at some distance from the seed and are later in relation to the above mentioned three-phase inclusions. At heating of inclusions in them in the beginning (to temperatures 200–220 °C) liquid drops of phase L3 in the basic oil-like phase L2 are dissolved (Fig. 1.17). Then, at 260–280 °C, a gas bubble (mainly methane) is dissolved in the same phase. The inclusion is transformed to a two-phase state with two liquids— oil-like L2 and water L1, with a ratio of L2 > L1. Further temperature increase up to 353–360 °C leads to complete dissolution of L1 phase in oil-like liquid L2 with formation of homogeneous essentially hydrocarbon fluid. The share of dissolved in it water phase reaches 15–20 vol% The cooling of inclusions in them in the reverse order completely restores the primary phase composition and volume ratios of the initial phases. This indicates a rapid establishment of equilibrium inclusions and stability of their phase composition.

Fig. 1.16 Predominantly oil water-hydrocarbon inclusions with volumetric ratios of phases L2 > L1 ≥ G > L3 > SB formed at hydrothermal solutions interact with oil shale. Inclusion forms in the overgrow layer of quartz near its boundary with the seed bar due to the massive precipitation of the smallest drops of oil saturated with dissolved water

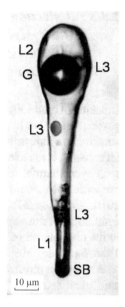

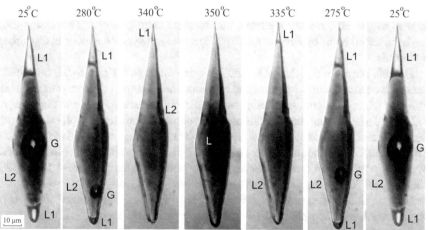

Fig. 1.17 The thermogram of predominantly oil inclusions shows reversible processes of sequential dissolution of the L3 phase (composition not yet installed) in the bulk of liquid hydrocarbons (L2), then dissolution of the gas bubble in the L2 phase and complete dissolution in the L2 phase of the aqueous phase and then transition of the fluid to the homogeneous state. As the temperature decreases, fluid heterogenization occurs with the sequential appearance of all previously dissolved phases

1.3.3 Reflection of Oil Cracking in Synthetic Water-Hydrocarbon Inclusions in Quartz and Assessment of Possible Maximum Depths of Its Location in the Earth's Interior

Cracking of crude oil and its heavy fractions is the main industrial process of obtaining gasoline and other hydrocarbon fuels, as well as raw materials for petrochemical and construction industries. At the same time, it has been established that to some extent oil cracking is manifested in natural conditions, especially when it is in direct contact with hydrothermal solutions and silicate (mica) and carbonate rocks with catalytic properties (Rokosova et al. 2001; Simoneit 1995; Huang and Otten 2001; Ping et al. 2010; Xiao et al. 2010; Zhao et al. 2008). If oil cracking really takes place in natural conditions, it can explain the vertical zoning observed in many oil and gas basins in the distribution of different types of hydrocarbons (Petrov 1984; Andreev et al. 1958; Samvelov 1995; Yeremenko and Botneva 1998).

Among the physico-chemical factors, the most significant changes in the vertical geological section of the Earth's strata undergo thermobaric parameters. It seems likely that their increase as the depth of oil and gas fields should inevitably lead to the cracking of oil and increase in its proportion of light and medium fractions, as well as gas HC, mainly methane, and residual solid bitumen. This assumption has been tested by us by conducting experiments using synthetic water-hydrocarbon inclusions.

For this purpose, polished ZX- and ZY-orientation plates of 0.5–2.0 mm thickness with captured water-hydrocarbon inclusions were prepared from quartz crystals grown in water-oil solutions at temperatures of 240, 280 and 310 °C and pressures of 5, 7 and 20 MPa. Using the methods of thermobarogeochemistry, the phase composition, state and behavior at temperatures up to 400 °C were determined in inclusions. Then the plates with the studied inclusions were placed in autoclaves with the volume of 30 ml, poured with clean water and maintained at the temperatures of 300, 320, 350 and 380 °C and the pressure of 100 MPa for 7–14 days.

In total, about 30 quartz crystals have been re-treated and studied. All grown crystals contained two-phase (L1 + G) and three-phase (L1 + L2 + G) inclusions. Under room conditions, the fractions of the water (L1) and gas (G) phases in two-phase inclusions were in the volume ratios corresponding to the autoclave filling coefficients taking into account the reduction of their free volume by the amount of additives in aqueous crude oil solution. In three-phase inclusions, the volume ratios of phases L1/G/L2 varied widely—from L1 > G ≫ L2 to L2 > L1 > G, regardless of the share of liquid hydrocarbons in the initial water-oil mixtures. This indicated the heterogeneous state of mother fluids during the capture of inclusions by quartz crystals. No hydrocarbon gases were detected in the gas bubble of inclusions. When the inclusions are heated, they disappear first. The two-phase inclusions become homogeneous and the three-phase inclusions become liquid two-phase, with practically the same volumetric L1/L2 phase ratios as before the disappearance of the gas phase. In the process of further heating liquid two-phase inclusions are stored up to 365–405 °C (depending on the density of the solution) with an insignificant (1.5–2.0 vol%) decrease in the proportion of L2 phase. At higher temperatures, most of these inclusions lose their hermetic state, often with an explosion.

After the autoclave heat treatment at 350 °C and 150 MPa, the two-phase (L1 + G) inclusions remain identical to the initial inclusions. At the same time, the phase composition and volumetric ratios of phases in three-phase inclusions of the composition (L1 > L2 > G) undergo significant changes at temperatures above 300 °C, while the inclusions, heat-treated at lower temperatures do not bear any changes.

In the gas phase of three-phase inclusions, according to the data of the local FT-IR, at 320 °C there appears methane with an insignificant admixture of propane and ethane, the content of which increases with increasing temperature and fraction inclusions of liquid hydrocarbons. Spherical releases of SB solid bitumen always occur in droplets and the liquid hydrocarbon trimming at the boundary of water and gas phases, and the share of gasoline fraction increases in the liquid hydrocarbon phase according to chromatogram data (Fig. 1.18). This is also confirmed by the beginning of boiling of the liquid hydrocarbon phase at 90 °C. Boiling continues up to 280 °C up to disappearance (dissolution) of L2 phase and transition of inclusion into two-phase state (L1 + G). Further heating of the inclusion leads to a gradual reduction of the gas bubble and its disappearance at 350 °C with the transition to a homogeneous state of inclusion. Solid bitumen SB emissions are maintained virtually

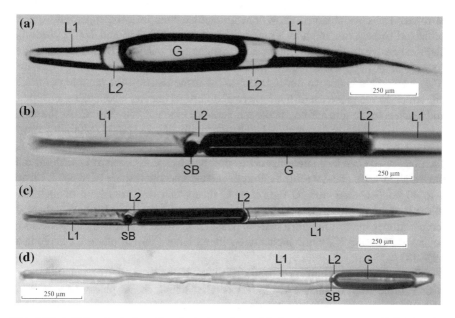

Fig. 1.18 Initial water-hydrocarbon inclusions formed in heterogeneous water-oil fluids 350–380 °C and 100 MPa, SB—absent (**a**), inclusions after heat treatment at 350 °C (**b–d**) with the appearance of solid bitumen, methane and with increase proportion of light fractions. The change in the composition of heat treated inclusions is result of oil cracking

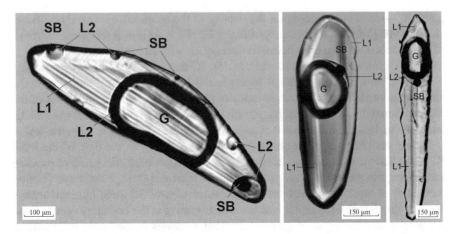

Fig. 1.19 Initial water-hydrocarbon inclusions in quartz grown at temperatures of 350–370 °C and pressure ~90 MPa. The phase composition of the inclusions is identical to the composition of low-temperature inclusions heat-treated at 350–380 °C and 100 MPa

unchanged. When cooling to 335 °C, a gas bubble appears in the inclusion, indicating the beginning of its heterogenization. At the further cooling the phase composition and volumetric ratios of water phase, liquid hydrocarbons and gas phase were fully restored. The data obtained show that fluid inclusions in low-temperature quartz subjected to heat treatment at temperatures of 300, 320, 350 and 380 °C become identical in terms of phase composition, states and behavior during heating and cooling of primary inclusions in quartz grown at similar and close temperatures (Fig. 1.19). This is due, in our opinion, to the same reason—the cracking of crude oil, which in the presence of hydrothermal solutions becomes noticeable at 320 °C and reaches maximum activity in the temperature range of 350–380 °C. At the same time, the thermometry of individual inclusions with different phase volumetric ratios made it possible to find out at what temperatures (and corresponding pressures) such inclusions can be in heterogeneous and homogeneous state. It turned out that during short-term (60–90 min) heating of inclusions with a share of phase L2 from 3 to 40 vol% at 266–308 °C the phase G disappears with transition of inclusion into a liquid two-phase state with a ratio of phases L1 > L2. At further increase of temperature in inclusions with a share of phase L2 less than 4–5 vol% homogenization occurs at 350–357 °C, while inclusions with a higher share of phase L2 explode at 355–385 °C.

Other behaviour at short-term heating is characterized essentially by oil inclusions with volume ratio of phases L2 ≫ L1 ≈ G. At first, the G phase disappears at 270–285 °C and the inclusion becomes two-phase with a phase ratio of L2 ≫ L1. This state is maintained up to the temperature of 345–355 °C, above which the aqueous phase (L1) is completely dissolved in the phase of liquid hydrocarbons (L2) with the transition to a homogeneous state. Decrease in temperature leads to heterogenization of inclusion with appearance at 335–350 °C of phase L1 and further, at 265–270 °C—gas, basically methane phase G.

Inclusions formed at temperatures from 335 to 495 °C produce solid bitumen, methane and, less frequently, propane, ethane and CO_2, while the L2 phase is substantially (up to 70%) enriched with light gasoline-kerosene fractions. In the highest temperature inclusions on the periphery of L2 phase secretions the appearance of one or two phases of liquid hydrocarbons—L3 and L4—is recorded. They disappear (dissolve) in phase L2 at 95 and 126 °C accordingly.

Inclusions captured by quartz at 335–355 °C are characterized by a phase ratio of L1 ≥ G > L2 > SB. The share of L2 phase in different inclusions varies from one hundredths to 30%. At heating of inclusions with a share of phase L2 to 4 vol% it disappears at 160–180 °C with transition of a fluid in a two-phase condition (L1 + G). Homogenization of such inclusions occurs at 287–322 °C. At higher fractions of phase L2, the increase in temperature up to 308–315 °C leads to disappearance of phase G, and then, at 340–360 °C inclusions explode, not having reached homogenization.

In water-hydrocarbon inclusions formed at 420–490 °C, the share of L2 phase in inclusions corresponds to its volume in initial solutions, changing from 5 to 35 vol%. Short-term heating of inclusions with the ratio of phases L1 ≥ G > L2 > SB has allowed to establish that phase L2 (at a share of phase L2 10–12 vol%) are disappears (dissolves) at 250–285 °C with transition of inclusion in a two-phase condition with a parity of phases L1 ≥ G; at the further heating to 370 °C the phase G disappears with transition of inclusion in a homogeneous condition. Hard bitumen is retained without change. Cooling of inclusions leads to the appearance of all these phases in reverse order at temperatures that are 12–15 °C below their disappearance temperatures. Increasing the share of L2 phase up to 25 and 36 vol% increases the temperature of its disappearance up to 315 and 335 °C, respectively. Complete homogenization of such inclusions occurs at 380–395 °C.

In experiments at 650/670 °C, the autoclaved oil captured in the inclusions is completely transformed into graphite and methane (Fig. 1.20).

As noted above, the nature of oil transformations at high temperatures largely depends on the volume ratios in the fluid of the oil, water and gas phases. When the phase volumetric ratios in high-temperature inclusions are subject to the inequality of L1 ≥ L2 > G and the fluids are in a homogeneous state, the oil is stored in them, at least up to 500 °C. This is indicated by the very presence of oil in the primary fluid inclusions captured by quartz at temperatures of 335–500 °C (Fig. 1.21), and its preservation in the secondary inclusions after repeated treatment in homogeneous solutions of supercritical fluids for 30–40 days. In our opinion, this is due to the fact that oil hydrocarbons do not exist as individual compounds with their inherent properties when dissolved in aqueous solution. This is indicated by the disappearance in the high-temperature (400 °C) FT-IR spectra of typical oil absorption bands near 2972, 2949 and 2287 cm^{-1} (Fig. 1.22), as well as the characteristic greenish-blue fluorescent glow in UV light. At cooling and heterogenization of inclusions all disappeared phases, their volume ratios, typical absorption bands in FT-IR spectra and characteristic fluorescent glow are completely restored.

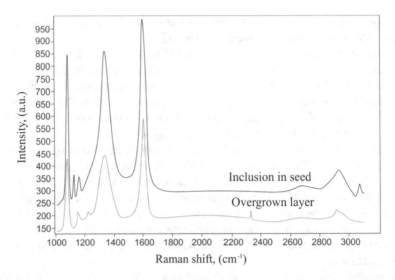

Fig. 1.20 Raman spectra of the inclusion in the seed and spectra the overgrown layer of quartz testifies to the transformation of water-hydrocarbon fluid into fine crystalline graphite

Essentially oil inclusions with a phase ratio of L2 $\gg$ L1 $\approx$ G behave differently during heat treatment. In the local high-temperature FT-IR spectra, when the homogenization of inclusions with the dissolution of gas G (270–285 °C) and water L1 (350–357 °C) phases is achieved, the specified absorption bands typical of oil (Fig. 1.23) are completely preserved, as well as its characteristic fluorescent glow. When the inclusions are cooled down, the composition and volume ratios of the phases are fully restored. After 14 days of heat treatment at 300 °C, the composition and volume ratios of the phases inclusions, as noted above, remain unchanged. However, already at 320 °C residual solid bitumen appears in them and the share of light oil fractions increases by 10–15%. The increase in temperature up to 350 and especially 380 °C leads to complete conversion of oil into solid bitumen, methane and, to a lesser extent, carbon dioxide (Fig. 1.24).

Thus, the experimental data obtained testify to the heterogeneous (three-and two-phase) state of water-oil systems at relatively low temperatures (up to 240–310 °C) and pressures (up to 30–50 MPa). Even in the longest (up to 120 days) experiments at the specified *PT*-parameters the oil remains stable without being subjected to any changes. But already at 320 °C and especially in a range of temperatures 350–380 °C it undergoes essential changes. The nature of these depends largely on the volume ratios of the oil, water and gas phases. Under conditions of water solution predominance over oil, dissolved in homogeneous, including supercritical fluids, oil hydrocarbons lose their ability to split due to the formation of true solutions with water. At least under such conditions, the oil present to the gas condensate of fluid inclusions remains stable up to 490 °C. However, in cases where oil prevails over aqueous solution, HC in its composition when reaching a homogeneous state are preserved as such and continue to be cracked until irreversible transformation into

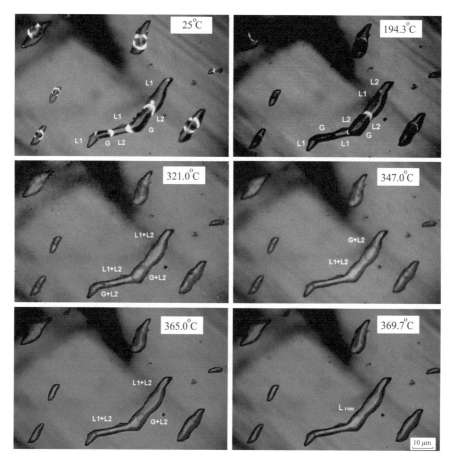

Fig. 1.21 Fragment of the thermogram of water-hydrocarbon inclusion in quartz. Quartz growth conditions: 5 wt% Na_2CO_3 + 10 vol% oil, 390/400 °C, ~120 MPa

methane and residual solid bitumen, up to graphite. All this makes it possible to consider that those oil deposits, where oil prevails over the water phase, can spread taking into account the actual thermal gradients of oil and gas bearing strata of 2–30 °C/km up to the depth of 12–14 km. But if the water component exceeds the share of oil in the reservoirs, such water-oil fluids can reach a depth of 18–20 km. In this case, the safety of oil is ensured by finding it together with associating aqueous solutions in sub and supercritical states.

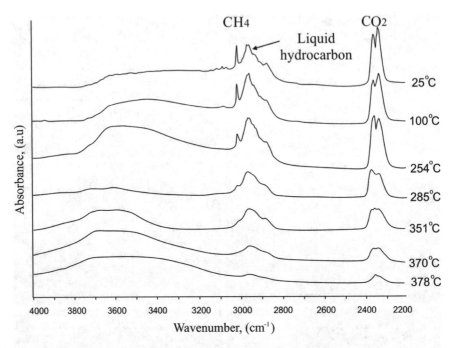

Fig. 1.22 FT-IR spectra of phase absorption of water-hydrocarbon inclusion in the temperature range 25–378 °C. An increase in temperature leads to the disappearance of the absorption bands of methane (3016 cm^{-1}) and liquid hydrocarbons (2800–3000 cm^{-1}). The band associated with the presence of CO_2 (2450 cm^{-1}) almost completely disappears

1.4 Conclusions

1. Modified method of hydrothermal growth of quartz crystals allows to form fluid inclusions of stimulated and spontaneous nucleation. These inclusions are essentially microsamples of fluids formed during the interaction of hydrothermal solutions with bituminous rocks and crude oil.
2. Numerous experiments and especially data on in situ study of fluid inclusions by thermobarogeochemical methods testify to fundamentally different behavior and phase states of water-hydrocarbon fluids formed in temperature ranges: 240–320 °C, 320–380 °C and 380–500 °C at pressures of saturated steam and higher. These differences are due to the cracking of crude oil (or, more precisely, its heavy fractions) in hydrothermal solutions at temperatures above 330 °C with the formation of predominantly light and medium fractions.
3. Water-hydrocarbon fluids formed at temperatures up to 300–320 °C and pressures above saturated steam (experimentally traced 90 MPa) turn into liquid two-phase oil-water fluids without free gas when heated. Their existence in this state is traced up to the depressurization of inclusions at 365–405 °C. It is possible that at higher temperatures and pressures such fluids may become homogeneous. But

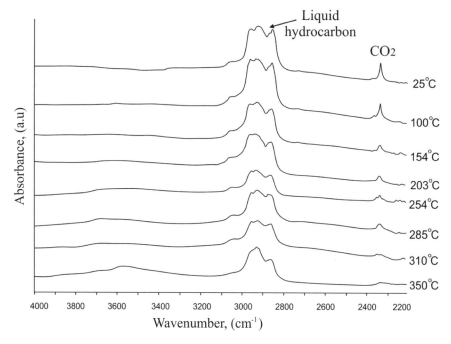

Fig. 1.23 FT-IR spectra of phase absorption of predominantly oil water-hydrocarbon inclusion in the temperature range 25–378 °C. The spectra shows a stable existence of oil up to 350 °C

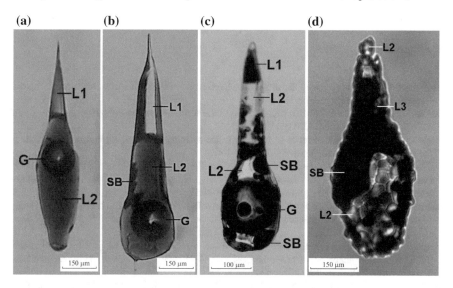

Fig. 1.24 Predominantly oil water-hydrocarbon inclusion after autoclave treatment at **a**—300 °C, **b**—320 °C, **c**—350 °C, **d**—380 °C and 100 MPa for 14 days

oil should be enriched with light and medium fractions due to cracking and other specific hydrocarbons and other compounds.

4. Water-hydrocarbon fluids formed at temperatures of 330–500 °C behave differently when heated: first, they pass into two-phase gas-liquid state with light oil fractions dissolved in gas hydrocarbons, and then, when the critical point of water is exceeded, they transform into homogeneous, probably supercritical fluids.

5. Different ratios of the same phases and their different compositions in the captured inclusions indicate the heterogeneous state of aqueous-hydrocarbon fluids formed during the interaction of hydrothermal solutions with bituminous rocks and crude oil. And only fluids, in which the share of dissolved hydrocarbons at given thermobaric parameters is lower than their solubility, are homogeneous.

6. Oil content in hydrothermal slightly alkaline and alkaline solutions, established by the visual method, in the temperature range of 260–400 °C and pressure up to 90 MPa increases from the hundredth and first tenths of shares up to 8–10 vol. At the same time, the total content of hydrocarbons formed as a result of oil cracking at 380–450 °C and pressure up to 110 MPa, is about 10–20% in the same solutions.

Acknowledgements The authors are grateful to J. Pironon, A. Randy and O. Barres (GeoResources Laboratory of the Nancy University, France), G. V. Bondarenko and S. V. Penteley for the cooperation in carrying out the research. The study was support by the IEM RAS project № AAAA-A18-118020590150-6

References

Andreev PF, Bogomolov AI, Dobryansky AF, Kartsev AA (1958) Oil transformation in nature. Gostoptehizdat, Leningrad, p 416

Balitskaya OV, Balitsky VS (2010) Mechanisms of the formation and morphogenetic types of fluid inclusions. Dokl Earth Sci 435(1):1442–1445

Balitsky VS (1965) On occurrence of bitumen in the mercury and mercury-antimony deposits in the Northwestern Caucasus. Sov Geol 3:144–150

Balitsky VS, Balitskaya LV, Bublikova TM, Borkov FP (2005) Water-hydrocarbon inclusions in synthetic quartz, calcite, and fluorite crystals grown from oil-bearing hydrothermal solutions (experimental data). Dokl Earth Sci 404(1):1050–1053

Balitsky VS, Prokof'ev VY, Balitskaya LV, Bublikova TM, Pentelei SV (2007) Experimental study of the interaction of mineral-forming hydrothermal solutions with oil and their coupled migration. Petrology 15(3):211–223

Balitsky VS, Penteley SV, Balitskaya LV, Novikova MA, Bublikova TM (2011) Visual in-situ monitoring of the behavior and phase states of water-hydrocarbon inclusions at high temperatures and pressures. Petrology 19(7):653–674

Balitsky VS, Bondarenko GV, Pironon J, Penteley SV, Balitskaya LV, Golunova MA, Bublikova TM (2014) The causes of vertical zonation in the distribution of hydrocarbons over the Earth's interior: Experimental evidence of the cracking of crude oil in high-temperature water-hydrocarbon fluids. Russ J Phys Chem B 8(7):901–918

Balitsky VS, Penteley SV, Pironon J, Barres O, Balitskaya LV, Setkova TV (2016) Phase states of hydrous–hydrocarbon fluids at elevated and high temperatures and pressures: study of the forms and maximal depths of oil occurrence in the earth's interior. Dokl Earth Sci 466(2):130–134

Bazhenova OK, Lein AY (2002) Geochemistry of carbonaceous rocks in modern hydrothermal systems. In: Proceedings of Russian conference on organic mineralogy. SPb Gos. Univ., St. Petersburg, pp 95–96

Beskrovnyi NS (1967) Oil bitumen and hydrocarbon gases accompanying hydrothermal activity. VNIGNI, Leningrad, p 209

Chukanov NV, Pekov IV, Sokolov SV, Nekrasov AN, Ermolaeva VN, Naumova IS (2006) On the problem of the formation and geochemical role of bituminous matter in pegmatites of the Khibiny and Lovozero alkaline massifs, Kola Peninsula, Russia. Geochem Int 44(7):715–728

Ermakov NP (1972) Geochemical systems of mineral inclusions. Nedra, Moscow, p 376

Florovskaya VN, Zaraiskii GP, Zezin RB (1964) Kerites and other carbon compounds at the Komsomol'skoe sulfide deposits, Southern Urals. Dokl Akad Nauk SSSR 157(5):1131–1134

Huang WL, Otten GA (2001) Cracking kinetics of crude oil and alkanes determined by diamond anvil cell-fluorescence spectroscopy pyrolysis: technique development and preliminary results. Org Geochem (32):817–830

Ivankin PF, Nazarova NI (2001) Deep fluidization of the earth's crust and its role in petro-uregenesis, salt and oil formation. TSNIGRI, Moscow, p 206

Lapidus AL, Strizhakova YuA (2004) Oil shale—an alternative raw material for chemistry. Bull Russ Acad Sci 74(9):823–829

Mel'nikov FP, Prokof'ev VY, Shatagin NN (2008) Thermobarogeochemistry. Academic project, Moscow, 244 p

Naumov VB, Balitsky VS, Khetchikov LN (1966) On relations between the capture, homogenization, and decrepitation temperatures of gas-liquid inclusions. Dokl Akad Nauk SSSR 171(1):146–148

Naumov GB, Ryzhenko BN, Khodakovskii IL (1971) Thermodynamic data handbook. Atomizdat, Moscow, p 240

Ozerova NA (1986) Mercury and endogenous ore-forming processes. Nauka, Moscow, p 232

Petrov AA (1984) Hydrocarbons of oil. Nauka, Moscow, p 263

Pikovskii YuI, Karpov GA, Ogloblina AI (1987) Polycyclic aromatic hydrocarbons in the products of Uzon oil in Kamchatka. Geokhimiya 6:869–876

Ping H, Chen H, Song G, Liu H (2010) Oil cracking of deep petroleum in Minfeng sag in north Dongying depression, Bohai Bay basin, China: evidence from natural fluid inclusions. J Earth Sci 21(4):455–470

Prokof'ev VY, Balitsky VS, Balitskaya LV, Bublikova TM, Borkov FP (2005) Study using IR Spectroscopy of Fluid Inclusions with hydrocarbons in synthetic quartz. In: Proceedings of 15th Russian conference on experimental mineralogy. Geoprint, Syktyvkar, pp 194–196

Roedder E (1984) Fluid inclusions. Rev. Mineral. Mineralogical Society of America, 12, 644p

Rokosova NN, Rokosov YuV, Uskov SI, Bodoev NV (2001) Simulation of transformations of organic matter into hydrothermal petroleum (a review). Pet Chem 41:221–233

Samoilovich LA (1969) Relations between the pressure, temperature, and density of aqueous salt solutions. VNIISIMS, Moscow, p 48

Samvelov RG (1995) Features of hydrocarbon pools formation at greater depths. Geol Neft Gaz (9):5–15

Simoneit BRT (1995) Organic geochemistry of aqueous systems at high temperatures and elevated pressures: hydrothermal petroleum. Main research avenues in geochemistry. Nauka, Moscow, pp 236–259

Teinturier S, Elie M, Pironon J (2003) Oil-cracking processes evidence from synthetic petroleum inclusions. J Geochem Explor 78:421–425

Xiao QL, Sun YG, Zhang YD (2010) The role of reservoir mediums in natural oil cracking: preliminary experimental results in a confined system. Chin Sci Bull 55(33):3787–3793

Yeremenko NA, Botneva TA (1998) Hydrocarbon deposits at great depth. Geol Oil Gas 1:6–11

Zaidelson MI, Vaynbaum SY, Koprova NA (1990) Formation and oil and gas content of domanicoid
    formations. Nauka, Moscow, 79 p
Zhao WZ, Wang ZY, Zhang SC (2008) Cracking conditions of crude oil under different geological
    environments. Sci China Ser D Earth Sci 51:77–83

Chapter 2
Experimental Studies of Hydrothermal Fluid

G. V. Bondarenko and Y. E. Gorbaty

Abstract Structural features of pure water and aqueous solutions of the electrolytes NaCl, NaCO$_3$ and Zn[NO$_3$]$_2$ were comprehensively studied using the X-ray diffraction, infrared absorption spectroscopy and Raman spectroscopy methods. Spectral in situ measurements of the test samples at pressures up to 1000 bar and temperatures 25–500 °C were performed with the use of the high-temperature high-pressure cells with transparent sapphire windows. It was experimentally determined a temperature and pressure impact on the infinite clusters of the hydrogen-bound molecules that is inherent to the liquid water. Therewith, the critical isotherm of the liquid water is defined as the conditional boundary of the percolation threshold below which the clusters of finite sizes can exist only. An evaluation of the effect of the electrolytes, dissolved in water, onto the hydrogen bonds and structural features of water of the aqueous solutions as well as determination of the polyatomic anions stability under high temperatures and pressures was also performed. The vibrational spectra of NaCl are not revealed a significant difference between the properties of the aqueous solution and pure water under the test conditions.

Keywords Supercritical water · Fluids · Critical isotherm · Percolation threshold · Hydrogen bonds · Correlation functions · Molecular spectroscopy

2.1 Introduction

Hydrothermal fluid is a necessary participant of the most important geological processes and shows highly active in the formation of the Earth's crust, volcanic activity, transport and concentration of ore-forming components in the deposits. Due to its high mobility and solubility, it makes a decisive contribution to the processes of heat and mass transport in the earth's crust and mantle. In geological terms, hydrothermal

G. V. Bondarenko (✉) · Y. E. Gorbaty
D.S. Korzhinskii Institute of Experimental Mineralogy, Russian Academy of Science,
Academician Osipyan Street 4, Chernogolovka, Moscow Region, Russia 142432
e-mail: bond@iem.ac.ru

Y. Litvin and O. Safonov (eds.), *Advances in Experimental and Genetic Mineralogy*,
Springer Mineralogy, https://doi.org/10.1007/978-3-030-42859-4_2

fluid is a hot, highly compressed water solution of many components, which exists, as a rule, at supercritical temperatures and pressures. The basis of the hydrothermal fluid is supercritical water (scH_2O), which is directly involved in the most important geological processes.

The studies of water under supercritical conditions are also of interest in connection with the possibility of important practical applications in environmentally safe chemical technologies, such as oxidation in supercritical water (Bermejo and Cocero 2006). Indeed, the scH_2O easily dissolves oxygen and organic compounds and facilitates their interaction (Wang et al. 2018). Mixed with other substances, the scH_2O can be used not only for oxidation processes but also for hydrogenation (Fedyaeva et al. 2013), various chemical reactions (Akiya and Savage 2002) to produce nanoparticles of oxidies (Cabanos and Poliakoff 2001; Viswanathan and Gupta 2003) and many other practical applications (Purkarová et al. 2018).

Due to the enormous role of the scH_2O in geological and technological processes, we will try to understand what the physical state of water is at supercritical temperatures and pressures, to summarize the available data on the intermolecular interaction and the nature of near-ordering in liquid and supercritical water, i.e. what is commonly referred to as "structure". On the one hand, due to the specific interaction between water molecules caused by hydrogen bonds, experimental data (especially vibrational spectra) are difficult to interpret. However on the other hand, it is this interaction that can be used as a kind of probe to look inside this complex, continuously changing system. Exactly hydrogen bonds determine the high critical temperature of water.

At normal pressure, we can heat water only to 100 °C. In this range of temperatures, changes in the network of hydrogen bonding that bonds water molecules are noticeable poorly. But at high pressures, we can heat the water to very high temperatures, keeping it liquid or liquid-like. At the same time, radical changes occur in the network of hydrogen bonds, which are well noticeable in vibrational spectra and paired correlation functions. But before we begin to consider these phenomena, we must remember how the substance goes into a supercritical state and what it is.

There is no qualitative difference between gas and liquid. They differ only in a greater or lesser degree of interaction between molecules. Gas and liquid can be differed from each other only when they exist simultaneously. At temperatures and pressures above critical, there are no various phases, and we cannot characterize the supercritical state as a gas or liquid. Thus, the term "fluid" is used to describe the supercritical phase. In fact, when using this term, we mean that the supercritical phase of high density should approach the properties of a liquid. A phase with a lower density at a small supercritical pressure and at a high temperature is close in properties to a gas. Therefore, it is usually to talk about the "liquid-like" and "gas-like" states of the supercritical phase.

Numerous attempts have been made to define the approximate boundaries of the "liquid-like" and "gas-like" supercritical states in the phase diagram, in particular of water. For example, as a possible boundary between the "liquid-like" and "gas-like" states of the supercritical fluid, critical isochore and maximum lines of some thermodynamic parameters were proposed. (such as thermal expansion coefficient, maximums C_p and C_v.). However, the fact that the lines of the maxima of the thermodynamic parameters above the critical point diverge shows that there can be no continuation of the two-phase equilibrium curve to the supercritical region (Gorbaty and Bodarenko 2007). Another assumption shared by many scientists is that supercritical fluid is simply, or at least technically a gas (Poliakoff and King 2001).

Hydrogen bonds exist in supercritical water, but the infinite network of hydrogen bonds disappears. Based on the experimental data, we will try to clarify two important questions: under what conditions does the infinite network of hydrogen bonds disappear and what is the form of the final clusters formed by remaining hydrogen bonds.

2.2 Experimental

The specific feature of the approach to the study of water and systems modeling hydrothermal fluid was the simultaneous application of several methods of investigation of the same object, if it is possible under the same thermodynamic conditions. The X-ray scattering intensity was measured. The data obtained were supplemented by the results of vibrational spectroscopy studies (Infrared Absorption and Raman spectroscopy). For each of the experimental methods, a unique high-temperature high-pressure equipment was created and special methods of measurement and processing of the obtained data were developed.

X-ray scattering intensity measurements were performed by the energy dispersion method (Giessen and Gordon 1968). The advantage of this method is that it significantly simplifies the task of creating a high-temperature high-pressure cell. The description of the cell used in our experiments and the procedure of data processing were discussed in detail (Gorbaty and Okhulkov 1994). At a constant pressure of 1000 bar, paired correlation functions of the studied systems were obtained in the temperature range of 25–500 °C.

Infrared absorption spectra (IR-absorption) were obtained at temperatures of 20–550 °C and pressures up to 1000 bar. To obtain absorption spectra, a high-temperature high-pressure cell with a variable absorption layer thickness was produced (Gorbaty and Bondarenko 1999). The use of such a cell avoids a number of errors associated with the distortion of the optimal geometry of the incident radiation due to the small linear and angular aperture of the cell, the loss of light for reflection and absorption in the windows, the absorption of water vapor and carbon dioxide in the atmosphere, the lens effect. In addition, the measurement method automatically takes into account the own radiation of the cell heated to a high temperature.

A miniature cell with 180° geometry was used for Raman spectroscopy (Gorbaty et al. 2004; Gorbaty 2007). In some cases, it was more convenient to use another cell equipped with an internal vessel to isolate the sample from the pressure transfer medium (Gorbaty and Bondarenko 1995). Raman spectra were obtained at a constant pressure of 1000 bar up to 500 °C. The advantage of the Raman spectroscopy over IR absorption is that the materials most usually used for optical windows, such as sapphire or diamond, are transparent across the entire spectrum. On the other hand, Raman spectroscopy, unlike absorption spectroscopy, is not a quantitative method. It does not work with absolute scattering intensities, except for specially designed experiments. Only the relative intensity of the lines in the spectrum can be measured using an internal or external standard. In addition, the intensity of Raman spectra depends on many factors that are very difficult, or not impossible, to control and correct.

2.3   Results and Discussion

2.3.1   Supercritical State of Water

The specific feature of the water structure is the tetrahedral nearest-ordering, arising due to the hydrogen bonds between the water molecules. Each molecule can be join four nearest molecules by means of hydrogen bonds. The main indicator of the tetrahedral structure is the peak at ~4.5 Å in the paired correlation function (Bernal and Fowler 1933). The peak corresponds to a distance between tops of more or less ideal tetrahedron. The paired correlation functions of water g(r) obtained at temperatures up to 500 °C and a constant pressure of 1000 bar are shown in Fig. 2.1 (Gorbaty and Bondarenko 2007). Only part of the g(r) function is presented to focus on behavior of peak at 4.5 Å. With increasing temperature the peak amplitude decreases and near the critical temperature the peak almost disappears. It is clear that with increasing temperature the number of relatively strong hydrogen bonds decreases, the bonds themselves become longer and more distorted. However, at temperatures above the critical one it reappears again and tends to grow with a further increase of temperature. The same is observed for aqueous solutions (Okhulkov and Gorbaty 2001).

This peak behavior consists in the fact that molecules enveloped by hydrogen bonds tend to unite into aggregates with tetrahedral ordering. An important role in the formation of such aggregates is played by the of effect cooperativity of hydrogen bonds (Barlow et al. 2002; Yukhnevich 1997). This effect consists in increasing the average hydrogen bond energy in the H-bound aggregate with each new molecule attached to it. Therefore, striving to minimize the free energy, the molecules are combined into clusters of finite size. This is the reason for the appearance of a peak at 4.5 Å in the pair correlation function, which indicates a tetrahedral ordering above the critical isotherm.

Calculations of MD (Kalinashev and Churakov 1999) and MC (Krishtal et al. 2001) did not reveal the appearance of tetrahedral ordering at temperature above critical. One of the reason is that these methods cannot yet take into the effect of cooperativity of hydrogen bonds. The results of such calculations are close to the statistical distribution of molecules with different numbers of hydrogen bonds. This

Fig. 2.1 The pair correlation functions of water at constant pressure 1000 bar

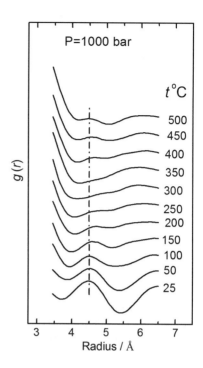

leads to the conviction that, at high supercritical temperature, molecules with four and three hydrogen bonds practically absent.

But the experimental functions of g(r) (Fig. 2.1) show that this conclusion does not reflect the real situation. It is quite clear that it is impossible to explain the behavior of experimental paired correlation functions using theoretical or computational methods. To explain the behavior of experimental paired correlation functions, an imaginary speculative two-dimensional model is proposed (Gorbaty and Bondarenko 2017).

The temperature dependence of probability of hydrogen bond formation P_b constructed from experimental datas, is shown in Fig. 2.2 (Gorbaty and Kalinuchev 1995). The value P_b decreases throughout the studied temperature range, and at critical temperature is 0.34 ± 0.03. This value is quite close to percolation threshold P_c calculated for a diamond-like crystal lattice (Stanley and Teixeira 1980).

This observation gives rise to an assumption that the line of percolation threshold for the network of hydrogen bonding in water is close to a critical isotherm. Above the percolation threshold, the substance is in a liquid state when the molecules are joined into a three-dimensional network, forming an infinite cluster. The infinity of the cluster means that one can always mentally "walk" through the connections from one wall of the vessel to another, no matter how big the vessel. But as the temperature rises, the number of hydrogen bonds decreases, until such a time as any path we choose is inevitably cut off. This means that the system has crossed the percolation threshold, below which there are clusters of only finite sizes. In the

Fig. 2.2 The temperature
dependence of probability of
OH-bonding in water
obtained with different
methods

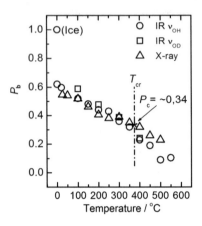

supercritical state, an infinite cluster of hydrogen bound molecules can no exist. The destruction of an infinite cluster leads to the appearance of free (inertial) rotating monomers and water molecules aggregates bound by remaining hydrogen bonds.

There is some evidence of specific water behavior near the critical isotherm. For example, the temperature derivative of the sound velocity in water has minima on the critical isotherm in a wide pressure range (Frank 1987). The isotope distribution coefficient undergoes a jump near the critical temperature (Driesner 1996).

Another example showing the special role of the critical isotherm shows in Fig. 2.3.

Fig. 2.3 The temperature
dependencies of
concentrations for the
component of aqueous
solution of sulfurs

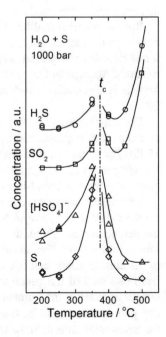

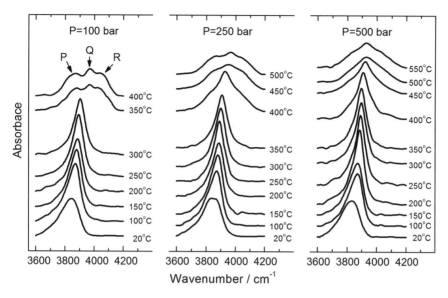

Fig. 2.4 The IR spectra of ($v_2 + v_3$) D_2O at isobaric heating

These data were obtained with the use Raman spectroscopy directly at high temperatures and a constant pressure of 1000 bar (Bondarenko and Gorbaty 1997). A piece of sulphur was placed in an inner sapphire vessel filled with water in a high-temperature high-pressure cell for Raman spectroscopy. Temperature dependences of concentrations of the components of the solution reach their maximum values close to the critical temperature even at a pressure much higher than the critical one.

Free rotation of monomers and aggregates of water molecules, covered by hydrogen bonds, is well visible in the spectra of infrared absorption obtained at temperatures of 20–550 °C for three isobars 100, 250 and 500 bar. Figure 2.4 shows the IR spectra of ($v_2 + v_3$) combination mode of the D_2O molecule. Unbound molecules are involved in free (inertial) rotation, when the molecule has time to make several turns between the collisions. This is manifested in the occurrence of the envelope P-, Q- and R- branches of the vibrational-rotational spectrum. The envelopes P, Q and R of the branches are particularly well visible in the spectra of gas phase at 350 and 400 °C, obtained at a pressure of 100 bar. At higher pressures, rotational branches are poorly resolved, but there is a sharp increase in the bandwidth above critical isotherm (Gorbaty and Bondarenko 1998).

Thus, the infinite cluster of hydrogen-bound molecules inherent in liquid water cannot exist in a supercritical state, but the probability of hydrogen binding is still high. Critical isotherm is a formal boundary separating qualitatively different states of the substance. Near the critical isotherm, the probability of hydrogen binding in the water is close to the percolation threshold.

2.4 Influence of Electrolytes on Hydrothermal Fluid Properties

Water solutions of electrolytes: NaCl, Na_2SO_4, Na_2CO_3, $NaNO_3$, $Zn[NO_3]_2$ and orthophosphoric acid H_3PO_4 were studied by molecular spectroscopy at high temperatures and pressures. The most discussed question is the effect of ions on hydrogen bonded network of water molecules. The effect of some salt on Raman spectra of uncoupled OD vibration mode HDO is shown in Fig. 2.5. The important feature of spectra is a weak high frequency shoulder in the counter of band. This shoulder usually assigned OD to the non-bound or weakly bound groups. The weak shoulder is seen in spectra of HDO, NaCl, Na_2CO_3 and maybe Na_2SO_4. The shoulder turns into a peak in spectrum of $NaClO_4$ solution. From viewpoint of structure-making/breaking it may assume that ion ClO_4^- is structure-weakening (chaotropic) what then the other ions are structure-strengthening (kosmotropic) (Bondarenko and Gorbaty 2011).

Stretching mode of water molecule is very sensitive to the formation of hydrogen bonds. Especially strong is the dependence of the intensity of absorption on the energy of hydrogen bonds. Due to the strong overlap of symmetrical and antisymmetrical stretching modes H_2O, it is more productive to study the modes OD and OH of HDO molecules, which are approximately 1000 cm^{-1} apart.

The most complete results were obtained from the water-NaCl system, which was studied at the same parameters of the state of vibrational spectroscopy (IR and Raman), X-ray diffraction and computer simulation (Bondarenko et al. 2006). The concentration of the electrolyte was chosen so that 100 water molecules per two formulae units of salt (about 1.1 M). It is believed that the nearest environment of ions includes from 4 to 8 water molecules. Thus, the hydrate shell of ions should contain at least 25% of all water molecules. This seems to have a very strong affect

Fig. 2.5 The OD band in the Raman spectra of some aqueous solutions. The concentration of all the salts is 5 mol%

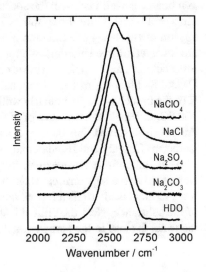

Fig. 2.6 Infrared spectra of HDO/H$_2$O mixture at isobaric heating (**a**), the same of NaCl solution (**b**)

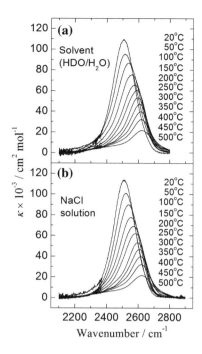

on the extend of coverage of water molecules by hydrogen bonds and have to be well visible in the vibrational spectra. Only a few of the large number of results will be presented here.

The infrared spectra of band of absorption of ν_{OD} HDO for pure solvent (H$_2$O + HDO) and for 1.1 M solution NaCl are not very different (Fig. 2.6). So only a careful analyzed of the shape of observed bands one can help to understand whether ions of dissolved salt influences the character of hydrogen bonding in water. Indeed the integrated intensity (Fig. 2.7a) and position of the band maximum (Fig. 2.7b) for pure solvent and solution, which are very sensitive to the extent of hydrogen bonding are practically in the limit of experimental error. Nevertheless the temperature trends show that most likely hydrogen bonds are stronger in water at low temperatures while at high temperatures they are stronger in solution. The temperature dependence of the band half-width in the infrared spectra of the solution and solvent is shown in Fig. 2.8. The increase in half-width at supercritical temperatures can easily be explained. It has been shown (Gorbaty and Bondarenko 1998) that above the critical temperature free (inertial) rotation of water molecules becomes possible. This leads to an increase half-width of the stretching band with increasing temperature and, at lower density, to the appearance of unauthorized P, Q and R-rotational branches. The maximum half-width observed at 100–150 °C seems to reflect the change in the energy distribution between the different species of the structure. In fact, the maximum half-width is a purely quantitative consequence of the monotonic decreasing of the degree of hydrogen bonding with the increase in temperature.

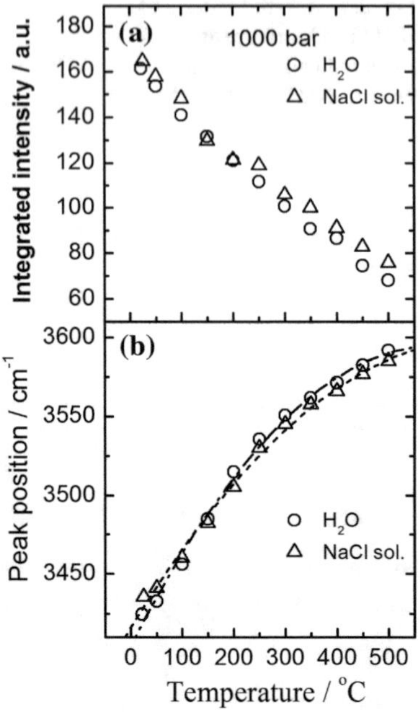

Fig. 2.7 Temperature trend
of the integrated intensity of
absorption for the solution
and solvent at isobaric
heating (**a**). Temperature
trend of the maximum peak
position (**b**)

Fig. 2.8 Temperature trends
of the IR spectral band
half-widths for solvent and
NaCl solution

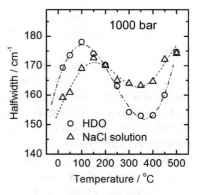

In Raman spectra of NaCl aqueous solution at the same parameters as in the absorption spectra, the quantitative parameters of solution and water spectra practically do not differ. One can only notice that the half-width of the solution spectra in the Raman scattering spectra as well as in the absorption is somewhat smaller than in spectra of the solvent, which probably means a higher degree of orderliness.

The results of the study of vibrational spectra of aqueous NaCl solution show that a large number (not less than a quarter) of all molecules in the nearest environment of ions have practically no or very little influence on the state of hydrogen bonds. There are two possible explanations for this unexpected fact.

1. Energy of interaction of a water molecule, in particular, with chlorine ion is close to energy of normal hydrogen bonding.
2. Chlorine and sodium ions do not form clearly defined hydrate shells, but are simply built into the cavities between the bonded molecules, with little or no effect on the near order in such a matrix.

Water solutions of salts $NaNO_3$, $Zn[NO_3]_2$ and orthophosphoric acid H_3PO_4 were investigated by the method of Raman spectroscopy. These studies were mainly aimed at determining the stability of multi-atomic anions at high temperatures and pressures. All obtained spectra were reduced to the intensity of bending mode of the water, taken as an internal standard, taking into account the change in the density of the solution.

Figure 2.9 shows the behavior of aqueous solution $Zn[NO_3]_2$ at heating up to 450 °C at constant pressure 1000 bars. Narrow intensive line at 980 cm^{-1} belongs to anion $[NO_3]^-$. It can be clearly seen that at a temperature of 350 °C the intensity of the line begins to decrease rapidly. At 450 °C, it disappears completely, which indicates the complete decomposition of the anion. The appearance of a weak line at 1850 cm^{-1} may indicate the formation of NO. The same result was obtained for $NaNO_3$ aqueous solution

Raman spectra of H_3PO_4 aqueous solution with concentration of 2.91 mol% were obtained at a constant pressure of 1000 bar in the temperature range of 25–400 °C. The results are shown in Fig. 2.10. It can be assumed that the spectra change little between 25 and 250 °C, and only at 300 °C does the spectrum change dramatically and especially the scattering intensity. In fact, the composition of the solution changes continuously, starting from the lowest temperatures. It is quite clear that at 300 °C, a new compound appears in the solution, which has not yet been confidently identified, and the concentration of anion $[PO_4]^-$ is greatly reduced. It is known that for ions

Fig. 2.9 Raman spectra of $Zn[NO_3]_2$ aqueous solution at isobaric heating up to 450 °C

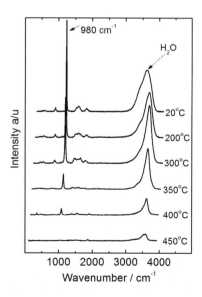

Fig. 2.10 Raman spectra of
H₃PO₄ aqueous solution at
isobaric heating up to 300 °C

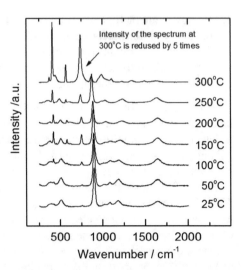

containing phosphorus, there is a strong temperature dependence of the forms of their existence. In addition, the phosphorus-containing compounds in the aqueous solution are more prone to polymerization. All this, along with the fact that phosphates are a valuable mineral raw material and the basis for some semiconductor compounds, makes such an object extremely interesting.

Near the critical water temperature, the solution content falls into the sediment. The solution becomes opaque and it is no longer possible to obtain the spectrum at 400 °C. A microscopic study of the sediment deposited on a sapphire window shows that it is mainly composed of rounded particles and needle crystal fragments. This is likely to indicate that, at high temperatures, not one but several compounds are formed in the solution. At supercritical temperatures, the solubility of phosphorus-containing compounds in the hydrothermal fluid decreases dramatically.

Acknowledgements The study was support by the draft AAAA-A18-118020590149-0

References

Akiya N, Savage PE (2002) Roles of water for chemical reaction in high temperature water. Chem Rev 102(8):2725–2750

Barlow SJ, Bondarenko GV, Gorbaty YE, Yamaguchi T, Poliakoff M (2002) An IR study of hydrogen bonding in liquid and supercritical alcohols. J Phys Chem A 106(43):10452–10460

Bermejo MD, Cocero MJ (2006) Supercritical water oxidation: a technical review. AIChE J 52(11):3933–3951

Bernal J, Fowler RH (1933) A theory of water and ionic solution with particular reference to hydrogen and hydroxyl ions. Chem Phys 1:515–548

Bondarenko GV, Gorbaty YE (1997) In situ Raman spectroscopic study of sulphur-saturated water at 1000 bar between 200 and 500 °C. Geochim et Cosmochim Acta 61(7):1413–1420

Bondarenko GV, Gorbaty YE (2011) Hydrogen bonding in aqueous solution of NaClO₄. Mol Phys 109(5):783–788

Bondarenko GV, Gorbaty YE, Okhulkov AV, Kalinichev AG (2006) Structure and hydrogen bonding in liquid and supercritical aqueous NaCl solutions at a pressure of 1000 bar and temperature up to 500 C: a comprehensive experimental and computational study. J Phys Chem A 110(11):4042–4052

Cabanas A, Poliakoff M (2001) The continuous hydrothermal synthesis of nano-particulate ferrites in near critical and supercritical water. J Mater Chem 11:1408–1416

Driesner T (1996) The effect of pressure on deuterium-hydrogen fractionation in high temperature. Science 277:791–794

Fedyaeva ON, Vostricov AA, Sokol MY, Fedorova NI (2013) Hydrogenezation of bitumen in supercritical water flow and the effect of zinc addition. Russ J Phys Chem B7(7):820–828

Franck EU (1987) Fuids at high pressure and temperature. Pure Appl Chem 59(1):25–29

Giessen BC, Gordon GE (1968) X-ray diffraction new high-speed technique based on X-ray spectrography. Science 159(3818):973–975

Gorbaty YE (2007) Spectroscopic techniques for studying liquids and supercritical fluids at high temperature and pressure. Sverhcriticheskie fluidy Teoriya Pract 2(1):40–53

Gorbaty YE, Bodarenko GV (1995) High-pressure high-temperature Raman cell for corrosive liquids. Rev Sci Instrum 66(8):4347–4349

Gorbaty YE, Bodarenko GV (1999) Experimental technique for quantitative IR studies of highly absorbing substances at high temperature and pressure. Appl Spectroscopy 53(8):908–913

Gorbaty YE, Bodarenko GV (2007) Water in supercritical state. Sverhcriticheskie fluidy Teoriya Pract 2(2):5–18

Gorbaty YE, Bodarenko GV (2017) Transition of liquid water to the supercritical state. J Mol Liq 239:5–9

Gorbaty YE, Bondarenko GV (1998) The physical state of supercritical fluids. J Supercrit Fluids 14:1–8

Gorbaty YE, Kalinichev AG (1995) Hydrogen bonding in supercritical water. 1. Experimental results. J Phys Chem 99(15):5336–5340

Gorbaty YE, Okhulkov AV (1994) High-pressure x-ray cell for studying the structure of fluids with the energy-dispersive technique. Rev Sci Instrum 65(7):2195–2198

Gorbaty YE, Bodarenko GV Venardow E, Barlow S, Garsia-Verdugo E, Poliakoff M (2004) Experimental spectroscopic high-temperature high-pressure techniques for studying liquids and supercritical fluids. Vibr Spectrosc 35:97–101

Kalinashev AG, Churakov SV (1999) Size and topology of molecular clusters in supercritical water: a molecular dynamic simulation. Chem Phys Lett 302(5–6):411–417

Krishtal S, Kiselev M, Puhovski Y, Kerdcharoen T, Hannongbua S, Heinzinger K (2001) Study of hydrogen bond network in sub- and supercritical water by molecular dynamics simulation. Z Naturforsch 56a:579–584

Okhulkov AV, Gorbaty YE (2001) The pair correlation functions of 1.1 M NaCl aqueous solution at constant pressure of 1000 bar in the temperature range 20–500 °C. J Mol Liq 93:39–42

Poliakoff M, King P (2001). Phenomenal fluids. Nature 412:125–125

Purcarova E, Ciahotny K, Svab M, Skoblia S, Beno Z (2018) Supercritical water gasification of wastes from the paper industry. J Supercrit Fluids 135:130–136

Stanley HE, Teixeira J (1980) Interpretation of the unusual behavior of H_2O and D_2O at temperatures: test of a percolation model. J Chem Phys 73(7):3404–3422

Viswanathan R, Gupta RB (2003) Formation of zinc oxide nanoparticles in supercritical water. J Supercrit Fluids 27(2):187–193

Wang J, Zhang Y, Zheng W, Chou I-M, Pan Z (2018) Using Raman spectroscopy and a fused quartz tube reactor to study the oxidation of o-dichlorobenzene in hot compressed water. J Supercrit Fluids 140:380–386

Yukhnevich GV (1997) The mechanism of occurrence of cooperative properties in conjugate hydrogen-bonds. Spectrs Lett 30(5):901–914

Chapter 3
Influence of Silicate Substance on Pyrochlore and Tantalite Solubility in Fluoride Aqueous Solutions (Experimental Studies)

A. R. Kotelnikov, V. S. Korzhinskaya, Z. A. Kotelnikova, and N. I. Suk

Abstract Experimental results of behavior of natural minerals of pyrochlore and tantalite in solutions KF, NaF and LiF in the presence of quartz (granite) at T = 550°–850 °C, P = 1 kbar are presented. The considerable influence of silicate substance presence on mineral solubility in water solutions of alkali metal fluorides in hydrothermal conditions is shown. The study of fluid inclusions in quartz showed that under experimental conditions (heterogeneous state of the fluid) the reactions of high-temperature hydrolysis KF: $KF + H_2O = KOH + HF$, with the separation of acid and alkaline components between the immiscible phases of the fluid are intensive. In this case, the interaction of alkaline components with quartz occurs: $SiO_2 + 2KOH = K_2SiO_3 + H_2O$, with the formation of a phase of silicate glass (aqueous solution-melt). This phase of silicate alkaline melt enriched with fluorine concentrates tantalum and niobium (up to 8% of Ta_2O_5 and 16% of Nb_2O_5) and can serve as a phase—a concentrator of ore elements in the formation of tantalum-niobate deposits at the last low-temperature stages of crystallization of rare-metal granites.

Keywords Niobium-tantalum deposits · Experimental modeling · Hydrothermal fluoride fluid · Pyrochlore solubility · Tantalite solubility · Alcaline-silicate melt-concentrator

3.1 Introduction

In the 50s of the last century in our country a new type of rare-metal deposits of Ta, Nb, Be, W, Sn, timed to the domes of albitized and greisenized granite was discovered and studied in detail (Beus et al. 1962). Typical representatives of such fields are

A. R. Kotelnikov (✉) · V. S. Korzhinskaya · N. I. Suk
D.S. Korzhinskii Institute of Experimental Mineralogy, Russian Academy of Science
Academician, Osipyan Street 4, Chernogolovka, Moscow Region, Russia
e-mail: kotelnik@iem.ac.ru

Z. A. Kotelnikova
Institute of Geology of Ore Deposits, Petrography, Mineralogy, and Geochemistry, Russian
Academy of Sciences, Staromonetny St., 35, Moscow, Russia

© The Editor(s) (if applicable) and The Author(s), under exclusive license to Springer
Nature Switzerland AG 2020
Y. Litvin and O. Safonov (eds.), *Advances in Experimental and Genetic Mineralogy*,
Springer Mineralogy, https://doi.org/10.1007/978-3-030-42859-4_3

Orlovskoye and Etykinskoye fields of Ta, Nb, which are located in the Eastern Trans-baikalia. The development of the zones of albitization and greisenization, contain-ing own ore minerals, took place on specific lithium-fluorine granites. This term for lithium and fluorine enriched granite rocks was first introduced by Kovalenko (1977). These rocks contain initially magmatic lithium- and fluorine-containing minerals—lithium mica and topaz. Discovery of this type of deposits activated the search for similar objects in other parts of our planet. These include Voznesenskoye, Pogranich-noye, Snezhnoye fields in Russia and a number of similar fields in China, South-East Asia, Egypt and France. Lithium-fluorine granites and their subvolcanic analogues—ongonites—are more frequent. As a rule, all ongonites are enriched with Ta, Nb, Li, Rb, Cs, Be, Sn, W. Almost all lithium-fluorine granites produce tantalum and tanta-lum mineralization ore manifestations. Lithium-fluorine ore-bearing granites belong to the alkaline earth granitoid series. According to morphology, they are small dome-shaped bodies of 0.5–2 km in diameter ("stems"), formed in approximately equal proportions of quartz, albite and microcline (often amazonite), with a small number of lithium mica (lepidolite, tsinnvaldite) and topaz. Ore bodies are located in the upper parts of granite domes, usually covered by keratinized crystalline shale. The industrial mineralization is concentrated in the albitized and greisenized granites. The ores are relatively poor (100–350 g/t), and Ta_2O_5 reserves usually do not exceed 10–15 thousand tons (Zarayskiy 2004).

3.1.1   Magmatic Stage of Formation of Rare-Metal Deposits

All deposits of tantalum-niobium ores in lithium-fluorine granites are located near large biotite and leucocratic massifs and have a genetic connection with them. The initial melt of initially granite composition usually appeared under post-threshold activation conditions of stretching. Deep troughs were laid at the periphery of the previously consolidated collision areas. The source of such granite melts could be the anatectic or paling melting of the granitogneissic continental crust near the Conrad boundary. That is, the upper crust in the path structures was lowered to the area of high temperatures (up to the depth of 15–20 km), where the rocks were heated up. Additional heating occurred due to mantle fluids coming from deep faults in zones of tectonic-magmatic activation (Zarayskiy 2004). The granite melt rose along the deep faults and was introduced in the form of laccolithic intrusions. Laccolite and lenticular bodies of granites are located in a subhorizontal orientation on the boundary of the crystalline rocks of the pre-Cambrian basement and the Phanerozoic cover. Their sizes reach 20–30 km with a capacity of up to 5–8 km in their central parts (Zarayskiy 2004, 2009). Further evolution of melts at their crystallization differentiation resulted in formation of residual melts. These melts corresponded to the compositions of lithium-fluorine granites and were separated from the mother granites in their upper part, squeezing out the cracks. Lithium-fluorine granites represent the latest portions of melt introduction and form small bodies of 0.5–2 km in size and thickness up to 0.5 km. Mass ratios of mother and lithium—fluorine granite are about 500 ÷ 100:

1. According to experimental data (Chevychelov et al. 2005; Borodulin et al. 2009) partition coefficients in the system of lithium-fluorine granite melt—fluid is about $150 \div 350$ in favor of the melt (at 800–650 °C). Therefore, tantalum and niobium remain in the melt and crystallize in the form of micron inclusions of tantalum-niobates along the grain boundaries of rock-forming minerals when separating the fluid. Calculations of the model of crystallization of the ancestral Hangilaisky pluton (Zaraiskiy 2004) show that to enrich lithium-fluorine granites with tantalum up to 30–50 g/t it is necessary to crystallize 95% of the initial granite melt. It is difficult to expect complete compression and crack segregation of the entire volume of Li-F granites with a high degree of crystallization. Therefore, it is practically impossible to achieve ore concentrations of tantalum by direct crystallization fractionation.

3.1.2 Hydrothermal-Metasomatic Stage of Formation of Rare-Metal Deposits

Accumulation of such rare metals as Be, Sn, W, Mo in granite melts at fractional differentiation is much lower than their saturation concentrations. Therefore, they also cannot crystallize in the form of independent mineral isolations and dissipate in large volumes of crystallizing granites. However, unlike tantalum and niobium, they have significantly lower melt/fluid partition coefficients and can be converted into a separating fluid in significant amounts. According to a number of researchers (Chevychelov 1998; Rjabchikov 1975; Gramenitskiy et al. 2005), these coefficients are $\sim 2 \div 5$, and for molybdenum and tungsten up to 1. Therefore, the formation of deposits of these metals can occur hydrothermally-metasomatically.

For tantalum and niobium there is no possibility of extraction by the separating fluid because of the high affinity of Nb and Ta to melts. Therefore, the question of concentration and re-deposition of the scattered ore substance of tantalum-niobates at the hydrothermal stage of evolution is acute. It is known (Rub and Rub 2006; Aksyuk 2002, 2009) that lithium-fluorine granites producing tantalum-niobium ore deposits have potassium specifics. According to the data (Aksyuk 2002) obtained on the basis of the experimentally developed mica geofluorometer, at the transient magmatic-hydrothermal stage of evolution of ore-magmatic systems the concentration of HF in water fluids separating from the granite melt is about 0.01 mol/kg H_2O for copper-molybdenum porphyry deposits. It is an order of magnitude higher at Greisen deposits of W, Mo, Sn, Be, Bi of Akchatau type (0.1 mol/kg H_2O) and reaches the maximum values of 1.0 mol/kg H_2O at rare-metal deposits of tantalum in lithium-fluorine "apogranites" of the Orlovsky and Etykinsky rare-metal massifs. Thus, fluorine content in hydrothermal solutions can reach 1 mol/kg. It is clear that fluorine is contained in a postmagmatic fluid in the form of fluoride salts (LiF, NaF, KF). Moreover, taking into account the potassium specificity of rare-metal lithium-fluorine granites, potassium fluoride (KF) solutions will dominate. This fact is confirmed by the study of water extracts from lithium-fluorine granite rocks (Aksyuk 2009), which showed high content of potassium fluoride. In addition, the paper (Badanina et al. 2010) suggested that the final concentration of tantalum and niobium is carried out at the postmagmatic stage by a high concentration salt solution containing significant amounts of fluorine. Earlier it was shown that fluorine-containing salt fluid actively interacts with silicate substance (quartz), forming a phase of silicate glass (Kotelnikova and Kotelnikov 2002, 2008). Therefore, it was interesting to study the transport of tantalum and niobium in saline fluoride solutions in the presence of silicate substance.

We have carried out experiments on the solubility of minerals containing niobium and tantalum (pyrochlore and tantalite) in hydrothermal solutions of LiF, NaF, KF in the temperature range of 550–850 °C and pressures of 0.5–1 kbar. To study the influence of the presence of silicate substance (quartz or granite melt) on the solubility of pyrochlore and tantalite in fluoride solutions, experiments were carried out with the addition of quartz (granite) and without silicate substance.

3.2 Experimental Method

3.2.1 Starting Materials

As initial materials we used quartz (Perekatnoye deposit, Aldan), amorphous SiO_2; pyrochlore $(Ca, Na)_2(Nb, Ta)_2O_6(O, OH, F)$ from weathering zone of carbonatite deposit of Tatarka of following composition (wt%): Na_2O—7.61; CaO—14.28; Nb_2O_5—71.61; F—5.18; TiO_2—0.83; $Ta_2O_5 \leq 1$ (conversion to 4 cations including charge balance): $(Na_{0.92}Ca_{0.95}Sr_{0.06})_{1.93}(Ti_{0.04}Nb_{2.02})_{2.06}O_6[F_{1.02}(OH)_{0.18}]_{1.20}$ and tantalite $(Mn, Fe)(Ta, Nb)_2O_6$ from the Vishnyakovsky tantalum deposit in the rare-metal pegmatites of the East Sayan belt, which has a composition (wt%): Ta_2O_5—57.98; Nb_2O_5—21.93; MnO—15.59; WO_3—4.16. Recalculation of composition (by 6 atoms O) has shown the following formula: $Mn_{0.984}(Nb_{0.738}Ta_{1.172}W_{0.08})_{1.990}O_6$. For experiments with the presence of silicate substance we used leukocrat granite of Orlovka deposit of following composition (wt%): SiO_2—72.10; TiO_2—0.01; Al_2O_3—16.14; Fe_2O_3—0.68; MnO—0.09; CaO—0.30; MgO—0.01; Na_2O—5.17; K_2O—4.28; P_2O_5—0.02; F—0.32; H_2O—0.18.

The investigated area of physico-chemical parameters to study the solubility of these minerals was 550–850 °C at a pressure of 1 kbar. Concentrations of initial fluoride solutions of KF, NaF and LiF ranged from 0.08 to 1M, which were prepared from the corresponding high purity reagents on the basis of bidistilled water. It should be noted that the selected range corresponds to the real range of fluoride concentrations in natural postmagmatic fluids in deposits associated with granites.

In order to study the influence of the presence of silicate substance (quartz or granite melt) on the solubility of pyrochlore and tantalite in fluoride solutions, the composition of fluids was determined by solutions with concentrations: LiF (0.08M); NaF (1M); KF (0.5 and 1M).

The experiments were carried out on a high gas pressure vessel (HGPV-10000) of IEM design (at T > 650 °C). HGPV is characterized by high accuracy of setting and determination of temperature and pressure, high working volume, high productivity and the possibility of long-term experiments, which makes it indispensable in the study of magmatic processes. The heater with tungsten-rhenium or molybdenum wire as a heating element allows to conduct experiments at temperatures up to 1400 °C. Temperature was regulated and controlled by Pt-PtRh10 thermocouples with accuracy of ±2 °C, pressure up to 6 kbar was measured by a spring pressure gauge with an error of ±1%. The heater was calibrated by the melting point of gold. The gradient-free zone is 50 mm, the operating diameter of the furnace is 15 mm. Part of the experiments in the temperature range of 550–650 °C was conducted on a hydrothermal plant with a cold gate and external heating (IEM RAS design). Temperature control accuracy was ±5 °C, pressure ±50 bar.

3.2.2 Methodology of Experiments

Experiments were conducted using the ampoule method. The solubility of pyrochlore and tantalite was estimated by the composition of quenching solutions. Quantitative analysis of the solution required a sufficiently large amount of fluid, so the experiments were carried out in platinum ampoules with a diameter of 7 mm and a volume of about 2 mL. Ampoules were loaded with the pyrochlore or tantalite crystals and the solution was poured. For experiments with silicate matter, a quartz crystal or granite powder was added, as well as amorphous silica. The method of synthetic fluid inclusions was used to estimate the phase state of the fluid. For the synthesis of fluid inclusions (by healing cracks) we used prisms sawn from natural quartz crystals (Perekatnoye deposit, Aldan). Quartz prisms were subjected to a thermal shock by heating them up to 250–300 °C, and then discharging them into alcohol, thus forming a network of cracks in them. After drying in a drying closet and ignition at 1100 °C in a muffle furnace, a quartz crystal was placed in a platinum ampoule. In addition, amorphous silica was added for better healing of cracks in quartz and poured with the selected fluoride solution. Then the ampoule was hermetically sealed and weighed. After run (time of quenching was 2–5 min) the ampoules were weighed to check the tightness. Experiments on healing cracks in quartz were carried out both in the presence of ore minerals (pyrochlore and tantalite) and without them.

3.2.3 Methods of Analysis

After the experiment, the hardening solution was analyzed by ICP/MS and ICP/AES (mass spectra and atomic emission) methods on a number of elements (Nb, Ta, Na, Ca, Mn, Fe, Ti, W, etc.). For the analysis of solutions experimentally obtained, 0.2 mL of concentrated HCl (37% GR, ISO, Merck) and an internal standard (Cs) were added to the original aliquot. The solution was then diluted to 10 mL with water. This method was described in a previous article (Zarayskiy et al. 2010). No precipitation was observed in the solutions. This is due to the fact that transition metals such as Ta and Nb are not deposited at low pH values. The time between solution preparation and measurement did not exceed 4-8 h. The concentrations of elements in the solutions were determined by the quantitative method using standard solutions containing 1–500 μg/L of Nb, Ta, Mn, Ti, W and Sn and 100–10 000 μg/L of Mn, Fe and Ti. Error in element concentration depends on the setting spectrometer, purity of input system, conical state of mass spectrometer, etc. The content of elements was determined by atomic emission analysis (AES) for rock-forming elements (Mn, Fe) and mass spectral (MS) for impurity elements (Nb, Na). For standard solutions, the error of element concentration according to Mn (MS; AES) was 0.7 μg/L; Fe (AES)—9 μg/L; Nb (MC)—0.4 μg/L; and Ta (MC)—0.02 μg/L. After experience, the pH value of the quenching solution was measured.

Compositions of solid products after experiments were determined by the method of electron-probe X-ray spectra analysis using scanning electron microscope Tescan Vega II XMU (Tescan, Czech Republic), equipped with INCA Energy 450 X-ray microanalysis system with energy dispersive (INCAx-sight) and crystal diffraction (INCA wave 700) X-ray spectrometers (Oxford Instruments, England) and INCA Energy + software platform. The analysis conditions when using only the energy dispersive spectrometer were as follows: accelerating the voltage of 20 kV, the current of absorbed electrons at Co 0.3 nA, the analysis time at 70 s. When using the crystal diffraction spectrometer together with the energy dispersive conditions of the analysis were different: accelerating the voltage of 20 kV, the current of absorbed electrons at Co 20 nA, the total time of analysis at 170 s. The accuracy of quantitative X-ray analysis using an energy dispersive spectrometer is comparable to that of a crystal diffraction spectrometer with elemental content above 1 wt% (Reed 2005). The content of the elements to be determined in the samples under study in most cases exceeded 1 wt%. When using an energy dispersive detector to record X-ray radiation, the detection limits of the detectable elements are in the range of 0.1–1 wt%; when detecting elements with a crystal diffraction spectrometer, the detection limits of the elements are in the range of 0.02–0.7 wt%. The accuracy of the determination is 0.2 wt% of the element when using an energy dispersive spectrometer and 0.05 wt% of the element when using a crystal diffraction spectrometer. To prevent sodium losses (especially from water-bearing phases), the scanning mode was used: the analysis was performed from the area of 10×10 μ and more.

The samples were polished and sprayed with a thin conductive layer of carbon. Shooting of microphotographs was carried out in back-scattered electron mode with real contrast (BSE—back-scattered electrons) with magnifications from 8.5 to 2500 times.

3.3 The Results of the Experiments

3.3.1 Fluid Phase State

Data on the phase state of the fluid were obtained by studying synthetic fluid inclusions in quartz. If the fluid was homogeneous at the experimental parameters, all inclusions had the same phase composition and the same values at thermo- and cryometry. In the case of heterogeneous fluid, different inclusions capture portions of immiscible phases or mechanical mixtures of these phases, as a result of which two or more types of inclusions with different thermobarogeochemical characteristics are formed.

In experiments with LiF solution (concentration 0.08M) at the pressure of 1 kbar and temperatures of 550, 650 °C the fluid existed in heterogeneous state: two-phase gas + liquid (G + L) and three-phase gas + liquid + crystal (G + L+S) inclusions were found in the samples.

In experiments with NaF solution (concentration of 1M), unlike LiF solutions, at pressures of 0.5, 1 kbar and temperature of 550 °C there was a homogeneous fluid: only two-phase inclusions of G + L were found. With increasing temperature up to 650 °C, the interaction of the fluid with solid phases resulted in the deposition of a small number of solid phases. The course of homogenization of two-phase inclusions is near-critical, which testifies to the proximity of TPX parameters of the experiment to the critical point of the system.

In experiments with KF solution (concentrations of 0.5M and 1M) for 550 °C and 1 kbar at the starting solution concentration of 2.9 wt% (0.5M) the fluid is homogeneous: only G + L inclusions were found. If under the same PT conditions the concentration of the initial solution is increased up to 1M (5.8 wt%), the fluid is heterogenized and different types of inclusions are formed. By results of study of fluid inclusions in quartz it is possible to conclude that at parameters of experiments (550, 650, 750 and 850 °C and pressure of 1 kbar) 1M (5.8 wt%) solutions of potassium fluoride stratified on vapor (low concentrated) phase and a phase of a salt fluid (highly concentrated phase). At the temperature of 650 °C and higher (at P = 1 kbar) the phase of silicate glass is found in inclusions.

3.3.2 Solubility of Pyrochlore and Tantalite

Data on the solubility of natural minerals of pyrochlore and tantalite in the presence of quartz (granite) and without it are presented in Tables 3.1, 3.2 and Figs. 3.1, 3.2, 3.3, 3.4.

It has been experimentally established that at T = 550 °C and P = 1 kbar the presence of quartz at dissolution of pyrochlore mineral significantly reduces the content of Nb in KF solutions by more than 3 orders of magnitude; in 1M NaF—by half an order and in 0.08M LiF—quite insignificantly (Fig. 3.1) (Kotelnikov et al.

Table 3.1 Conditions and results of experiments on the solubility of pyrochlore in alkali metal fluoride solutions (T = 550–850 °C, P = 1 kbar)

Sample No.	Hitch	Initial solution	pH before run	pH after run	Phases	Composition of solution after run		
						Nb, mg/L	Na, mg/L	Ca, mg/L
550 °C								
Px239-1	46 mg *Pchl*	1745 mcl 0.5M KF	6.72	7.54	*Pchl*	17.5	112	2.3
Px239-2	62 mg *Pchl* + 367 mg *Qtz* + 92 mg amf. SiO$_2$	1777 mcl 0.5M KF	6.72	7.48	*Pchl* + *Qtz*	0.042	151	1.9
Px240-1	145 mg *Pchl*	1953 mcl 1M KF	7.32	8.42	*Pchl*	45.4	252	2.2
Px240-2	31 mg *Pchl* + 499 mg *Qtz* + 102 mg amf. SiO$_2$	1805 mcl 1M KF	7.32	8.10	*Pchl* + *Qtz*	0.045	290	2.4
Px232-1	32 mg *Pchl*	1983 mcl 0.08M LiF	6.31	6.25	*Pchl*	0.10	22.8	<DL
Px232-2	27 mg *Pchl* + 121 mg *Qtz* + 31 mg amf. SiO$_2$	1942 mcl 0.08M LiF	6.31	5.22	*Pchl* + *Qtz* + *Sil-gl*	0.16	15.9	<DL
Px231-1	49 mg *Pchl*	2088 mcl 1M NaF	6.33	7.21	*Pchl*	0.3	14,944	1.2
Px231-2	39 mg *Pchl* + 255 mg *Qtz* + 62 mg amf. SiO$_2$	1804 mcl 1M NaF	6.33	7.27	*Pchl* + *Qtz*	0.12	15,857	<DL
650 °C								
Px251-1	259 mg *Pchl*	693 mcl 1M KF	7.32	9.93	*Pchl*	2.7	1754	1
Px251-2	275 mg *Pchl* + 163 mg *Qtz* + 15 mg amf. SiO$_2$	699 mcl 1M KF	7.32	6.97	*Pchl* + *Qtz* + *Sil-gl*	14.8	2080	1.1
Px252-2	238 mg *Pchl* + 365 mg *Qtz* + 21 mg amf. SiO$_2$	543 mcl 1M NaF	6.33	8.25	*Pchl* + *Qtz* + *Sil-gl*	2.6	10,955	9.3
Px253-2	247 mg *Pchl* + 235 mg *Qtz* + 15 mg amf. SiO$_2$	601 mcl 0.08M LiF	6.31	–	*Pchl* + *Qtz*	1.5	907	7.9

(continued)

Table 3.1 (continued)

Sample No.	Hitch	Initial solution	pH before run	pH after run	Phases	Composition of solution after run		
						Nb, mg/L	Na, mg/L	Ca, mg/L
850 °C								
Px250-1	129 mg *Pchl*	638 mcl 1M KF	7.32	10.2	*Pchl*	22	3068	<DL
Px250-2	74 mg *Pchl* + 157 mg *Qtz* + 25 mg amf. SiO$_2$	601 mcl 1M KF	7.32	1.90	*Pchl + Qtz + Sil-gl*	222	189	22
Px255-2	75 mg *Pchl* + 157 mg granite + 22 mg amf. SiO$_2$	423 mcl 1M KF	7.45	6.23	*Pchl + Qtz + AlSil-gl*	166.8	442	54.2
Px256-2	119 mg *Pchl* + 167 mg *Qtz* + 17 mg amf. SiO$_2$	734 mcl 1M NaF	6.33	1.80	*Pchl + Qtz + Sil-gl*	175	1801	4

<DL—less than detection limit

2018a, b). At the same time pyrochlore solubility in KF solutions increases at the increase of concentration from 0.5 to 1M by 0.6 order of magnitude, in similar experiments in the presence of quartz this effect is not observed.

For temperatures of 650° and 850 °C, the presence of quartz in the system increases the equilibrium content of Nb in the solution by an order of magnitude: at 650 °C, the niobium content in 1M KF solution is without quartz—$2.91 * 10^{-5}$, and in the presence of quartz—$1.59 * 10^{-4}$ mol/kg H$_2$O; at 850 °C the content of Nb is without quartz—$2.38 * 10^{-4}$, and with quartz (or with granite melt)—$2.39 * 10^{-3}$ mol/kg H$_2$O (Fig. 3.2). Measurement of pH of solutions (Tables 3.1 and 3.2) before and after the experiments showed that for 550 °C the initial solution of 1M KF after the experiment is alkalized, which reduces the solubility of pyrochlore; for 650 °C—pH is shifted to the acidic area, but only by half an order; for 850 °C—pH after the experiment is shifted to the acidic area significantly: (before the experiment pH = 7.32 and after experiment pH = 1.901). It can be assumed that due to the acid-base interaction at T $\geq$ 650 °C, and especially at 850 °C, the concentration of HF in the fluid increases significantly. Note that at a temperature of $\geq$650 °C and P = 1 kbar the fluid is heterogenized to the high-salt and vapor phases. The effect of acidification of solutions, apparently, is associated with the formation of glass phase during the experiments and significant redistribution of potassium into the glass and high-salt phases. On the Fig. 3.3 the temperature dependence of equilibrium content of tantalum (Fig. 3.3a) and niobium (Fig. 3.3b) at dissolution of tantalum in 1M KF and 1M NaF depending on presence of quartz (granite melt) is presented.

Study of fluid inclusions in quartz, study of phase composition of experiments on pyrochlore and tantalite solubility showed that under the conditions of experiments

Table 3.2 Conditions and results of experiments on the solubility of tantalite in solutions of alkali metal fluorides (T = 550–850 °C, P = 1 kbar)

Sample No.	Hitch	Initial solution	pH before run	pH after run	Phases	Nb, mg/L	Ta, mg/L	Mn, mg/L	W, mg/L
550 °C									
KT104-1	34 mg *Tnt*	789 mcl 1M KF	7.32	10.5	*Tnt*	4.0	0.8	0.47	9.7
KT104-2	28 mg *Tnt* + 406 mg *Qtz* + 40 mg amf. SiO_2	789 mcl 1M KF	7.32	8.5	*Tnt + Qtz*	0.2	0.022	1.7	19.4
KT105-1	20 mg *Tnt*	518 mcl 1M NaF	6.33	–	*Tnt*	0.23	0.18	0.10	1.8
KT105-2	33 mg *Tnt* + 366 mg *Qtz* + 35 mg amf. SiO_2	544 mcl 1M NaF	6.33	8.0	*Tnt + Qtz*	0.89	0.29	4.3	1.6
650 °C									
KT109-1	22 mg *Tnt*	627 mcl 1M KF	7.32	10.3	*Tnt*	9	1.7	32.4	14.8
KT109-2	31 mg *Tnt* + 67 mg *Qtz* + 21 mg amf. SiO_2	645 mcl 1M KF	7.32	7	*Tnt + Qtz + Sil.gl*	4.4	0.04	9.3	16.9
KT111-1	32 mg *Tnt*	508 mcl 1M NaF	6.33	8.5	*Tnt*	0.54	0.042	6.8	1.3
KT111-2	58 mg *Tnt* + 117 mg *Qtz* + 12 mg amf. SiO_2	522 mcl 1M NaF	6.33	7	*Tnt + Qtz + Sil.gl*	2.0	0.018	6	0.72

(continued)

Table 3.2 (continued)

Sample No.	Hitch	Initial solution	pH before run	pH after run	Phases	Nb, mg/L	Ta, mg/L	Mn, mg/L	W, mg/L
750 °C									
263-2	86 mg *Tnt* + 350 mg *Qtz* + 88 mg amf. SiO$_2$	769mcl 1M KF	7.32	6	*Tnt* + *Qtz* + *Sil.gl*	55.6	0.16	66.5	42.5
265-1	146 mg *Tnt*	523 mcl 1M NaF	6.33	≤10	*Tnt*	0.63	0.96	2.1	8.3
265-2	97 mg *Tnt* + 148 mg *Qtz* + 14 mg amf. SiO$_2$	501 mcl 1M NaF	6.33	≈5	*Tnt* + *Qtz* + *Sil.gl*	82.7	0.038	7.1	23.7
850 °C									
KT107-1	59 mg *Tnt*	648 mcl 1M KF	7.32	7	*Tnt*	35.8	4.7	15.5	82.1
KT108-2	71 mg *Tnt* + 152 mg granite + 24 mg amf. SiO$_2$	538 mcl 1M KF	7.32	5.5	*Tnt* + *Qtz* + *Sil.gl*	53.0	0.93	21.5	4.2

A. R. Kotelnikov et al.

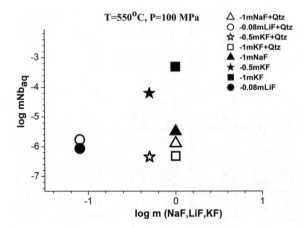

Fig. 3.1 The concentration dependence of the equilibrium content of niobium when pyrochlore dissolves in solutions of 0.08M LiF; 0.5 and 1M NaF; 0.5 and 1M KF at 550 °C and P = 1 kbar (in the presence of quartz and without it)

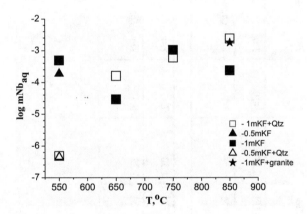

Fig. 3.2 Temperature dependence of the equilibrium niobium content when pyrochlore dissolves in 0.5 and 1M KF solutions at 550, 650, and 850 °C and P = 1 kbar (in the presence of quartz and without it)

the reactions of high-temperature hydrolysis of KF occur: $KF + H_2O = KOH\downarrow + HF\uparrow$, the interaction with quartz occurs: $SiO_2 + 2KOH = K_2SiO_3 + H_2O$, with the formation of a silicate glass phase (aqueous melt-solution) at $T \geq 650$ °C and P = 1 kbar. This phase of alkaline glass is a concentrator of Ta and Nb (Table 3.3). Contents of tantalum and niobium (Ta_2O_5 and Nb_2O_5) in the granite melt reach 6 wt% (850 °C). When dissolving pyrochlore in 1M KF ($T \geq 650$ °C) in the presence of quartz, a silicate melt with a content of Nb_2O_5 up to 8 wt% is formed.

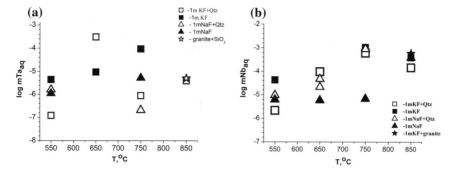

Fig. 3.3 Temperature dependence of the equilibrium content of tantalum (**a**) and niobium (**b**) when tantalite is dissolved in 1M KF and 1M NaF solutions, depending on the presence of quartz (granite). Filled shapes—without quartz; not shaded—with quartz; asterisk—1M KF + granite + SiO_2 amorphous

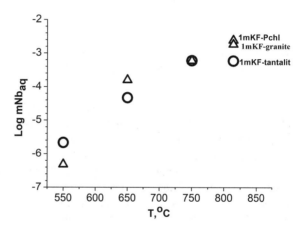

Fig. 3.4 Comparison of data on the content of niobium in solution in the presence of quartz by dissolving natural pyrochlore minerals (triangles) and tantalite (circles) in 1M KF solution

On the Fig. 3.4 the data of niobium content in solution in the presence of quartz at dissolution of natural minerals of pyrochlore and tantalite in 1M KF are compared. As can be seen, the dependences of the niobium content in the solution for both minerals are of the same type.

3.4 Discussion of Results

3.4.1 Pyrochlore Solubility

It is shown that pyrochlore solubility in KF solutions at 550 °C and P = 1 kbar depends on their concentration (Fig. 3.1). Thus, when the solution concentration

Table 3.3 The compositions of the products of experiments (wt%) on the solubility of pyrochlore in solutions of alkali metal fluorides

	650 °C, 1 kbar, 1M KF		750 °C, 1 kbar, 1M KF		
Sample No.	Px-251-2		Px-257-2		
Phases	*Pchl + Qtz + Sil* (melt)		*Pchl + Qtz + Sil* (melt)		
Oxide	*Pchl*	*Gl*	*Pchl*	*Rip**	*Gl*
F	4.82	0.98	4.92	–	2.70
Na_2O	7.55	0.36	7.11	–	0.72
Al_2O_3	–	–	–	–	–
SiO_2	–	71.35	–	37.67	56.51
K_2O	1.20	10.69	0.25	16.13	18.57
CaO	15.28	–	15.14	–	0.23
MnO	0.40	0.26	–	–	–
Nb_2O_5	72.11	8.16	70.06	44.19	12.44
Ta_2O_5	–	–	–	–	–
Sum	101.36	91.80	97.48	97.99	88.47

Rip*—$K_{1.023}Nb_{1.023}Si_{1.967}O_7$ (the crystal-chemical formula may be correspond to the mineral rippite: $K_2(Nb,Ti)_2Si_4O_{12}O(O,F)$ (Doroshkevich et al. 2016)

changes from 0.5 to 1M, pyrochlore solubility [lg(mNb)aq] increases by an order of magnitude (-4.3 to -3.3). The temperature dependence of pyrochlore solubility is insignificant; it can be described by the linear regression equation: [lg(mNb)aq] $= -3.745 + 0.0001*(T°C)$; (Sx $= 0.71$, Ex $= 1.1$). Based on this dependence at T $= 550$ °C and P $= 1$ kbar lg(mNb)aq $= -3.70$, and at T $= 850$ °C and P $= 1$ kbar it will be equal to -3.66. The presence of silicate substance in experiments at T $= 550$ °C and P $= 1$ kbar reduces solubility [lg(mNb)aq] to -6.3 (i.e. by about 3 orders of magnitude). In the presence of silicate substance (quartz or granite) there is a clear dependence of Nb solubility on temperature: [lg(mNb)aq] $= -12.44 + 0.0122 * (T°C)$; (Sx $= 0.52$, Ex $= 0.8$). According to this equation, the solubility of Nb increases from -5.8 to -2.1 with an increase in the temperature of experiments from 550 to 850 °C. The solubility of pyrochlore at T $= 550$ °C and P $= 1$ kbar in 1M KF solutions is almost two orders of magnitude higher than in 1M NaF (lg(mNb)aq values are -3.3 and -5.5, respectively).

3.4.2 Tantalite Solubility

Solubility of tantalite in relation to tantalum [lg(mTa)aq] differs from pyrochlore solubility (by niobium). Tantalite solubility experiments were carried out at P $= 1$ kbar in the temperature range from 550 to 850 °C in 1M NaF and 1M KF solutions. Presence of a silicate phase at T $= 550$ °C reduces solubility of tantalite in 1M

KF from -5.4 to -6.9 (by one and a half orders of magnitude). At the same time, at 650 °C, the presence of quartz increases the solubility of tantalite. However, in the temperature range from 550 to 850 °C the dependence of tantalite solubility on the temperature in the presence of silicate matter is small, the regression equation calculated from the data for temperatures of 550, 650, 750 and 850 °C as follows: $[\lg(mTa)aq] = -6.69 + 0.0017 * (T°C)$; $(Sx = 1.1, Ex = 1.4)$. The calculated values of $\lg(mTa)aq$ are -5.7 (550 °C) and -5.2 (850 °C). Without silicate substance dependence of tantalite solubility in 1M KF on temperature is described by the following linear equation $[\lg(mTa)aq] = -9.35 + 0.0070 * (T°C)$; $(Sx = 0.14, Ex = 0.4)$. The calculated solubility values are as follows: -5.5 (550 °C) and -3.4 (850 °C). In 1M NaF solutions, the solubility of tantalite at 550 °C is practically independent of the presence of quartz: -5.9 (without quartz) and -5.7 (with quartz). With increasing temperature, the solubility in the system without quartz increases to -5.2 (750 °C). In the presence of silicate substance in the system solubility of tantalite with increasing temperature decreases to -6.6 (750 °C). In the tantalite used by us there is a lot of niobium (up to 21.93 wt% Nb_2O_5), that is why solubility of tantalite in relation to niobium has also been studied. In silicate-containing systems, the solubility of tantalite with respect to Nb is similar to that of pyrochlore (Fig. 3.4). Solubility of Nb $[\lg(mNb)aq]$ for pyrochlore and tantalite depending on temperature was studied in 1M KF at a pressure of 1 kbar in the presence of silicate substance (quartz or granite). The temperature dependences of solubility $[\lg(mNb)aq]$ are described by polynomials of the 2nd degree: for pyrochlore:

$$[\lg(mNb)aq] = -33.904 + 0.0751 * (T°C) - 0.0000451 * (T°C)^2$$
$$(Sx = 0.22; \ Ex = 0.35);$$

for tantalite:

$$[\lg(mNb)aq] = -25.904 + 0.0548 * (T°C) - 0.0000330 * (T°C)^2$$
$$(Sx = 0.13; \ Ex = 0.21).$$

In general, we can conclude that the niobium component of the solid solutions of pyrochlore and tantalite in fluoride aqueous solutions is significantly more soluble than that of tantalum.

3.4.3 Silicate Melt Phase

As it was mentioned earlier, in experiments in the presence of silicate substance (quartz or granite) in the conditions of heterophase fluid silicate (aluminosilicate) melt was formed, which dissolved significant amounts of niobium and tantalum oxides. This melt can be considered to be saturated with respect to Nb_2O_5 and Ta_2O_5, because in the experiments in equilibrium with silicate melt there were pyrochlore or tantalite ("saturation phases"). The formation of silicate melt enriched with niobium and tantalum is observed both in experiments with 1M NaF fluid and for experiments with 1M KF fluid. Most of the experiments were performed with 1M KF solution (Tables 3.3 and 3.4). In experiments with pyrochlore the content of Nb_2O_5 in silicate glass varies from 8.16 wt% (650 °C) to 12.44 wt% (750 °C). In experiments with tantalite in silicate glass the concentration of Nb_2O_5 changed from 1.38 wt% (650 °C) to 1.9–3.2 wt% (850 °C). The content of Ta_2O_5 in the melt varied from 5.3 (650 °C) to 8.50 wt% (850 °C). In experiments with pyrochlore one can draw a conclusion about the incongruent dissolution of pyrochlore, while a significant share of such elements as Na, Ca, Mn was redistributed into the fluid phase, while the main part of niobium was distributed into the silicate melt phase. In experiments with tantalite, the picture is different. From the content of MnO, Nb_2O_5, Ta_2O_5 oxides in silicate melts it is possible to calculate the formula of hypothetical dissolved tantalite. On the average, the following formula (conversion to 6 atoms O) was obtained for experiments with fluid 1M KF: $Mn_{1.06}(Nb_{0.72}Ta_{1.26})_{1.98}O_6$. The original tantalite has a formula: $Mn_{0.98}(Nb_{0.74}Ta_{1.17}W_{0.08})_{1.99}O_6$. Within the error of analysis, these formulas are similar. One can note a relative increase in the tantalum content in the melt with a decrease in temperature. The molar fraction of tantalum $X_{Ta}^{Liq} = Ta/(Nb + Ta)$ depending on temperature is described by the following equation:

$$X_{Ta}^{Liq} = 0.945 - 0.000414 * (T°C); \quad S_x = 0.05.$$

Thus, according to this equation at 850 °C, the value of $X_{Ta}^{Liq} = 0.59$; with a decrease in temperature to 650 °C, $X_{Ta}^{Liq} = 0.68$.

In some experiments on solubility of tantalite in the presence of silicate melt small amounts of microlite—tantalum member of pyrochlore—microlite solid solution were formed (Fig. 3.5). Tantalites (saturation phase) and silicate melts are completely free of tungsten. According to the analysis of compositions of solutions coexisting with the melt, tungsten is redistributed into the fluid phase. This is consistent with the data (Marakushev and Shapovalov 1994; Shapovalov et al. 2019) on the predominant enrichment of tungsten salt phase. In experiments with pyrochlore (in the presence of 1M KF solution), rippite crystals ($KNbSi_2O_7$), an ore mineral from a number of niobium deposits, were found in the silicate glass phase (Doroshkevich et al. 2016). This phase is also a niobium concentrator. The pyrochlore itself is almost identical in composition to the original one, except for a small amount of potassium, apparently added from potassium-fluoride fluid.

Based on the experimental data obtained by us it is possible to conclude that a low-temperature high-alkaline melt is formed under heterogeneous fluid conditions at a pressure of 1 kbar. Thus, for aluminosilicate melt (experience of KT-108) the agpaitic coefficient is 2.5 ± 0.2. The formation of such an ultra-alkali melt is connected with the fluid-magmatic interaction of the high-temperature fluid under conditions of its heterogenization. You can write the following reaction equation: $KF + H_2O = KOH\downarrow + HF\uparrow$ (arrows indicate that HF enriches the vapor phase of the stratified fluid and KOH—the liquid phase). This saline alkaline phase reacts with the silicate phase

Table 3.4 The compositions of the products of experiments (wt%) on the solubility of tantalite in solutions of alkali metal fluorides

Sample No.	650 °C, 1 kbar, 1M KF		750 °C, 1 kbar, 1M KF			850 °C, 1 kbar, 1M KF			850 °C, 1 kbar, 1M KF		750 °C, 1 kbar, 1M NaF		
	KT-109-2		Px-263-2			KT-108-2			KT-103-2		Px-265-2		
Phases	Tnt + Qtz + Sil (melt)		Tnt + Qtz + Sil (melt)			Tnt + Mcl + granite (melt)			Tnt + Qtz + Sil (melt)		Tnt + Qtz + Sil (melt)		
Oxide	Gl	Tnt	Tnt	Gl1	Gl2	Mcl	Tnt	Gl	Tnt	Gl	Tnt	Mcl	Gl
F	2.80	–	–	1.15	2.24	4.21	–	5.53	–	1.10	–	4.91	3.92
Na$_2$O	–	–	–	–	–	5.66	–	2.62	–	0.40	–	7.42	9.63
Al$_2$O$_3$	0.46	–	–	–	–	–	–	8.83	–	–	–	–	3.58
SiO$_2$	69.15	–	–	63.84	60.62	–	–	49.74	–	63.53	–	–	60.49
K$_2$O	16.06	–	–	9.42	6.63	0.83	–	16.61	–	15.67	–	–	1.44
CaO	0.15	–	–	–	–	9.22	–	0.20	–	–	–	4.73	–
MnO	1.52	16.38	15.84	3.85	2.90	1.42	16.92	2.21	16.54	1.14	17.06	4.50	4.77
Nb$_2$O$_5$	1.38	25.49	20.45	3.99	4.23	10.54	25.27	3.24	23.54	1.88	24.97	20.82	3.30
Ta$_2$O$_5$	5.30	58.66	63.21	7.50	14.50	68.33	58.12	6.31	59.92	8.50	57.96	54.76	7.90
Sum	96.82	100.53	99.50	89.75	91.12	100.21	100.31	95.29	100.00	92.22	99.99	97.14	95.03

Fig. 3.5 Products of experiments on the solubility of tantalite in a solution of 1M KF in the presence of quartz at 850 °C, 1 kbar. Tnt—tantalite, Mcl—microlite

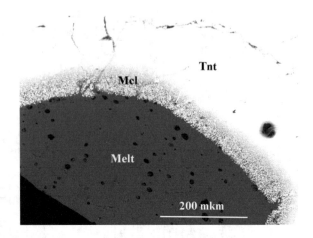

to form aqueous melt KSi_2O_5, with an excessive SiO_2 content. This is consistent with the stoichiometry of silicate melts from our experiments with quartz (Tables 3.3 and 3.4). This alkaline melt in experiments with granite dissolves in itself aluminum oxide, forming aluminosilicate melt. Such alkaline phase is a concentrator of tantalum and niobium—at a temperature of 650–850 °C the total content ($Nb_2O_5 + Ta_2O_5$) reaches from ~7.5 (650 °C) to ~10.5 (850 °C) wt%.

This alkaline low-temperature melt contains up to ~10 wt% volatile ($F + H_2O$) and can serve as a phase—concentrator of rare-metal elements (Ta, Nb) at the latest stages of crystallization of lithium-fluorine granites of potassium specialization. Due to the concentration of rare metals in this phase, there are ore accumulations of tantalum and niobium minerals in the upper parts of lithium-fluorine granite massifs—the so-called apogranites (Beus et al. 1962). Apogranites are formed during hydrothermal processing of ultra-alkali low-temperature melts. The fluids carry easily soluble alkaline components, and ore metals are concentrated in the form of iron and manganese tantalum-niobates.

3.5   Conclusions

1. The boundaries of heterogenization fields of fluoride fluids in the temperature range of 550–850 °C and pressure of 1 kbar were determined by the method of synthetic fluid inclusions.
2. The influence of silicate substance on pyrochlore and tantalite solubility in alkali metal fluoride solutions (LiF, NaF, KF) was studied. It is shown that the solubility of the above minerals is significantly affected by the presence of silicate (quartz) or aluminosilicate (granite) material in the experiments.

3. The temperature dependences of pyrochlore and tantalite solubility in aqueous solutions of alkali metal fluorides for the systems ($H_2O \pm LiF \pm NaF \pm KF \pm$ Pchl $\pm$ Tnt) in the range of 550–850 °C and pressure of 1 kbar are determined.
4. It is established that in the presence of fluoride heterophase fluid in silicate systems with ore minerals (pyrochlore, tantalite) for concentrations of solutions of 0.5 and 1M at temperatures of 650–850 °C and pressure of 1 kbar the formation of alkaline silicate melt, which is a concentrator of Nb and Ta, occurs.
5. This phase of highly alkaline silicate melt can serve as a phase that concentrates significant amounts of ore elements. This phase, when treated with low concentrated aqueous solutions, forms ore apogranites.

Acknowledgements This work was supported by the grant of the Russian Foundation for Basic Research № 15-05-03393-a.
Theme # AAAAA-A18-118020590151-3.

References

Aksyuk AM (2002) Experimentally established geofluorimeters and the fluorine regime ingranite-related fluids. Petrology 10(6):557–569

Aksyuk AM (2009) Fluorine regime in deep hydrothermal fluids and near-surface waters. Doctoral dissertation. Moscow, 166 p. (in Russian)

Badanina EV, Syritso LF, Volkova EV, Thomas R, Trumbull RB (2010) Composition of Li-F granite melt and its evolution during the formation of the ore-bearing Orlovka massif in Eastern Transbaikalia. Petrology 18(2):131–157

Beus AA, Severov EA, Sitnin AA, Subbotin KD (1962) Albitized and Greisenized Granites (apogranites). Akad Nauk SSSR, Moscow, 196 p. (in Russian)

Borodulin GP, Chevychelov VY, Zaraysky GP (2009) Experimental study of partitioning of tantalum, niobium, manganese and fluorine between aqueous fluoride fluid and granitic and alkaline melts. Dokl Earth Sci 427(1):868–873

Chevychelov V Yu (1998) The influence of the composition of granitoid melts on the behavior of ore metals (Pb, Zn, W, Mo) and petrogenic components in the melt—water fluid system. Experimental and theoretical modeling of mineral formation processes. Nauka, Moscow, pp 118–130

Chevychelov V Yu, Zarayskiy GP, Borisovskii SE, Borkov DA (2005) Effect of melt composition and temperature on the partitioning of Ta, Nb, Mn and F between granitic (alkaline) melt and fluorine-bearing aqueous fluid: fractionation of Ta and Nb and conditions of ore formation in rare-metal granites. Petrology 13(4):305–321

Doroshkevich AG, Sharygin VV, Seryotkin YV, Karmanov NS, Belogub EV, Moroz TN, Nigmatulina EN, Eliseev A, Vedenyapin N, Kupriyanov IN (2016) Rippite. IMA 2016-025. CNMNC Newslett (32):919: Min Mag 80:915–922

Gramenitskiy EN, Shchekina TI, Devyatova VN (2005) Phase relationships in fluorine-containing granite and nepheline-syenite systems and the distribution of elements between phases (experimental study). GEOS, Moscow, p 188

Kotelnikov AR, Korzhinskaya VS, Kotelnikova ZA, Suk NI, Shapovalov YuB (2018a) Effect of silicate matter on pyrochlore solubility in fluoride solutions at $T = 550$–850 °C, $P = 50$–100 MPa (experimental studies). Dokl Earth Sci 4(1):1199–1202

Kotelnikov AR, Korzhinskaya VS, Kotelnikova ZA, Suk NI, Shapovalov YuB (2018b) Tantalite and pyrochlore solubility in fluoride solutions at $T = 550\text{-}850°C$ and $P = 1$ kbar in presence of silicate material. Exp Geos 24(1):177–180

Kotelnikova ZA, Kotelnikov AR (2002) Synthetic NaF-bearing fluid inclusions. Geochem Intern 40(6):594–600

Kotelnikova ZA, Kotelnikov AR (2008) NaF-bearing fluids: experimental investigation at 500–800°C and P = 2000 bar using synthetic fluid inclusions in quarts. Geochem Intern 46(1):48–61

Kovalenko VI (1977) Petrology and geochemistry of rare metal granitoids. Nauka, Sibirian Devision Novosibirsk, p 207

Marakushev AA, Shapovalov YuB (1994) Experimental study of ore concentration in fluoride granite systems. Petrology 2(1):4–23

Reed SJB (2005) Electron microprobe analysis and scanning electron microscopy in geology. Cambridge University Press, Cambridge

Rub AK, Rub MG (2006) Rare metal granites of Primorye. VIMS, Moscow, p 86

Ryabchikov ID (1975) Thermodynamics of the fluid phase of granitoid magmas. Nauka, Moscow, p 232

Shapovalov YuB, Kotelnikov AR, Suk NI, Korzhinskaya VS, Kotelnikova ZA (2019) Liquid immiscibility and problems of ore genesis (according to experimental data). Petrology 27(5)

Zarayskiy GP (2004) Conditions of formation of rare-metal deposits associated with granitoid magmatism. Smirnovsky digest-2004 Moscow: Fund named after acad Smirnov, pp 105–192

Zarayskiy GP (2009) The relationship of magmatism, metasomatism and ore formation in rare metal deposits of apogranite type. In: Paragenesis and ore formation ekaterinburg: ural branch RAS, pp 105–109

Zarayskiy GP, Korzhinskaya VS, Kotova NP (2010) Experimental studies of Ta_2O_5 and columbite-tantalite solubility in fluoride solutions from 300 to 550 °C and 50 to 100 MPa. Miner Petrol 99(3–4):287–300

Chapter 4
Experimental and Theoretical Studies of the Viscosity of the Fluid Magmatic Systems in Conjunction with the Structure of Melts at the Thermodynamic Parameters of the Earth's Crust and Upper Mantle

E. S. Persikov and P. G. Bukhtiyarov

Abstract Paper provides a brief overview of the results of the established general regularities of the concentration, temperature, pressures and phase dependency of viscosity of the fluid-magmatic systems in connection with the anniversary of IEM RAS (50 years have passed since the establishment of this unique Institution). The study of the viscosity of such melts was carried out in the full range of compositions of natural magmas (acid-ultrabasic) in a wide range of fluid compositions (Ar, H_2O, $H_2O + HCl$, $H_2O + NaCl$, $H_2O + HF$, CO_2, $H_2O + CO_2$, H_2), and thermodynamic parameters of the earth's crust and upper mantle ($T = 800°–1950$ °C, $P = 100$ MPa–12.0 GPa, $P_{fl} = 10–500$ MPa). The study of the viscosity of such melts was carried out in the IEM RAS in conjunction with the study of structural features of melts. The features of the unique equipment and techniques developed in the IEM RAS for such original studies are briefly considered. The possibilities and advantages of the developed structural-chemical model of reliable predictions and calculations of viscosity of fluid-magmatic systems in the full range of magma compositions from acidic to ultramafic at thermodynamic parameters of the earth's crust and upper mantle are discussed. Some examples of successful application of the obtained experimental and theoretical results to natural processes are briefly considered.

Keywords Viscosity · Structure · Acidic-ultramafic melts · High temperatures and pressures · Activation energy · Crust · Mantle · Model · Sphere

4.1 Introduction

The processes of origin of magmatic melts in the earth's crust and upper mantle, kinetics and dynamics of their evolution in a variable range of temperatures and

E. S. Persikov (✉) · P. G. Bukhtiyarov
D.S. Korzhinskii Institute of Experimental Mineralogy, Russian Academy of Sciences, Academitian Osipyan Street 4, Chernogolovka, Moscow Region, Russia 142432
e-mail: persikov@iem.ac.ru

Y. Litvin and O. Safonov (eds.), *Advances in Experimental and Genetic Mineralogy*, Springer Mineralogy, https://doi.org/10.1007/978-3-030-42859-4_4

69

pressures, structural and textural features of magmatic rocks, the processes of differentiation, and partial melting are largely control by the viscosity of magma. The great diversity in the composition of main and volatile components, heterogeneity, a wide range of temperatures and pressures are the main features of the existence of magmatic melts in nature. Experimental study of viscosity of such systems at high temperatures and pressures is an important and at the same time complex scientific problem.

Significant results have been obtained of the experimental and theoretical study of the viscosity of model and magmatic melts in the process of intensive development of experimental methods of physical-chemical petrology and geochemistry over the past half a century. It should be noted that progress in this field of petrology and geochemistry has been achieved by many researchers in various laboratories of the world, for example: (USA—Shaw et al. 1968; Shaw 1972; Waff 1975; Mysen et al. 1979; Mysen 1991; Hui and Zhang 2007; Lange 1994; Kono et al. 2014; Wolf and McMillan 1995; Allwardt et al. 2007; and other); (France—Carron 1969; Bottinga and Well 1972; Richet 1984; Neuville and Richet 1991; Champaller et al. 2008; and other); (Germany—Dingwell 2006; Dingwell et al. 2004; Hess and Dingwell 1996; Reid et al. 2003; Liebske et al. 2005; Behrens and Schulze 2003; and other), (Japan—Kushiro 1980; Fujii and Kushiro 1977; Uhira 1980; Ohtani et al. 2005; Suzuki et al. 2005; and other); (Canada—Brealey et al. 1986; Scarfe 1986; Scarfe et al. 1987; Russell et al. 2002, 2012; and other); (Italy—Giordano and Dingwell 2003; Giordano et al. 2004, 2008; Vetere et al. 2010; Poe et al. 2006; and other); (England—Sparks et al. 2009; Dobson et al. 1996; and other); (Russia—Volarovich 1940; Lebedev and Khitarov 1979; Sheludyakov 1980; Ivanov and Stiegelmeyer 1982; Chepurov and Pokhilenko 2015; and others), including our long-term research in IEM RAS (Persikov 1984, 1991, 1998; Persikov and Bukhtiyarov 2004, 2009; Persikov et al. 1990, 2018a; and others). This paper provides a brief overview of the results of the established generalized regularities of temperature, pressure, concentration and phase dependences of the viscosity of fluid-magmatic systems in the full range of compositions of natural magmas (acid-ultrabasic) and thermodynamic parameters of the earth's crust and upper mantle ($T = 800$–$1950\,^\circ\mathrm{C}$, $P = 20\,\mathrm{MPa}$–$12.0\,\mathrm{GPa}$, $P_{fl} = 20$–$500\,\mathrm{MPa}$). The study of the viscosity of such melts is carried out in conjunction with the study of the structural features of the melts. The developed structural-chemical model of reliable predictions and calculations of viscosity of fluid-magmatic systems in the full range of magma compositions from acidic to ultramafic at thermodynamic parameters of the earth's crust and upper mantle are also discussed. Some examples of successful application of experimental and theoretical results to natural processes are briefly considered.

4.2 Experimental and Analytical Methods

Viscosity of model and magmatic melts without and with volatile components was investigated using four types of original equipment. Unique high gas pressures radiation viscometer, discussed in detail earlier (Persikov 1991), was used to study the viscosity of melts of different compositions at fluid pressures up to 500 MPa (developed in IEM RAS). Solid-state high pressures apparatuses were used to study the viscosity of melts at pressures up to 12 GPA: (1) The "anvil with a hole" type apparatus (developed in IEM RAS, e.g., Litvin 1991). (2) The multi-anvil devices of the "BARS" type (developed in IGM SB RAS, e.g., Sokol and Palyanov 2008). (3) "Kawaii" type apparatuses (developed at Tokyo University, Japan, e.g., Persikov et al. 1989). To conduct experiments on the synthesis of fluid-bearing glasses and subsequent experiments on measurements of melt viscosity at high temperatures and fluid pressures of almost any composition, a unique device has been developed, which is located inside the high gas pressure vessel (IHPV) of the radiation viscometer (Persikov and Bukhtiyarov 2002). The scheme of this device with the scheme of the internal heater of the IHPV is shown in Fig. 4.1. The internal volumes of the reactor (5) with a platinum or molybdenum ampoule with a starting sample (6) and a separator equalizer (9) under the piston (10) were filled with a fluid of the required composition at a pressure of 10 MPa using a special system. In experiments with water fluid the mentioned internal volumes of the device were filled with purified water at atmospheric pressure. The device assembled in this way, together with the internal heater (2), was placed inside the IHPV so that the ampoule with the sample (6) was in the gradient-free temperature zone of the heater. Due to the displacement of the piston (10), the gas pressure (Ar) in the IHPV during the experiment was always equal to the fluid pressure in the internal volume of the reactor (5). At the beginning of the experiment on the synthesis of fluid-bearing glasses, the pressure of Ar in the IHPV and, accordingly, the fluid in the reactor (5) was raised within one hour to the required value, for example 100 MPa. Further, the temperature of the experiment was raised to the required value. At these parameters, the samples were kept in automatic mode for the required time of experiments. After that, isobaric quenching was carried out with the internal heater turned off. At the same time, a sufficiently high melt quenching rate (~300 °C/min) was achieved, which ensured the production of quenched glasses with a fluid dissolved in them. After isobaric quenching, the device was removed from the IHPV, an ampoule with a sample (6) was extracted from the reactor for subsequent analysis of the glass composition and for experiments on measurements of melt viscosity. In experiments with oxidizing fluid, the inner surface of the reactor (5) was lined with platinum to prevent oxidation of the molybdenum reactor. The falling sphere method was used to measure the melt viscosity in all experiments. Platinum or platinum-rhodium spheres with a diameter of 0.13–0.21 cm, filled with radioactive isotope Co^{60} and manufactured in a special device, were used to measure the viscosity of relatively viscous polymerized melts on a radiation viscometer. At the same time, full safety of work was ensured, since the activity of each sphere was very low ($\leq 10^{-4}$ curie). The time of falling of the

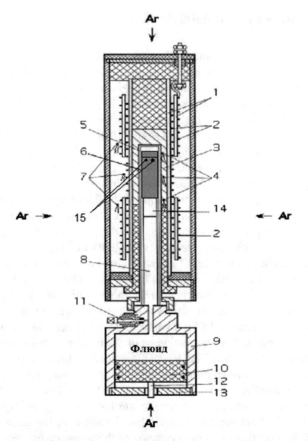

Fig. 4.1 Scheme of the unique internal device and heater of IHPV. 1, 3—insulators; 2—two windings of the heater; 4—three thermocouples to control gradient free zone along the Pt capsule with the sample; 5—molybdenum tube; 6—Pt ampoule with sample (melt) and spheres; 7—two thermocouples to control the temperatures of each windings of the heater; 8—sapphire cylinder; 9—vessel of the separator-equalizer; 10—piston of the separator-equalizer; 11—valves; 12—sensor to control of the piston position; 13—cover; 14—Pt ampoule with oxygen buffer; 15—two Pt60Rh40 spheres of different diameter

sphere in the melt of a certain distance between two collimators of the viscometer (the distance between the maxima of gamma radiation intensity) at the experiment parameters was measured in such kind experiments. To measure the viscosity of low-viscosity depolymerized melts on all types of apparatus, a quenching version of the falling sphere method was used. In this case, two platinum-rhodium spheres of small and different diameter (50–250 μk) were used as a rule. The spheres made by wire fusing with water quenching. The time of falling of the spheres in the melt was determined from the moment of reaching the required values of temperature and pressure of the experiment to the moment of isobaric quenching. The distances covered by the spheres in quenching experiments were determined in sections under a microscope. Naturally, the viscosity measurement error increased by about half (up to 25–30 rel. %) compared with the radiation method of viscosity measuring ($\leq$15 rel. %). The melt viscosity at given T and P was calculated according to the known Stokes law with the Faxen correction for the wall effect (e.g., Persikov 1991):

$$\eta = \frac{2gr^2\Delta\rho}{9V(1+3,3r/h)}[1-2,104\frac{r}{r_a}+2,09(\frac{r}{r_a})^3-0,95(\frac{r}{r_a})^5] \qquad (4.1)$$

where g is the gravitational constant, r is the radius of the falling sphere, r_a is the internal radius of the ampoule with the melt, $\Delta\rho$ is the density contrast between the sphere and the melt, V is the falling velocity, h is the height of the ampoule. The term in square brackets is the Faxen correction for the wall effect.

Calculation viscosity with Eq. (4.1) requires a knowledge of melt and sphere densities. The melt density was determined by the next way: the density of quenched melts (glasses) from runs at the different pressures was measured by the method of hydrostatic weighting. The correction for experimental temperatures was applied. The spheres density at normal conditions was measured by the same method of hydrostatic weighting and was used to calculate the melts viscosity by Eq. (4.1). The correction on temperature and pressure on spheres density was not used because the temperature and pressure have opposing effects (Persikov 1991; Liebske et al. 2005). In fact, the melt density can vary from 2.46 to 3.09 g/cm^3 without changing the calculated viscosity by more than $\pm$3% (Brearley et al. 1986; Persikov et al. 2018a). The relative error of viscosity measurements calculated was $\pm$15–30% rel. Temperature range (800–1950 °C) was chosen so that liquidus interval of viscosity measurements was provided over the whole range of pressures and for all studied compositions (in the limit of ~200°–250 °C above the corresponding liquidus temperatures). Platinum-rhodium or tungsten-rhenium thermocouples calibrated at the different pressures using melting points of Au, Al and Ag were used. Relative errors of temperature measurement did not exceed $\pm$5 °C on a radiation viscometer and $\pm$20 °C on solid-phase high pressure devices. As usual, the correction for pressure effect on e.m.f. of thermocouples was not taken into account for these devices. The relative error of pressure measurement on the radiation viscometer did not exceed $\pm$0.1% using a special pressure sensor. The error of pressure measurement on solid-phase high pressure devices did not exceed $\pm$0.1 GPa. A characteristic feature of the performed experiments is the wide range of fluid compositions used to study viscosity of the fluid-magmatic systems: Ar, H_2O, H_2O + HCl, H_2O + NaCl, H_2O + HF, CO_2, H_2O + CO_2, H_2. Homogeneous glasses prepared by melting of powders of natural minerals and rocks, as well as synthetic glasses obtained by melting stoichiometric mixtures of natural minerals at a temperature of 1400 °C and ambient pressure were used as starting materials in the experiments. A pressure of CO_2 equal to 100 MPa at a temperature of 1400 °C were used to prepare the starting materials for the experiments with carbonate—bearing model kimberlite and dunite melts.

Chemical composition of the starting glasses and run products were determined by microprobe analysis, on CamScan MV2300 (VEGA TS 5130 MM) digital electron microscope with INCA Energy 450 and Oxford INCA Wave 700 energy dispersive

X-ray spectrometry (EDS). Analyses were conducted at an acceleration voltage of 20 kV with a beam current of 0.3 nA and counting time was 50–100 s. The following standards were used: SiO_2 for Si and O, albite ($NaAlSi_3O_8$) for Na, microcline for K, wollastonite ($CaSiO_3$) for Ca, pure titanium for Ti, corundum for Al, pure manganese for Mn, pure iron for Fe, periclase for Mg. Date reduction was performed using the standard INCA Energy 200 and the INCA program of A. Nekrasov (IEM RAS). The concentration of water in the quenching melts (glasses) was determined by Karl-Fischer titration method using KFT AQUA 40.00 device and IR spectroscope. CHN-1 type gas chromatograph is used to determine the concentration of H_2 in glasses. The concentration of Cl in quenched melts (glasses) were determined by the method of microprobe analysis, the F concentration by the photometric method using analizaron of chelating agent (Persikov et al. 1990). The concentrations of the two forms of water (hydroxyl OH^- and molecular water H_2O) and carbon dioxide (CO_3^{2-} and CO_2) in quenched melts (glasses) were measured using the methods of Raman and IR spectroscopy. Raman spectroscopy was mainly used to qualitatively illustrate the dissolution of the carbonate form in melts (glasses). Raman spectra of the glasses were obtained using a RENIS/HAW-1000 spectrometer equipped with a Leica microscope. The slit width for spectra was 50 μk, and the spectra were processed using the standard GRAMS program. The results of microprobe analyses of some samples are shown in Table 4.1, which also shows the values of the structural-chemical parameter of the melts—the degree of depolymerization or basicity coefficient ($K = 100NBO/T$) used for comparative evaluation of the chemical composition and structure of magmatic melts (see below).

The water content in melts (glasses) for different types of dissolved water (OH^- groups and H_2O molecules) was determined by quantitative infrared microspectroscopy using the Bouguer-Beer-Lambert ratio (Stolper 1982; Persikov et al. 2010):

$$C = 100 \cdot 18.015 \cdot A/(\rho \cdot \delta \cdot \varepsilon_{H_2O}), \tag{4.2}$$

where C is the concentration of the corresponding forms of water (OH^- or H_2O) in the melt (glass) wt%, A is the peak height of the absorption, ρ is the glass density, g/L, δ—thickness of the plate, cm, ε_{H_2O}—molar absorption coefficient of the appropriate type of water, L/mol cm, 18.015 is the molecular weight of water, g/mol.

The thickness of the glass plates was determined by a micrometer with an error of ±2 μk. IR spectra of hydration samples were obtained using the infrared spectrometer Nicolet Magna-IT 860, equipped with a microscope Nicolet Continuum IR (Caltech, USA). The total concentration of water in the melt (glass) was determined by summing the concentrations of its two forms, determined by the reduced Eq. (4.2).

Table 4.1 The chemical composition of some melts (glasses, wt%) and their structural-chemical parameter (K = 100NBO/T)

Acidic-mafic melts and melts of some minerals

Samples	SiO_2	TiO_2	Al_2O_3	Fe_2O_3	FeO	MnO	MgO	CaO	Na_2O	K_2O	P_2O_5	H_2O	OH^-	Σ	K
Granite	73.23	0.19	13.6	2.56	0.48	0.20	0.17	1.69	3.78	4.11	0.13	–	–	99.94	0.3
Granite + H_2O	69.22	0.18	12.84	2.46	0.56	0.18	0.17	1.59	3.66	3.98	0.13	–	5.20	99.99	41.0
Andesite	58.56	0.64	18.98	3.95	3.90	0.15	3.48	6.17	3.24	0.92	–	–	–	99.84	17
Andesite + H_2O	55.92	0.62	18.14	1.72	6.00	0.14	3.3	5.86	3.08	0.9	0.10	–	4.60	100.1	63
Basalt	53.54	1.05	17.29	7.46	1.11	0.12	5.46	8.32	3.59	1.64	0.21	–	–	99.95	47
Basalt + H_2O	50.2	0.98	16.2	2.91	5.31	0.1	5.12	7.8	3.37	1.54	0.19	2.36	4.0	99.98	103
Albite	68.73	–	19.45	–	–	–	–	–	11.82	–	–	–	–	100	0
Albite + H_2O	64.21	–	17.94	–	–	–	–	–	10.95	–	–	0.5	6.4	100	50
Jadeite	59.45	–	25.22	–	–	–	–	–	15.33	–	–	–	–	100	0
Diopside	55.51	–	–	–	–	–	18.75	25.74	–	–	–	–	–	100	200
Diopside + H_2O	54.68	–	–	–	–	–	18.47	25.34	–	–	–	–	2.0	100	123

Model ultramafic melts

Samples	SiO_2	TiO_2	Al_2O_3	MgO	CaO	Na_2O	K_2O	P_2O_5	H_2O	OH-	CO_2	$CO_3{}^{2-}$	Σ	K
Kimberlite[a]	34.4	–	10.3	–	39.8	4.9	–	–	–	0.05	0.15	10.4	100	313
Kimberlite[a] + H_2O	33.06	–	9.91	–	38.24	4.85	–	–	2.04	1.86	0.32	10.0	100	247
Dunite[b]	39.85	–	2.98	34.11	21.68	1.38	–	–	–	–	–	0.2	100	340

Notes

Dash—no data

[a]Model kimberlite melt after melting the initial mixture ($Ab_{38}Cal_{62}$, wt%) at $T = 1300$ °C and at a CO_2 pressure of 100 MPa (Persikov et al. 2017, 2018a)

[b]Model dunite melt after melting the initial mixture ($Ab_5Di_{30}Mgz_{50}Cal_{50}$, mol%) at $T = 1400$ °C and at a CO_2 pressure of 100 MPa (Persikov et al. 2018b)

4.3   Results and Discussion

The generalized regularities of temperature, pressure, concentration and phase dependences of viscosity of fluid-magmatic systems in the full range of compositions of
natural magmas (acid-ultrabasic) and thermodynamic parameters of the earth's crust
and upper mantle were established during of long-term experimental and theoretical
studies of viscosity of such melts.

4.3.1   Temperature Dependence of Viscosity of Magmatic Melts

The temperature dependence of viscosity is the key parameter of melts. The knowledge of this dependence allows estimating quantitatively the ascent rates of magmatic melts and their heat and mass transfer in different geodynamic setting. Some
equations for temperature dependence of the viscosity in model silicate and natural
magmatic melts have been suggested earlier (Adam and Gibbs 1965; Shaw 1972;
Richet 1984; Whittington et al. 2000; Giordano and Dingwell 2003; Giordano et al.
2008; Russell et al. 2002; etc.). We have proved that the temperature dependence of
the viscosity of near-liquidus magmatic melts at ambient, moderate and high pressures, in a large range of viscosity $0.1–10^8$ Pa s (last one for anhydrous near-liquidus
granitic melts) fits quite well the fundamental Arrhenian-Frenkel-Eyring equation
(e.g., Persikov 1998; Persikov et al. 2018a):

$$\eta = \eta_0 \exp(E/RT), \tag{4.3}$$

where η_0 is the pre-exponential factor for the viscosity of melts at $T \to \infty$, ($\eta_0 =
10^{-4.5} \pm 10^{-0.1}$ Pa s); T is the absolute temperature in K; $R = 8.3192$ (J/mol K)
is the gas constant; E is the activation energy of viscous flow (J/mol), which is a
function of the pressure and melt compositions; and η is the melt viscosity at the
given temperature, in Pa s.

Figure 4.2 shows the temperature dependence of viscosity for melts of the albite-
diopside-water system and jadeite melts at atmospheric and high pressures. Figure 4.3
shows the temperature dependences of the viscosity of model ultrabasic kimberlite
melts at a moderate CO_2 pressure of 100 MPa and high lithostatic pressures (5.5
and 7.5 GPa), in comparison with the temperature dependence of the viscosity of the
basalt melts. According to Eq. (4.3), the temperature dependence of viscosity for all
studied melts (e.g., Fig. 4.2), is exponential, i.e. the viscosity of these melts decreases
exponentially with increasing temperature, and conversely, increases exponentially
with decreasing temperature, both at moderate and high pressures. It is necessary
to emphasize the following important result. The constancy of the pre-exponential
factor (η_0) of Eq. (4.3) and its independence from the melt composition and pressure
are experimentally established (e.g., see the equations given in Fig. 4.3a–c). It is

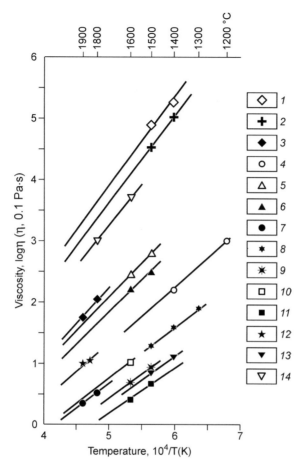

Fig. 4.2 Temperature dependence of the viscosity of some melts of the system albite–diopside–H_2O and jadeite melts at atmospheric and high pressures (Persikov et al. 1989; Persikov and Bukhtiyarov 2004, 2009). 1—Ab_{100} (model granite), P = 1 atm.; 2—Ab_{100}, Ptot. = 400 MPa; 3—Ab_{100}, Ptot. = 4 GPa; 4—$Ab_{75}(H_2O)_{25}$, P(H_2O) = 400 MPa; 5—$Ab_{80}Di_{20}$, P = 1 atm; 6—$Ab_{80}Di_{20}$, Ptot. = 500 MPa; 7—$Ab_{80}Di_{20}$, Ptot. = 4 GPa; 8—$Ab_{30}Di_{70}$, P = 1 atm; 9—$Ab_{30}Di_{70}$, Ptot. = 500 MPa; 10—$Ab_{30}Di_{70}$, Ptot. = 2.5 GPa; 11—Di_{100}, P = 1 atm. and Jd_{100}, Ptot = 9 GPa; 12—Di_{100}, Ptot. = 4 GPa; 13—$Di_{92}(H_2O)_8$, P(H_2O) = 500 MPa; 14—Jd_{100}, P = 1 atm. (melts compositions in mol%)

fully consistent with the theoretical findings (Frenkel 1975) and previously obtained data at moderate and high pressures (Persikov 1991; Russell et al. 2002; Persikov et al. 2018a). On this basis temperature-independent values of activation energies of the viscous flow of all studied fluid-magmatic melts were obtained for the first time with high accuracy (±1% rel.). Note that the activation energy of viscous flow is a very important energy and structural-chemical parameter of the melt, which has a strict physical meaning as a potential barrier or a change in free energy of the system

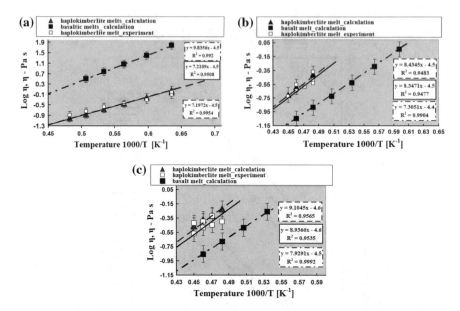

Fig. 4.3 Temperature dependence of the viscosity of haplokimberlite and basaltic melts. (uncertainties of experimental and calculating data are ±30% rel.; the calculated data on the temperature dependence of the viscosity of melts is obtained by our model, see text). **a** Pressure is 100 MPa, **b** pressure is 5.5 GPa, **c** pressure is 7.5 GPa

during shearing motion. Activation energy reflects intermolecular interaction, and consequently, a molecular structure and melt composition, and its value is close to the heat of compound formation (Adam and Gibbs 1965; Frenkel 1975; Richet 1984; Persikov 1998). The simple equations were suggested (Persikov 1998; Persikov and Bukhtiyarov 2009) to calculate the values of the activation energy for magmatic melts using the constant pre-exponential factor value ($\log \eta_0 = -4.5 \pm 0.14$).

$$E = 19.284 \, T \, (\log \eta_T^P + 4.5), \tag{4.4}$$

where: η_T^P is the melt viscosity at the given temperature and pressure, in Pa s; E is the activation energy of viscous flow, in J/mol; factor $19.284 = R/0.4314$; $R = 8.3192$ J/mol K is the gas constant. For example, the following values of the activation energy of viscous flow were obtained for some melts: model dunite melts, E = 137 ± 1.37 kJ/mol ($P = 100$ MPa) and E = 171 ± 1.7 kJ/mol ($P = 7.5$ GPa); basalt melts, E = 186 ± 1.86 kJ/mol ($P = 100$ MPa), and E = 153 ± 1.53 kJ/mol ($P = 7.5$ GPa). It is obvious that such high accuracy of determination of the activation energies of viscous flow (±1.0% rel.) is unattainable for many studies of the viscosity of magmatic melts in which (η_0) in Eq. (4.3) was assumed to be non-constant. For example, the viscosity of a model peridotite melt was measured at a wide pressure range of 2.8–13.0 GPa using a multi-anvil device combined with a synchrotron X-ray source (Liebske et al. 2005). The constant of the activation energy of the viscous flow of peridotite melts in a wide range of pressures (0.1–8 GPa) has been suggested in this work. It is obvious that the constancy of activation energy of the viscous flow (E) for peridotite melts in such a wide range of pressures contradicts our data (e.g. Persikov et al. 2018a), and the theoretical premises mentioned above.

4.3.2   The Concentration Dependence of Viscosity of Magmatic Melts

The viscosity of magmatic melts complex depends on composition of the melt and its structure, which causes serious difficulties of development of models for calculation and prediction the concentration dependence of viscosity. Viscosity is highly sensitive to the change of composition of magmatic melts, which is proved by the following data: the viscosity of near-liquidus melts in the granite-dunite series decreases very much, approximately by a factor of 10^7 times. There is certain progress in solving the problem of concentration dependence of the viscosity of such melts in an empirical way (Carron 1969; Bottinga and Weill 1972; Shaw 1972; Sheludyakov 1980; Hess and Dingwell 1996; Giordano and Dingwell 2003; Hui and Zhang 2007; Giordano et al. 2008; Russell et al. 2002). We propose quite a different approach (Persikov 1998; Persikov and Bukhtiyarov 2009). We have used an adequate criterion, which reflects the structure and composition of magmatic melts correctly and with proper sensitivity. It is a structural chemical parameter of the melt $K = 100(NBO/T)$, depolymerization degree or coefficient of relative basicity (Mysen et al. 1979; Mysen 1991; Persikov 1991). The calculation of this parameter is made from chemical composition of the melt in wt% of oxides, according to a simple technique (Persikov 1998) with the use of the following equation:

$$K = 200\,(O - 2T)/T = 100\,(NBO/T), \tag{4.5}$$

where T is total amount of gram-ions of the network-forming cations (Si^{4+}, Al^{3+}, Fe^{3+}, Ti^{4+}, P^{5+}, B^{3+}), which are in tetrahedral coordination over oxygen and enter the anionic part of the melt structure; O is a total amount of gram-ions of oxygen in the melt; NBO/T is the total amount of gram-ions of non-bridging oxygen in the melt; Note that B. Mysen sometimes used a 100NBO/T parameter too (Mysen 1991, Figs. 26–28).

Note that the isothermal comparison of viscosity in the full range of magma compositions has often no physical—chemical meaning that because of essential differences in liquidus temperatures of magmatic rocks as, for example, the viscosity of granite-dunite melts. It is more correctly to consider the activation energies of the viscous flow as temperature-independent structural—chemical and rheological parameters. The new comparable values of activation energies of viscous flow of the melts for all the published experimental data on the viscosity of near-liquidus silicate, aluminosilicate and magmatic melts have been obtained, based of the proven

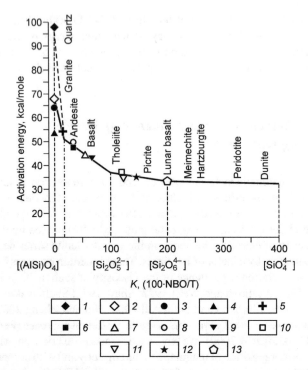

Fig. 4.4 Generalized structural-chemical dependence of activation energies of a viscous flow of model magmatic melts [bracketed are the basic terms of the melt anion structure (Persikov 1991, 1998; Persikov and Bukhtiyarov 2004). *1*, Q_{100}; *2*, Ab_{100} (model granite); *3*, Jd_{100}; *4*, $Nepf_{100}$; *5*, $Ab_{93}(H_2O)_7$; *6*, $Ab_{85}(H_2O)_{15}$; *7*, $Ab_{75}(H_2O)_{25}$; *8*, $Ab_{80}Di_{20}$; *9*, $Ab_{57}Di_{43}$; *10*, $Ab_{30}Di_{70}$; *11*, $Di_{92}(H_2O)_8$; *12*, $Di_{96}(H_2O)_4$; *13*, Di_{100} (melts composition in mole %)]

constancy value of ηo in Eq. (4.3). Moreover, a generalized structural—chemical dependence of activation energies of viscous flow on melts compositions have been suggested (Fig. 4.4). The diagram was analyzed in detail earlier (e.g., Persikov 1998), therefore, we would like to only emphasize some important features. The represented diagram has been corrected compared to the previously published one. Recently obtained experimental data have been taken into account (Dingwell 2006; Dingwell et al. 2004; Giordano and Dingwell 2003; Giordano et al. 2004, 2008; Hess and Dingwell 1996; Whittington et al. 2000, 2001; Vetere et al. 2010; and references therein). The inflection points in the dependence $E = f(K)$ characterize the changes in the melt structure and are interpreted in terms of the silicate melt theory. For a three—dimensional framework structure the break down is completed at $K = 17$, and at $K = 100$, 200, and 400 formation of di-, meta-, and ortho-silicate structures of melts respectively is brought to an end (see Fig. 4.4). A wide field of natural magmas is limited by vertical lines in Fig. 4.4, which was defined as follows: at an average water concentration in acidic magmas (0.67 wt% in granites, and 1.1 wt% in the rhyolites) their degree of depolymerization is $K \leq 5$, depolymerized ultramafic magmas are characterized by the value K = 200–400, for example, for dunite melts K = 380.

4.3.3   Pressure Dependence of Viscosity of Magmatic Melts

Pressure is the most important thermodynamic parameter of magmatic systems. Most of magmatic melts exist of a great depth, under high pressures, except volcanic melts. The influence of high pressure on viscosity of natural magmatic and model silicate melts has been largely studied (Waff 1975; Kushiro 1980; Fujii and Kushiro 1977; Brearley et al. 1986; Scarfe et al. 1987; Persikov et al. 1989, 2017, 2018a; Mysen 1991; Persikov 1991; Wolf and McMillan 1995; Reid et al. 2003; Liebske et al. 2005; Suzuki et al. 2005; Allwardt et al. 2007; Karki and Stixrude 2010; etc.). Note that all theories of liquids presume that viscosity increases exponentially with increasing pressure (Brush 1962; Scarfe 1986; Persikov 1998; Wolf and McMillan 1995; Reid et al. 2003; Liebske et al. 2005; Persikov et al. 2018b):

$$\eta^P = \eta_0 \exp(E + PV)/RT, \tag{4.6}$$

where η^P is the viscosity at given pressure P; V is the activation volume; the rest designations are the same as in Eq. (4.4).

An increase of melts viscosity with increasing pressures of pyroxene, peridotite, dunite, and kimberlite melts ($200 < 100 NBO/T < 400$), as well as more polymerized ($0 < 100 NBO/T < 200$) melts, observed after the minimum values (Fig. 4.5, for basaltic melts), is in full agreement with the theoretical predictions (Eq. 4.6).

These results are consistent with free volume theory (e.g. Frenkel 1975; Persikov 1998; Liebsky et al. 2005). According to these theories, an increase in the viscosity of a liquid with increasing pressure is associated with an increase in the potential

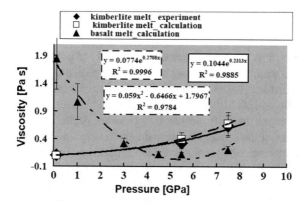

Fig. 4.5 Isothermal (1800 °C) dependence of the viscosity of haplokimberlite and basaltic melts on pressure (uncertainties of experimental and calculating data are ±30% rel., see text)

Table 4.2 Parameters and uncertainties[a] obtained from the calculation of Eq. 4.6 for data on the viscosity of haplokimberite and basaltic melts at moderate and high pressures

	Haplokimderlite melts			Basaltic melts		
Pressures (GPa)	0.1	5.5	7.5	0.1	4.0	7.5
Activation energy E (kJ/mol)	138 (1.4)	163 (1.6)	172 (1.7)	177 (1.8)	144 (1.4)	153 (1.4)
Activation volume V (cm^3/mol)	2.0 (5)	2.0 (5)	2.0 (5)	−2.0 (5)	−2.0 (5)	1.5 (4)
Logη_0 (η_0 − Pa s)	−4.5 (14)	−4.5 (14)	−4.5 (14)	−4.5 (14)	−4.5 (14)	−4.5 (14)

[a]Uncertainties at the 1σ level are given in parenthesis in terms of the last units cited

barrier, i.e. respectively, the activation energy of the viscous flow (E in Eq. 4.6) with increasing pressure. A change in the activation volume of the liquid (V in Eq. 4.6) is in the first approximation a positive and constant value in the measured pressure range for a positive viscosity dependence (kimberlite and dunite melts, basaltic and more polymerized melts after a minimum). Accordingly, for a negative dependence of the viscosity of liquids (basaltic and more polymerized melts up to a minimum of this dependence), the activation energy (E in Eq. 4.6) decreases with increasing pressure. A change in the activation volume of the liquid (V in Eq. 4.6) will be, in the first approximation, a negative and constant value in the measured pressure range. Table 4.2 presents the results of calculations of these parameters for the melts studied, which were obtained from Eq. (4.6) using new experimental data on the viscosity and activation energies of the viscous flow of anhydrous kimberlite and basaltic melts over a wide range of temperatures and pressures. The results obtained fully correspond to the above theoretical presumes. Note that this result does not agree with the result of Liebsky et al. (2005), who reported that the calculated activation energy of the viscous flow for peridotite melts (198 ± 23 kJ/mol) does not depend on the pressure in a wide range of pressures ($P = 0$–8 GPa, Table 5, Liebsky et al. 2005). The constancy of activation energy of the viscous flow (E) for peridotite melts in such a wide range of pressures also contradicts the theoretical premises mentioned above.

An anomalous pressure dependence of the viscosity of polymerized (up to basalt) magmatic melts, which decreases with increasing pressure has been reported in several works (Kushiro 1980; Brearley et al. 1986; Scarfe et al. 1987; Persikov 1998; Mysen 1991; Wolf and McMillan 1995; McMillan and Wilding 2009; and references therein). It was experimentally proven that pressure dependence of viscosity and activation energies of the model silicate and natural magmatic melts are inverse and has extremes (Persikov 1998; Reid et al. 2003; Liebske et al. 2005; Persikov et al. 2018a). As an example, Fig. 4.6 shows a diagram of dependence of melts viscosity on lithostatic pressure for the melts of the system albite-diopside as well as for jadeite melt. The results obtained indicate that viscosity and activation energy of viscous flow markedly decrease with lithostatic pressure all over the range of compositions (up to basalts) and the pressure at the minimum of melts viscosity and activation

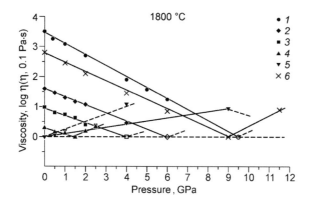

Fig. 4.6 Isothermal dependence of the viscosity of melts in the albite (Ab)–diopside (Di) join (Brearley et al. 1986; Kushiro 1980; Persikov and Bukhtiyarov 2004; Reid et al. 2003) and jadeite (Jd) melt on pressure (Persikov et al. 1989). *1*, Ab_{100} (model granite); *2*, $Ab_{80}Di_{20}$; *3*, $Ab_{57}Di_{43}$ (model basaltic melt); *4*, $Ab_{30}Di_{70}$; *5*, Di_{100}; [dashed line shows extrapolation from the model (Persikov 1998), and solid line shows extrapolation in this paper from the improved model (Persikov et al. 2015 and this study)]; *6*, Jd_{100}. The composition of melts is in mole %, and open symbols show calculated values

energies strongly depends on melt composition (Liebske et al. 2005; Persikov and Bukhtiyarov 2009).

The nature of this anomaly is still under discussion. The more probable mechanism of this anomaly can involve the change of melts structure under isochemical conditions, i.e. the transformation of Al and Si cations in the melts from four- to six-fold oxygen coordination with increasing pressure. It means that Al and Si transformation from network-forming to modifier cations is followed by an increase in depolymerization degree of melts (100NBO/T). This idea was first proposed by (Waff 1975) and has been supported by a wealth of experimental and spectroscopic data (Kushiro 1980; Wolf and McMillan 1995; Persikov 1998; Reid et al. 2003; Allwardt et al. 2007; Persikov et al. 2018a; and references therein). Note that the assumption on structural changes of Al and Si at high pressures in melts with pressure perfectly fits the phase correspondence principle, according to which the temperature increase expands the stability fields of polymerized phases and on the contrary, pressure increase expands the stability fields of depolymerized phases (Persikov 1998; Liebske et al. 2005; Karki and Stixrude 2010).

4.3.4 Viscosity of Heterogeneous Magmas

Experimental study of viscosity of subliquidus heterogeneous magmatic melts (liquid + crystals + bubbles) is the most urgent problem of magmatic rheology, because such state is natural for them in the Earth's crust depth and, probably, in the mantle.

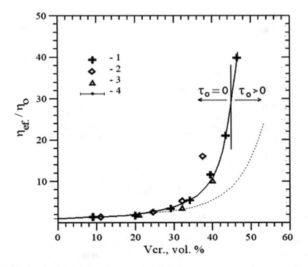

Fig. 4.7 Dependence of effective viscosity of model system (liquid + crystal) on volume content of crystals. 1, $\eta o = 3.8$ Pa s, crystal size 0.2–0.4 mm; 2, $\eta o = 0.68$ Pa s, crystal size 0.2–0.4 mm; 3, $\eta o = 0.68$ Pa s, crystal size 0.16–0.2 mm; 4, experimental errors; dash curve is consist to equation of Rosco (Persikov 1991, 1998; Persikov and Bukhtiyarov 2009)

Weak knowledge of this problem is caused not only by technical and methodological difficulties of experimental study of viscosity of such systems, especially at high pressures. To these are added the complexity of theoretical calculations, as such melts can exhibit the entire range of melt flows, referred to in rheology abnormally viscous (non-Newtonian, viscoplastic or Bingham, pseudoplastic). The separate account of the influence of crystalline phase and bubbles on viscosity of subliquidus magma can be estimated as the first approximation. As an example, Fig. 4.7 shows our results of viscosity measurements of the model (silicone) liquid + corundum crystals. Analysis of these results showed that the effective viscosity of the mixture increases significantly with increasing volume content of crystalline phase ($V_{cr.}$). At $V_{cr.} < 0.1$ dependence character $\eta = f(V_{cr.})$ is well approximated by the theoretical Einstein equation, and in the interval $0.1 < V_{cr.} < 0.30$ by the Roscoe empirical equation. However, with $V_{cr.} > 0.30$ the effect of the crystalline phase on the effective viscosity of the mixture is much greater (Fig. 4.7), which corresponds to the following empirical equation (Persikov 1998):

$$\eta_{ef.} = \eta_0(1 - V_{cr.})^{-3.35}, \quad 0 < V_{cr.} < 0.45 \tag{4.7}$$

where ηo is the viscosity of the liquid phase.

The Newtonian flow behavior of melts was observed at $V_{cr.} \leq 0.45$. At a further growth of crystal content, the Newtonian behavior of melts changed to the Bingham one with the limit of fluidity equal to 2900 din/cm^2.

As for a separate effect of bubbles on the effective viscosity of the mixture, then the data obtained by us agree well with the results from (Uhira 1980; Pal 2003) and at $V_{b.} \leq 0.45$ are described by the following empiric equation:

$$\eta_{ef.} = \eta_o(1 - 1.5V_{b.})^{-0.55}, \quad 0 < V_{b.} \leq 0.45 \tag{4.8}$$

where $V_{b.}$—volume fraction of bubbles in the liquid.

In this connection it is useful to note that the maximum content of crystal and fluid phases in magmas, reliably taken into account in our model was 45 vol%, because according to experimental data, exceeding of this limit leads to a fundamental change in the rheology of magmas—Newtonian flow regime, observed for almost all natural magmas, transforms into non-typical Bingham flow regime with yield stress increasing with increasing the content of these phases (Shaw et al. 1968; Pal 2003; Persikov and Bukhtiyarov 2009; and others). According to (Sparks et al. 2009; Russell et al. 2012; McMillan and Wilding 2009), the content of mantle and crustal xenoliths in basaltic and kimberlite magmas (including diamonds) during their evolution is not more than ~45 vol%, which is close to the threshold value of 50–60 vol%, when magma eruption becomes impossible (Kavanagh and Sparks 2009; Russell et al. 2012).

4.3.5 Structural-Chemical Model of Calculations and Prediction of Magmatic Melts Viscosity at High and Super High Pressures

A few models to predict and calculate temperature and concentration dependences of viscosity of magmatic melts at ambient pressure have been published (Giordano and Dingwell 2003; Hui and Zhang 2007; Giordano et al. 2008) after the pioneer works of (Bottinga and Weill 1972; Shaw 1972). The positive feature of these models is a very large temperature range to predict viscosity of silicate and magmatic melts: (700–1600 °C, Giordano et al. 2008); 300–1700 °C, (Hui and Zhang 2007). However, the essential disadvantages typical for all of them restrict the possibility to use these models for realistic prediction of magma viscosity. The main drawbacks of those models are as follows (see Table 4.3): (1) empiric approach with numerous fitting parameters: 285 (Bottinga and Weill 1972), 10 (Giordano and Dingwell 2003; Giordano et al. 2008), 37 (Hui and Zhang 2007); (2) low accuracy; (3) the effect of pressure, crystal and bubble content is ignored; (4) the effect of Al and Si ratio in melts (5) and the influence of oxygen fugacity are not considered either; (6) the effect of volatile content in melts is not considered in (Bottinga and Weill 1972; Giordano and Dingwell 2003); (7) only the effect of total water content on viscosity of silicate and magmatic melts is included into a few models (Shaw 1972; Hui and Zhang 2007; Giordano et al. 2008); (8) the amphoteric nature of dissolved water, and its polymerized effect in kimberlite and ultrabasic melts (see Table 4.1) is not

Table 4.3 Comparison of models suggested for calculating and predicting the viscosity of magmatic melts

Model	Shaw (1972)	Bottinga and Weill (1972)	Giordano et al. (2008)	Hui and Zhang (2007)	Persikov and Bukhtiyarov (2009; this study)
Basic principle	Empirical	Empirical	Empirical	Empirical	Structural-chemical, theoretical
Composition range of melts	Rhyolite-basalt	Rhyolite-basalt	Rhyolite-peridotite	Rhyolite-peridotite	Rhyolite-peridotite
Temperature range, °C	800–1200	1000–1400	700–1600	300 (?)–1600	Near-liquidus melts (T = Tm. ± 0.2 Tm.)
Volatiles effect	H_2O, up to 5 wt%	No	No	H_2O, up to 5 wt%	OH^-, H_2O, F^-, Cl^-, CO_2, CO_3^{2-}, up to any real contents
Pressure effect	No	No	No	No	Up to 12 GPa
Ration of $Al^{3+}/(Al^{3+} + Si^{4+})$	No	No	No	No	Yes
Ration of $Fe^{2+}/(Fe^{2+} + Fe^{3+})$	No	No	No	No	Yes
Ration of $Al^{3+}/(Na^+ + K^+ + Ca^+ + Mg^+ + Fe^{2+})$	No	No	No	No	Yes
Volume content of crystals and bubbles	No	No	No	No	Up to 45 vol%
Error, ±rel. %	50–100	50–100	250	200	30

Note Experimental uncertainty of the viscosity measurements is 15–30% rel. at ambient and high pressures

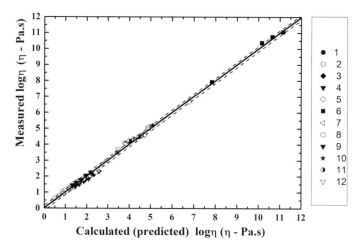

Fig. 4.8 Comparison of experimental viscosity values with calculated (predicted) values using a new model (T is the temperature in K). **1**—basanite (NIQ), T = 1573, (Whittington et al. 2000); **2**—tephrite, T = 1720 ÷ 1464, (Whittington et al. 2000); **3**—phonolite, T = 1816–1618, (Whittington et al. 2001); **4**—trachyte, T = 1929–1777, (Whittington et al. 2001); **5**—model rhyolite (HPG8), T = 1453, (Dingwell et al. 1996); **6**—model rhyolite (HPG8Na, HPG8K), T = 1073–927, (Hess et al. 1995); **7**—model rhyolite (AOQ), T = 1673–1423, C_{H_2O} = 0, 03–5, 0 мас. %, (Schulze et al. 1996); **8**—andesite, T = 1978–1650, (Richet et al. 1996); **9**—andesite, T = 1867–1670, (Neuville et al. 1993); **10**—rhyolite, T = 1173–973, C_{H_2O} = 8, 8 мас. %, (Burnham 1964); **11**—dacite, T = 1473–1375, (Giordano et al. 2008) **12**—phonolite, T = 1473–1326, (Romano et al. 2003)

considered either. Our new model is an advanced version of the previously published models (Persikov and Bukhtiyarov 2009) which has no the mentioned above disadvantages (see Table 4.3). Figure 4.8 shows the comparison of all the recently published experimental data on viscosity of magmatic melts and calculated values obtained using our model. The field of uncertainty for predicting data on magma viscosity using the new model ($2\sigma = 0.228$ in log scale or $\pm 30\%$ rel.) is presented in Fig. 4.6. It is necessary to emphasize high accuracy of calculations and prediction of viscosity of silicate and magmatic melts by the suggested model as compared to those suggested earlier (see Table 4.3). Moreover, the developed structural-chemical model allows to predict and calculate viscosity of near liquidus silicate and magmatic melts as a function of the following parameters: P; Pfl; T; melt composition, including H_2O, OH^-, CO_2, CO_3^{2-}, F^-, Cl^-; the cation rations: $Al^{3+}/(Si^{4+} + Al^{3+})$, $Al^{3+}/(Na^+ + K^+ + Ca^{2+} + Mg^{2+} + Fe^{2+})$, $Fe^{2+}/(Fe^{2+} + Fe^{3+})$, and a volume content of crystals and bubbles (up to 0.45).

Thus the characteristic features of that model are as follows: (1) structural chemical approach; (2) ultimate simplicity of analytical dependences; (3) high accuracy of prediction (± 30 rel. % for viscosity, and ± 1.0 rel. % for activation energy), which is consistent with the uncertainties of experimental data on viscosity, especially at high pressures.

4.3.6　Example of Successful Application of the Obtained Experimental and Theoretical Results to Natural Processes

A new version of our model allows to predict characteristic features of viscosity of basaltic and kimberlite magmas in the course of their origin in the mantle, evolution during ascent to the crust, and volcanic eruptions. The T and P values are taken from (Wyllie 1980; Yoder 1976; Dalton and Presnall 1998; Dasgupta and Hirschmann 2006; Kavanagh and Sparks 2009; Mitchell 2008; Russell et al. 2012; Sparks et al. 2009). Figure 4.9 shows the dynamics of change in the viscosity of kimberlite and basaltic magmas with different water contents under the T, P-conditions of their origin in the upper mantle and the subsequent evolution during the ascent of such magmas from mantle to crust. The data obtained with a high degree of correlation (Fig. 4.9) show that the viscosity of kimberlite magma in the course of its origin, evolution, and ascent decreases by a factor of ~4 despite the considerable decrease in the temperature of kimberlite magma (by ~150 °C) (Kavanagh and Sparks 2009; Sparks et al. 2009) during its ascent to the surface and partial crystallization and

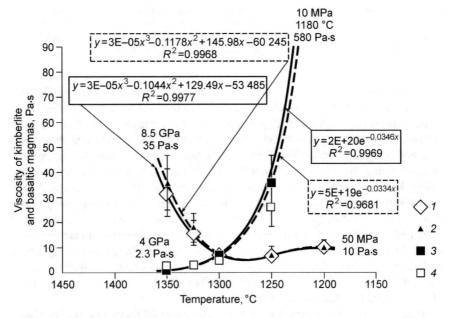

Fig. 4.9 Dynamics of change in the viscosity of kimberlite and basaltic magmas during their origin and ascent to the crust as well as volcanic eruptions. *1, 2*, kimberlite magma: *1*, water content in magma $X(H_2O) = 0.5$ wt% (solid curve); *2*, X $(H_2O) = 8.0$ wt% (dashed curve); *3, 4*, basaltic magma: *3*, water content in magma X $(H_2O) = 0.5$ wt% (solid curve); *4*, X $(H_2O) = 8.0$ wt% (dashed curve) (Pressures used in the diagram: 8.5 GPa–50 MPa for kimberlite magma and 4 GPa–10 MPa for basaltic magma; the error of the calculations based on the model ±30 rel. % is shown on the diagram)

degassing. The viscosity of the formed kimberlite melt is equal to ~40 Pa s at $P = 8.5$ GPa and $T = 1350$ °C, during the partial melting of mantle carbonated peridotite (degree of melting <1 wt%) at the depth of 150–300 km (Wyllie 1980; Brey et al. 2009; Dalton and Presnall 1998; Dasgupta and Hirschmann 2006; Mitchell 2008; Sparks et al. 2009). The volume fractions of the crystal and bubbles in the melts are Vcr. = 20 and Vb. = 5 vol%, respectively and water content in the melt is ~8 wt%, $(X(OH^-)$ ~2.0 wt%, and molecular water $X(H_2O)$ ~ 6.0 wt%. However, during the formation of kimberlite pipes, dikes, and sills, the viscosity of kimberlite melt near the surface is considerably lower. It is equal to ~10 Pa s at $P = 50$ MPa and $T = 1180$ °C, the volume fractions of the crystal and bubbles in the melts are Vcr. = 35 vol% and Vb. = 5 vol%, respectively, and water content in magma is $X(OH^-)$ ~ 0.5 wt% (Fig. 4.9).

The viscosity of basaltic magma during its ascent from mantle to crust increases by more than two orders of magnitude (the high degree of correlation is shown in Fig. 4.9). During the origin of basaltic magmas in the asthenosphere, at depths of ~100 km (Yoder 1976), the formed basaltic melt has minimum viscosity (~2.3 Pa s) at $P = 4.0$ GPa and $T = 1350$ °C, water content in the melt is $X(OH^-) = 3.0$ wt% and $X(H_2O) = 5.0$ wt%. B u t at the final stage of evolution (e.g., during the eruption of basaltic volcanoes), the viscosity of basaltic melt is much higher (~600 Pa s) at $P = 10$ MPa, $T = 1180$ °C, Vcr. = 30 vol%, Vb. = 15 vol%, $X(OH^-) = 0.5$ wt% (Fig. 4.9). The lower viscosity of basaltic magma (at 3.0–4.0 GPa), as compared to ultramafic (kimberlite) magma (at 4.0–8.5 GPa), at the temperatures 1325–1350 °C can be mainly explained by the reverse effect of pressure and chemically dissolved water (OH^-) on the viscosity of melts: the viscosity of basaltic melts decreases, whereas that of kimberlite melts viscosity increases with pressure and (OH^-) content. Such a seemingly anomalous effect is also amplified by the difference between the content of the crystal and bubbles in the magmas under the given T, P-conditions: Vcr. = 10–15 vol% and Vb. = 1.0 vol% in the forming basaltic magma and Vcr. = 20 vol% and Vb. = 5 vol% in parental kimberlite magma (Fig. 4.9). The analysis of the results on the dynamics of change in the viscosity of kimberlite and basaltic magmas (Fig. 4.9) in the course o f their origin in the mantle and the subsequent evolution during their ascent to the crust leads to an important conclusion on the possible role of dissolved water. According to the data obtained, the dissolution of water (up to ~8 wt%) in kimberlite and basaltic magmas cannot have a significant influence on the dynamics of change in the viscosity of those magmas within the error (~30 rel.%) of prediction calculations (Fig. 4.9) whereas pressure and the volume content of crystals and bubbles have a considerably greater effect on the viscosity of those magmas as compared t o the effect of dissolved water.

The influence of the second important fluid component of kimberlite and basaltic magmas (CO_2) on their viscosity was also taken into account. According to scarce experimental data, summarized, e.g., in (Lange 1994), the dissolution of CO_2 has almost no influence on the viscosity of magmatic melts at ~2.5 GPa. That conclusion is true for basaltic magmas, which are characterized by the low solubility of CO_2. However, a considerable amount of CO_2 can be dissolved in kimberlite magma at 2.5–8.5 GPa due to: (1) the reaction between carbonatite melts and mantle peridotite, with the dissolution of mainly orthopyroxene (Brey et al. 2009; Chepurov et al. 2013; Kavanagh and Sparks 2009; Russell et al. 2012; Sparks et al. 2009); (2) the possible reaction of decarbonatization at ~2.5 GPa carbon dioxide, which forms during those reactions, may partly dissolve in kimberlite melt in carbonate form, but, predominantly, it forms abundant fluid (CO_2) micro-bubbles in the melt. The volume of this fluid phase increases with the subsequent decrease in pressure and temperature. However, the viscosity of subliquidus kimberlite magma (liquid + crystals + CO_2 bubbles) rather decreases under the decisive influence of pressure (Fig. 4.9) than increases.

4.4 Conclusions

1. The generalized regularities of temperature, pressure, concentration and phase dependences of viscosity of fluid-magmatic systems in the full range of compositions of natural magmas (acidic-ultramafic) and thermodynamic parameters of the earth's crust and upper mantle were established in conjunction with the structural features of the melts. The range of viscosity values for the studied melts is extremely large: from $10^{6.3}$ Pa s (water bearing granitic melt (0.5 wt%) at atmospheric pressure, temperature 1100 °C and 100NBO/T = 4.9) to 10^{-1} Pa s (water free dunite melts at $P = 100$ MPa, temperature 1800 °C and 100NBO/T = 340).

2. It is now reliably established that from a wide range of fluid composition of the magma (H_2O, CO_2, HCl, NaCl, HF, NaF, H_2S), water and CO_2 dissolved in magmas in two forms, has a decisive influence on the viscosity of such magmas. Chemical dissolved water (OH^-) and carbon dioxide (CO_3^{2-}) strongly reduces the viscosity, significantly increases the degree of depolymerization of the melt. Physical ones in the forms of molecular H_2O and CO_2 weakly reduces the viscosity of the magma, without changing its degree of depolymerization, and pseudo-binary system magma-H_2O–CO_2 with a good degree of approximation simulates the rheological behavior of such magmas in the entire range of depths of the earth's crust and upper mantle. Temperature dependence of viscosity for all studied melts is exponential at ambient pressure, moderate and high fluid and lithostatic pressures. The new reliable values of the activation energies of the viscous flow for all studied fluid-magmatic melts at ambient, moderate and high pressures were obtained for the first time with high accuracy (±1% rel.). The range of values of the activation energies of the viscous flow for the studied melts ranged from 284 ± 2.8 kJ/mol (water bearing granitic melts (0.5 wt%) at ambient pressure) to 137 ± 1.4 kJ/mol (water free dunite melts at $P = 100$ MPa).

3. Pressure dependence of viscosity and activation energies of viscous flow for all studied melts are inverse and has extremes. Viscosity and activation energy markedly decrease with lithostatic and water pressure all over the range of compositions of polymerized melts (from acidic to mafic), except for depolymerized melts, such as diopside, kimberlite and dunite melts for which viscosity and

activation energy increase with lithostatic and water pressure. Pressures at the minimum viscosity and activation energy strongly depends on melt composition decreasing with increasing the degree of depolymerization.

4. An improved structural-chemical model is proposed to predict and calculate the viscosity of silicate and magmatic melts as functions of the following parameters: P_{total}; Pfl; T; melt composition, including volatiles (H_2O, OH^-, CO_2, CO_3^{2-}, F^-, Cl^-); cation ratios: $Al^{3+}/(Si^{4+} + Al^{3+})$, $Al^{3+}/(Na^+ + K^+ + Ca^{2+} + Mg^{2+} + Fe^{2+})$, $Fe^{2+}/(Fe^{2+} + Fe^{3+})$, and a volume content of crystals and bubbles (up to 0.45, as applied to magma viscosity). The new model is specified by: (1) structural chemical approach; (2) ultimate simplicity of analytical dependences; (3) high accuracy of the prediction ($\pm 30\%$ rel. for viscosity, and $\pm 1.0\%$ rel. for activation energy), which is consistent with the uncertainties of the experimental data on viscosity, especially at high pressures.

5. Prediction results on the ascent of diamonds carrier kimberlite magmas from mantle to the crust with the appreciable acceleration is an unique in that the viscosity of kimberlite magmas will decrease by more than 4 times during this process despite of the decreasing magmas temperature by more than ~150 °C. On the contrary, the viscosity of basaltic magmas will increase by more than 2.5 orders of magnitude during their origin, evolution and the ascent from mantle to the crust up to volcanic eruption.

Acknowledgements The work was carried out under the theme NIR AAAA-A18-118020590141-4 of the IEM RAS and was supported by the program № 8 of the Presidium of RAS. We thank A. N. Nekrasov (IEM RAS) for his generous help during electron microprobe analysis of samples, G. V. Bondarenko (IEM RAS) and S. Newman (Caltech, USA) for their help during FTIR and Raman spectroscopy study of samples. We thank A. G. Sokol from IGM SB RAS for his help during experimental study of the viscosity of kimberlite and basalt melts at high pressures. We are grateful to D. M. Sultanov from IEM RAS for his help in the preparation of high quality drawings.

References

Adam G, Gibbs JH (1965) On the temperature dependence of cooperative relaxation properties in glass-forming liquids. J Chem Phys 43:39–146

Allwardt JR, Stebbins JF, Terasaki H, Du LS, Frost DJ, Withers AC, Hirschmann MM, Suzuki A, Ohtani E (2007) Effect of structural transitions on properties of high-pressure silicate melts: Al-27 NMR, glass densities, and melts viscosities. Am Mineral 92:1093–1104

Behrens H, Schulze F (2003) Pressure dependence of melt viscosity in the system NaAlSi$_3$O$_8$-CaAlSi$_2$O$_6$. Amer Min 88:1351–1363

Bottinga Y, Weill DF (1972) The viscosity of magmatic silicate liquids: a model for calculation. Am J Sci 272:438–475

Brearley M, Dickinson JE Jr, Scarfe M (1986) Pressure dependence of melt viscosities on the join diopside-albite. Geochim Cosmochim Acta 30:2563–2570

Brey GP, Bulatov VK, Girnis AV (2009) Influence of water and fluorine on melting of carbonated peridotite at 6 and 10 GPa. Litos 112(1):249–259

Brush SG (1962) Theories of liquid viscosity. Chemic Rev 62:513–548

Burnham CW (1964) Viscosity of a H_2O rich pegmatite melt at high pressure (Abstract). Geol Soc Am Special Paper 76:26

Carron JP (1969) Vue d'ensemble sur la rheology des magmas silicates naturels. Bull Soc Franc Min Crestallogr 92:435–446

Champallier R, Bystricky M, Arbaret L (2008) Experimental investigation of magma rheology at 300 MPa: from pure hydrous melt to 75 vol. % of crystals. Earth Planet Sci Lett 267:571–583

Chepurov AI, Pokhilenko NP (2015) Experimental estimation of the kimberlite melt viscosity. Dokl Earth Sci 462(2):592–595

Chepurov AI, Zhimulev EI, Agavonov IV et al (2013) The stability of ortho-and clinopyroxenes, olivine, and garnet in kimberlitic magma. Russ Geol Geoph 54(14):533–544

Dalton JA, Presnall DC (1998) The continuum of primary carbonatite-kimberlitic melt compositions in equilibrium with lherzolite data from the system $CaO-MgO-Al_2O_3$- SiO_2-CO_2 at 6 GPa. J Petrol 39(11–12):1953–1964

Dasgupta R, Hirschmann MM (2006) Melting in the Earth's deep upper mantle caused by carbon dioxide. Nature 440(7084):659–662

Dingwell DB (2006) Transport properties of magmas: diffusion and rheology. Elements 2:281–286

Dingwell DB, Romano C, Hess KU (1996) The effect of water on the viscosity of a haplogranitic melt under P-T-X conditions relevant volcanism. Contrib Min Petrol 124:19–28

Dingwell DB, Copurtial P, Giordano D, Nichols ARL (2004) Viscosity of peridotite liquid. Earth Planet Sci Lett 226:127–138

Dobson DF, Jones AP, Rabe R, Sekine T, Kurita K et al (1996) In-situ measurement of viscosity and density of carbonate melts at high pressure. Earth Planet Sci Lett 143:207–215

Frenkel Y (1975) The kinetic theory of liquids. USSR Academy Press, Moscow, 415 p. (in Russian)

Fujii T, Kushiro I (1977) Density, viscosity, and compressibility of basaltic liquid at high pressures. Carneqie Inst Year Book 76:419–424

Giordano D, Dingwell DB (2003) Non-Arrhenian multicomponent melt viscosity: a model. Earth Planet Sci Lett 208:337–349

Giordano D, Romano C, Papale P, Dingwell DB (2004) The viscosity of trachytes, and comparison with basalts, phonolites, and rhyolites. Chem Geol 213(1–3):49–61

Giordano D, Russel JK, Dingwell DB (2008) Viscosity of magmatic liquids: a model. Earth Planet Sci Lett 271:123–134

Hess KU, Dingwell D (1996) Viscosities of hydrous leucogranitic melts: a non-arrhenian model. Amer Miner 81:1297–1300

Hess KU, Dingwell DB, Webb SL (1995) The influence of excess alkalis on the viscosity of a haplogranitic melt. Amer Mineral 80:297–304

Hui H, Zhang Y (2007) Toward a general viscosity equation for natural anhydrous and hydrous silicate melts. Geochim Cosmochim Acta 71:403–406

Ivanov OK, Stiegelmeier SV (1982) The viscosity and temperature of crystallization of melts of ultramafic rocks. Geochimiya 3:330–337 (in Russian)

Karki BB, Stixrude L (2010) Viscosity of $MgSiO_3$ liquid at mantle conditions: implications for early magma ocean. Science 96:740–742

Kavanagh JL, Sparks RSJ (2009) Temperature changes in ascending kimberlite magma. Earth Planet Sci Lett 286:404–413

Kono Y, Park C, Kenney-Benson C, Shen G, Wang Y (2014) Toward studies of liquids at 605 high pressures and high temperatures: combined structure, elastic wave velocity, and viscosity 606 measurements in the Paris-Edinburgh cell. Phys Earth Inter 228:269–280

Kushiro I (1980) Viscosity, density and structure of silicate melts at high pressures, and their petro-logical applications. In: Hargraves RB (ed) Physics of magmatic processes. Princeton University Press, New Jersey, pp 93–120

Lange RA (1994) The effect of H_2O, CO_2, and F on the density and viscosity of silicate melts. In: Carrol MR, Holloway JR (eds) Reviews in mineralogy. Volatiles in magmas MSA, Washington, vol 30, pp 331–369

Lebedev EB, Khitarov NI (1979) Physical properties of magmatic melts. Nauka, Moscow, 200 p. (in Russian)

Liebske C, Schmickler B, Terasaki H, Poe BT, Suzuki A, Funakoshi KI, Ando R, Rubie DC (2005) The viscosity of peridotite liquid at pressures up to 13 GPa. Earth Planet Sci Lett 240:589–604

Litvin YA (1991) Physical and chemical studies of deep melting of the Earth. Nauka, Moscow, 314 p. (in Russian)

McMillan PF, Wilding MC (2009) High pressure effects on liquid viscosity and glass transition behaviour, polyamorphic phase transitions and structural properties of glasses and liquids. J. Non Crystall Solids 355:722–732

Mitchell RH (2008) Petrology of hypabyssal kimberlites: relevance to primary magma compositions. J Volcanol Geochem Res 174(1–3):1–8

Mysen BO (1991) Relation between structure, redox equilibria of iron, and properties of magmatic liquids. In: Perchuk LL, Kushiro I (eds) Physical chemistry of magmas, vol 9. Adv Phys Geochem. Springer, New York, pp 41–98

Mysen BO, Virgo D, Scarfe CM (1979) Viscosity of silicate melts as a function of pressure: structural interpretation. Carnegie Inst Washington Yearb 78:551–556

Neuville DR, Richet P (1991) Viscosity and mixing in molten (Ca, Mg) pyroxenes and garnets. Geochim Cosmochim Acta 55(4):1011–1019

Neuville DR, Courtial P, Dingwell DB, Richet P (1993) Thermodynamic and rheological properties of rhyolite and andesite melts. Contrib Min Petrol 113:572–581

Ohtani E, Suzuki A, Audo A, Funakoshi A, Katayama Y (2005) Viscosity and density measurements of melts and glasses at high pressure and temperature by using the multi-anvil apparatus and synchrotron X-ray radiation. In: Wang J et al (eds) Advances in high-pressure technology for geophysical application. Elsevier, Amsterdam, pp 195–210

Pal R (2003) Rheological behavior of bubble-bearing magmas. Earth Planet Sci Lett 207:165–179

Persikov ES (1984) Viscosity of magmatic melts. Nauka, Moscow, 160 p. (in Russian)

Persikov ES (1991) The viscosity of magmatic liquids: experiment, generalized patterns; a model for calculation and prediction; application. In: Perchuk LL, Kushiro I (eds) Physical chemistry of magmas, vol 9. Adv Phys Geochem. Springer, New York, pp 1–40

Persikov ES (1998) Viscosities of model and magmatic melts at the pressures and temperatures of the Earth's crust and upper mantle. Russ Geol Geophys 39(12):1780–1792

Persikov ES, Bukhtiyarov PG (2002) Unique high gas pressure apparatus to study fluid—melts and fluid—solid—melts interaction with any fluid composition at the temperature up to 1400 °C and at the pressures up to 5 kbars. J Conf Abs 7(1):85

Persikov ES, Bukhtiyarov PG (2004) Experimental study of the effect of lithostatic and aqueous pressures on viscosity of silicate and magmatic melts. In: Zharikov VA, Fedkin VV (eds) A new structural-chemical model to calculate and predict the viscosity of such melts. Experimental mineralogy. Some results on 540 the Century's frontier. Nauka, Moscow, issues 1, pp 103–122. (in Russian)

Persikov ES, Bukhtiyarov PG (2009) Interrelated structural chemical model to predict and calculate viscosity of magmatic melts and water diffusion in a wide range of compositions and T-P parameters of the Earth's crust and upper mantle. Russ Geol Geophys 50(12):1079–1090

Persikov ES, Kushiro I, Fujii T, Bukhtiyarov PG, Kurita K (1989) Anomalous pressure effect on viscosity of magmatic melts. In: DELP international symposium phase transformation at high pressures and high temperatures: applications to geophysical and petrological problems. Misasa, Tottori-ken, Japan, pp 28–30

Persikov ES, Zharikov VA, Bukhtiyarov PG, Pol'skoy SF (1990) The effect of volatiles on the properties of magmatic melts. Eur J Min (2):621–642

Persikov ES, Newman S, Bukhtiyarov PG, Nekrasov AN, Stolper EM (2010) Experimental study of water diffusion in haplobasaltic and haploandesitic melts. Chem Geol (276):241–256

Persikov ES, Bukhtiyarov PG, Sokol AG (2015) Change in the viscosity of kimberlite and basaltic magmas during their origin and evolution (*prediction*). Russ Geol Geophys 56:883–892

Persikov ES, Bukhtiyarov PG, Sokol AG (2017) Viscosity of hydrous kimberlite and basaltic melts at high pressures. Russ Geol Geophys 58:1093–1100

Persikov ES, Bukhtiyarov PG, Sokol AG (2018a) Viscosity of haplokimberlite and basaltic melts at high pressures. Chem Geol 497:54–63

Persikov ES, Bukhtiyarov PG, Sokol AG (2018b) Viscosity of depolymerized dunite melts under medium and high pressures. Geochem Int 56(12):1148–1155

Poe BT, Romano C, Liebske C, Rubie D, Terasaki H, Suzuki A, Funakoshi K (2006) High-temperature viscosity measurements of hydrous albite liquid using in-situ falling-sphere viscometry at 2.5 GPa. Chem Geol 229:2–9

Reid JE, Suzuki A, Funakoshi KI, Terasaki H, Poe BT, Rubie DC, Ohtani E (2003) The viscosity of $CaMgSi_2O_6$ liquid at pressures up to 13 GPa. Phys Earth Planet Int 139:45–54

Richet P (1984) Viscosity and configurational entropy of silicate melts. Geochim Cosmochim Acta 48:471–483

Richet P, Lejeune AM, Holt F, Roux J (1996) Water and the viscosity of andesite melts. Chem Geol 128:185–197

Romano C, Giordano D, Papale P, Mincione V, Dingwell DB, Rosi M (2003) The dry and hydrous viscosities of alkaline melts from Vesuvius and Phlegrean Field. Chem Geol 202:23–38

Russell JK, Giordano D, Dingwell DB, Hess KU (2002) Modelling the non-Arrhenian rheology of silicate melts: numerical consideration. Eur J Miner 14:417–427

Russell JK, Porritt LA, Lavallee Y, Dingwell DB (2012) Kimberlite ascent by assimilation- fuelled buoyancy. Nature 481:352–357

Scarfe GM (1986) Viscosity and density of silicate melts. In: Scarfe GM (ed) Silicate melts mineral association of Canada short course handbook, vol 12. Canada, pp 36–56

Scarfe CM, Mysen BO, Virgo D (1987) Pressure dependence of the viscosity of silicate melts. In: Mysen B (ed) Magmatic processes: physico-chemical principles. Cheochem Soc Spec Publ 1:59–68

Schulze F, Behrens H, Holtz F, Roux J, Johannes W (1996) The influence of H_2O on the viscosity of a haplogranitic melt. Amer Min 81:1155–1165

Shaw HR (1972) Viscosities of magmatic silicate liquids: an empirical method of prediction. Amer J Sci 272(11):870–893

Shaw HR, Wright TL, Peck DL, Okamura R (1968) The viscosity of basaltic magma: an analysis of field measurements in Makaopuhi lava lake, Hawaii. Amer J Sci 266:225–264

Sheludyakov LN (1980) Composition, structure and viscosity of silicate and aluminosilicate melts. Science, Alma-ATA, 157 p. (in Russian)

Sokol AG, Palyanov YN (2008) Diamond formation in the system MgO-SiO_2-H_2O-C at 7.5 GPa and 1600 °C. Contrib Min Petrol 121:33–43

Sparks RSJ, Brooker RA, Field M, Kavanagh J, Schumacher JC, Walter MJ, White J (2009) The nature of erupting kimberlite melts. Lithos 112(S):429–438

Stolper EM (1982) Water in silicate glasses: an infrared spectroscopic study Contr Miner Petrol (81):1–17

Suzuki Akio, Ohtani Eiji, Terasaki Hidenori, Funakoshi Kenichi (2005) Viscosity of silicate melts in $CaMgSi_2O_6$-$NaALSi_2O_6$ system at high pressure. Phys Chem Min 32:140–145

Uhira K (1980) Experimental study on the effect of bubble concentration on the effective 706 viscosity of liquids. Bull Earth Res Inst 56:857–871

Vetere F, Behrens H, Holtz F, Vilardo G, Ventura G (2010) Viscosity of crystal-bearing melts and its implication for magma ascent. J Miner Petrol Sci 105:151–163

Volarovich MP (1940) Investigation of the viscosity of molten rocks. Zap All-Russ Min Soc 69(2–3):310–313 (in Russian)

Waff HS (1975) Pressure-induced coordination changes in magmatic liquids. Geophys Res Lett 2:193–196

Whittington A, Richet P, Holtz F (2000) Water and the viscosity of depolymerized aluminosilicate melts. Geochim Cosmochim Acta 64:3725–3736

Whittington A, Richet P, Holtz F (2001) The viscosity of hydrous phonolites and trachytes. Chem Geol 174:209–223
Wolf GH, McMillan PF (1995) Pressure effects on silicate melt structure and properties. In: Stebbins JF et al (eds) Reviews in mineralogy. Structure, dynamics and properties of silicate melts, vol 32. MSA, Washington, pp 505–561
Wyllie PJ (1980) The origin of kimberlite. J Geophys Res 85:6902–6910
Yoder HS (1976) Generation of basaltic magmas. National Academy of Sciences, Washington D.C., 265 p

Chapter 5
Crystallization of Cpx in the Ab-Di System Under the Oscillating Temperature: Contrast Dynamic Modes at Different Periods of Oscillation

Alexander G. Simakin, Vera N. Devyatova, and Alexey N. Nekrasov

Abstract The evolution of the texture of a partially crystallized melt at a temperature oscillating around liquidus has been experimentally investigated. The model pseudobinary albite—diopside system was used, in which the clino-pyroxene is the only phase formed under the experimental conditions P = 200 MPa, T = 1050–1150 °C, C_{H2O} = 3.3 wt%. All products of the experiments were characterized by the Crystal Size Distributions (CSD) measured by BSE image analysis. Due to the high nucleation rate of clinopyroxene and the slow cooling rate of about 100 °C/min a large volume of quenching crystals was formed with residual glass approaching pure albite in composition. The quench tails were identified, and parts of the CSD formed prior to quenching were restored. In the CSD recovery procedure, we used an estimate of the equilibrium volume of Cpx, based on an experimentally verified melting diagram. We also assume that the crystal growth rate at quenching can be approximated as a size-independent parameter. In the experiment yielding the largest crystals, the Cpx cores formed prior to quenching were identified using microprobe observations, and their sizes were estimated from BSE images. The measured size distribution of the cores of Cpx crystals turned out to be close to the restored CSD. In our experiments, the variable period of temperature oscillations in the same range of 1135 < T < 1155 °C was a parameter controlling the texture. With a large period of oscillation (about an hour), all crystals dissolve at the hot stage and reprecipitate in a cold one, which leads to the formation of log-linear CSDs. The shortest oscillation period of 4–10 min results in a gradual maturation of the texture with time with a decrease in the number density of crystals, and CSD evolved to a log-normal shape with a maximum. The growth rate independent of the crystal size is essential for effective ripening with temperature oscillations. Simple calculations show that with diffusion control of both dissolution and growth, only the first temperature oscillation affects CSD.

A. G. Simakin (✉) · V. N. Devyatova · A. N. Nekrasov
D.S. Korzhinskii Institute of Experimental Mineralogy, Russian Academy of Science, Osipyana 4, Chernogolovka, Moscow Region 142432, Russia
e-mail: simakin@ifz.ru

A. G. Simakin
O.Yu. Schmidt Institute of Physics of the Earth, Russian Academy of Science, Bolshaya Gruzinskaya, 10-1, Moscow 123242, Russia

© The Editor(s) (if applicable) and The Author(s), under exclusive license to Springer Nature Switzerland AG 2020
Y. Litvin and O. Safonov (eds.), *Advances in Experimental and Genetic Mineralogy*, Springer Mineralogy, https://doi.org/10.1007/978-3-030-42859-4_5

Keywords Melt · Crystal growth · Nucleation · CSD

5.1 Introduction

Shortly after the birth of a new Academic cluster (settlement, village) in Chernogolovka in 1969 a small research center for Earth Science was created, called the Institute of Experimental Mineralogy. Mark Epelbaum was close to the founding fathers of IEM and from the very beginning he headed igneous studies, aimed at the investigation of silicate melts and especially their dynamic properties, such as diffusion (Chekhmir et al. 1991), viscosity (Epel'baum 1980) and crystallization kineticks. As a man who began his career in the field of industrial glass and ceramics, M. B. Epel'baum realized the importance of understanding how crystalline phases nucleate and grow from a melt for the manufacture of glass and silicate materials. Over the next decades, magmatism lab researchers learned how the dynamics and kinetics of crystallization can affect natural processes: in volcanoes, small sills and dykes and large intrusions. Important results were obtained on the experimental measurement of the growth rates of magmatic minerals: feldspar from granite (Simakin and Chevychelov 1995), plagioclase (Simakin and Salova 2004), and clinopyroxene (Simakin et al. 2003) from hawaiitic melts. Experimental studies of magma crystallization caused by degassing were initiated by our early work with the Pietro Armienti (Simakin et al. 1999). The first results of numerical simulation of the growth and deposition of crystals in the kinetic approach using the CSD functions were obtained in Simakin et al. (1998).

In episodes of fast solidification caused by decompression and degassing, convective mixing of magmas contrasting in composition or temperature, flushing of hydrous magmas with carbonic fluid—deviations from equilibrium of the melt and magmatic minerals are high on a large scale and the kinetics of nucleation and growth becomes important. Spatially homogeneous sets of nucleating and growing crystals can be described with crystal size distribution (CSD) functions. Existing models (e.g., Brandeis et al. 1984; Toramaru 1991; Hort 1998), which neglect the effect of crystal settling are actually extended forms of the Avrami-Kolmogorov equation (Avrami 1939) and do not require the explicit use of CSD. The crystal size distribution becomes especially important for modeling the crystal settling in solidifying intrusions with a Stokes rate proportional to the square of the crystal radius. When solidification with crystal settling is simulated (Simakin et al. 1998), CSD gives an extra dimension (crystal size) and the problem generally turns from 3D into 4D ($F(t, x, y, z, R)$).

There have been attempts to measure and use CSD of magmatic minerals as such in petrological practice. Decades ago Marsh (1988) proposed a model that pretended to extract some useful information from the parameters of log-linear CSDs typical for quickly solidified magmatic rock. In this model it is assumed that magma enters the chamber, crystallizes, being perfectly mixed, and is permanently evacuated with periodic eruptions. Then, at constant growth and nucleation rate, parameters of CSD are related to the average residence time of the magma in a chamber. This model had been arbitrarily applied to a closed system, such as the lava lake (Cashman and Marsh 1988). The inadequacy of the model became apparent when Vona et al. (2011), Vona and Romano (2013) tried to use it to interpret experimental data on the dynamic melt crystallization in a rotational viscometer. There were two generations of crystals in their experiments: one formed in the main stage of experiment and second at quenching. In the first publication (Vona et al. 2011) the authors considered only those parts of CSDs that were formed during quenching, since they had log-linear form and could be interpreted in a conventional manner (Marsh 1988). In the second paper, based on these data, Vona and Romano (2013) dropped the quenching parts and get reasonable estimates of growth rates from the rest of the CSDs based on the results of viscosity measurement.

About decade ago, Simakin and Bindeman (2008) theoretically investigated effect of a series of dissolution and precipitation events in the system crystals-melt. Formation of logarithmic normal like shape of CSD typical for zircon and quartz phenocrysts in some large volume rhyolites was explained. Assumed periodic rapid heating and gradual slow cooling is a common mode for the long-lived magma chambers under volcanoes. The parameters of CSD can be semi-quantitatively interpreted with this model to estimate the number and strength of such events. Several experimental studies have focused on the evolution of CSD with periodic oscillations in temperature (Cabane et al. 2005; Mills and Glazner 2013). An experimental study conducted by Mills and Glazner (2013) did find ripening of CSD of olivine in a basalt melt with the applied temperature oscillations with a decrease in the volume number content of crystals.

Later, model of Simakin and Bindeman (2008) was questioned in Bindeman and Melnik (2016) based on the results of the simplified numerical simulation of the dissolution and growth of zircon crystals. The main difference in the models used in (Simakin and Bindeman 2008; Bindeman and Melnik 2016) is representation of the crystal growth rate as size independent or diffusion controlled (inversely proportional to size) processes, respectively. Direct observations of the crystallization of olivine from basalt melt in a moissanite cell lead Ni et al. (2014) to conclude that the rate is proportional to size, which is opposite to the dependence expected in diffusion control. Due to the complexity of this problem, experimental observations can make a valuable contribution to its solution.

This paper presents the results of an experimental study of clinopyroxene crystallization from the model Ab-Di melt with temperature oscillations. In experiments conducted by Mills and Glazner (2013), the temperature always oscillated below the liquidus of the phases under consideration. In our experiments, we applied oscillations with a different period when the temperature rises above the liquidus as in the periodically replenished magma chambers. We have shown that diffusional dissolution time of the largest crystals t_{dissol} becomes a parameter controlling the contrasting dynamic modes of textural evolution. We precisely measured CSD of clinopyroxene and show that the approximation of the size independent growth is applicable. Theoretically, we also demonstrated that, size-independent crystal growth rate leads to the maturation of the CSD with experimentally observed rates.

5.2   Experimental Method

Experimental strategy. For our research, we chose a simple pseudo-binary Ab-Di system. At $P_{H2O} = 2$ kbar it is characterized by a melting diagram with an essentially single diopside + melt (Di + L) field (Yoder 1966; Pati et al. 2000). It is pseudo-binary because of the presence of Ca in plagioclase and enstatite in diopside. The plagioclase + clinopyroxene + melt field exists near the solidus temperature $T_s = 820$–830 °C and a tiny phase field of plagioclase in equilibrium with melt occurs near the albite edge (see Fig. 5.1). Such an elementary diagram simplifies interpretation of the experimental data and the CSD measurements. When the fluid pressure rises from $P_{H2O} = 1$ ($C_{H2O} \approx 4$ wt%) to 2 Kbar ($C_{H2O} \approx 5.6$ wt%), T_l decreases insignificantly as reflected in the position of the Cpx liquidus for $P_{H2O} = 1$ Kbar approximately plotted in Fig. 5.1 as a line connecting corresponding diopside and albite melting points. This is explained by the fact that the first percentages of water dissolved mainly as hydroxyl formed by a reaction with an aluminosilicate melt, thereby greatly reducing the temperature of the liquidus (T_l) of the diopside. With a larger content, the water dissolves mainly in molecular form and has a smaller effect on the melting

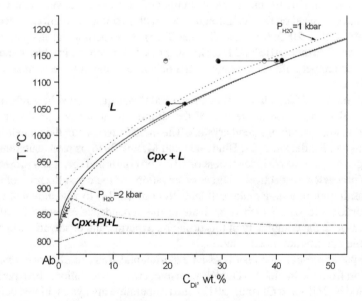

Fig. 5.1 The pseudo-binary melting diagram of Ab-Di system; the liquidus curve of Cpx at $P_{H2O} = 2$ Kbar (Yoder 1966) is shown by a short-dashed line, liquidus at $P_{H2O} = 1$ Kbar is approximately plotted on the basis of the melting temperatures of diopside and albite, since there are no direct experimental data. For $P_{H2O} = 2$ Kbar additional fields Pl + L and Cpx + Pl + L are shown. The points on the plot depict our results with the starting compositions in the Ab-Di system with an excess of silica (see text) and $C_{H2O} = 3.3$ wt%; semi-filled circles correspond to a pure melt, pair of small filled circles (some superimposed) display experiments with seeds, reflecting the original and diopside saturated compositions linked by an arrow. Our approximation of the Cpx liquidus is shown by a solid line

Table 5.1 Composition of the starting materials

	SiO_2	Al_2O_3	MgO	CaO	Na_2O	K_2O
$Ab_{80}Di_{20}$ (13)[a]	67.03 (0.34)[b]	15.50 (0.22)	3.99 (0.17)	4.35 (0.15)	8.90 (0.26)	0.22 (0.05)
$Ab_{71}Di_{29}$ (8)	66.09 (0.56)	14.15 (0.23)	5.64 (0.30)	5.94 (0.07)	7.9 (0.36)	0.14 (0.04)
$Ab_{55}Di_{45}$ (8)	65.54 (0.28)	10.49 (0.32)	9.27 (0.12)	8.50 (0.21)	6.06 (0.12)	0.14 (0.06)
diopside (seed. 11)	56.36 (0.33)	0.21 (0.05)	17.71 (0.11)	26.28 (0.11)	0.12 (0.07)	0.09 (0.04)

Electron microprobe analyses of glasses are normalized to 100%
[a] number of analyses, [b] one standard deviation

diagram. Thus chosen composition demonstrated low sensibility to the local water concentration variations caused by the probable inhomogeneity of the starting glass and induced by crystallization.

The experimental apparatus The experiments were performed in an internally heated pressure vessel (IHPV) with free volume 262 cm^3 made of a stainless steel, held vertically. Temperatures during the experiment were measured by an external unsheathed Pt-Rh thermocouple, thermal gradient free zone is 40–50 mm. The thermocouples were calibrated against the melting of Au at 1000 bar. Pressure was measured with transducers with accuracy of 5%. The temperature was maintained with a precision of ± 2.5 °C and recorded digitally. At the end of experiments the furnace was turned off and the temperature was lowered to T = 600 °C level for a few minutes at the cooling rate about 150°/min along with the vessel cooled with running water.

Starting materials. We prepared the starting mixtures using synthetic gels of albite and diopside compositions (see Table 5.1). Mixtures of powdered and carefully mixed components were loaded into a Pt capsule. Glasses with different albite contents were prepared by heating to 1450 °C in a high-temperature furnace at an ambient pressure for 5 h. Dry glasses were studied using SIM-EDS and X-ray methods. The synthesized glasses (see Table 5.1) contained a small amount (5–6 wt%) of excess SiO_2 in comparison of the target glasses of Ab-Di composition and were practically crystal free. During the next step, hydrous glasses were prepared in an IHPV at P = 1 kbar and T = 1250 °C during 24 h. In this synthesis large Pt capsule was loaded with 1 g of the dry glass and 45 mg of water. The water content in the homogeneous glasses without bubbles was determined by KFT method. Water content demonstrated systematic increase with rising albite content and comprises: Ab_{55}—3.33 wt% Ab_{60}—3.35% Ab_{70}—3.42 wt%, Ab_{80}—3.69 wt%. The later value approaches water solubility in the albite melt of about 4 wt% at P = 1 kbar and T = 1250 °C (McMillan and Holloway 1987). At P = 2 kbar the residual melt becomes water—saturated only when practically all Cpx crystallizes upon quenching as follows from the mass balance calculation that excludes vesiculation. All dynamic crystallization experiments were conducted with the starting glass Ab_{55}. Glasses with other compositions were used in the preliminary experiments on the Cpx liquidus determination.

Experimental regimes. Pt capsules with diameter 3 mm were loaded hydrous glass powder, welded and placed in the temperature gradient free zone of the IHPV furnace. In experiments the temperature oscillated with different periods from T = 1155 °C slightly above to T = 1135 °C moderately below the liquidus (see Table 5.2). The diopside crystals (Ioko-Dovyren intrusion skarns, North Transbaikal Territory) were used in some preliminary runs aimed on the Cpx liquidus determination.

5.3   Analytical Methods

Phase compositions of the each run products were studied. Samples were prepared by mounting the products of the runs in polystyrene and polishing one face of a polystyrene cylinder. The textural and chemical analyses were performed using a CamScan MV2300 and Tescan Vega TS5130MM SEMs with an energy-dispersive spectrometer (INCA Energy 450) at the Institute of Experimental Mineralogy, Chernogolovka, Moscow region, Russia. The spectrometer was equipped with semi-conductive Si(Li) detector INCA PentaFET X3. All the phases were analyzed at an accelerating voltage of 20 kV, the current of the absorbed electrons on the Co sample was 0.1–0.2 nA. The smallest beam diameter was 0.2 μm for point phase analysis, the glasses were analyzed using a rectangle scanning area with a width of up to 50–80 μm. The measurements results were processed by the software package INCA Energy 200. Qualitative X-ray analysis of the run products was performed on a Bruker "D2 Phaser" desktop diffractometer with a copper anode for X-ray powder diffraction.

The water content was determined with the Karl Fischer titration (KFT) method on a KFT Aqua 40.00 with a HT 1300 high-temperature solids module. Detectable water weight (m_W) range: 1–100 mg, reproducibility: ±3 μg at m_W 1–1000 μg, 3% at the m_W > 1 mg. With a common glass sample weight 20 mg and water content in the range 1–3 wt% precision is more than 1.5% relative.

5.4   CSD Measurements

We studied the CSD by analyzing backscattered electron images (BSE) of the surfaces of our samples. First, the images were analyzed using *ImageJ* 1.48v. Then the CSDs were calculated using the *CSD-Corrections* 1.40 software (Higgins 2000, 2002). In the chosen system the number density of crystals was high, and the typical value of counts for each measured sample ranged from 1200 to 2000, with the exception of 428 counts for the 72 experimental run, which involved the largest crystals (see Table 5.3). The shape of the crystals for each experiment was estimated by fitting the observed distribution of the width/length (*W/L*) ratio of crystals cross-sections

Table 5.2 Conditions of the experiments and observed phases

Run #	Initial comp.	Seed	Number of cycles	T_1 (°C)	t_1 (min)	T_2 (°C)	t_2 (min)	Products
Liquidus estimation								
10	$Ab_{55}Di_{44}$	–	0	1250	1440	1140	1440	Glass
13	$Ab_{80}Di_{20}$	–	0	1250	1440	1140	1440	Glass
18	$Ab_{55}Di_{44}$	+	0	1250	30	1140	1440	Glass + Cpx
14	$Ab_{55}Di_{44}$	+	0	1250	30	1100	1440	Glass + Cpx
16	$Ab_{80}Di_{20}$	+	0	1250	30	1100	1440	Glass
21	$Ab_{74}Di_{26}$	+	0	1250	30	1060	240	Glass + Cpx
27	$Ab_{80}Di_{20}$	+	0	1250	20	1040	240	Glass + Cpx
28	$Ab_{55}Di_{44}$	+	0	1250	20	1040	240	Glass + Cpx
Crystallization at the oscillating temperature								
72	$Ab_{55}Di_{44}$		3	1155	60	1135	60	Glass + Cpx
60	$Ab_{55}Di_{44}$		2	1155	45	1135	45	Glass + Cpx
54	$Ab_{55}Di_{44}$		4	1155	45	1135	45	Glass + Cpx
61	$Ab_{55}Di_{44}$		8	1155	20	1135	20	Glass + Cpx
56	$Ab_{55}Di_{44}$		5	1155	20	1135	20	Glass + Cpx
86	$Ab_{55}Di_{44}$		3	1155	20	1135	20	Glass + Cpx
49	$Ab_{55}Di_{44}$		5	1155	5	1135	5	Glass + Cpx
48	$Ab_{55}Di_{44}$		1	1155	5	1135	5	Glass + Cpx
50	$Ab_{55}Di_{44}$		5	1155	2	1135	2	Glass + Cpx
52	$Ab_{55}Di_{44}$		1	1155	120	1135	120	Glass + Cpx
51	$Ab_{55}Di_{44}$		2	1155	90	1135	90	Glass

Table 5.3 Results of CSD measurements

Run #	Number of measured crystals	Relative area ε_s	Crystals habitus: ellipsoid axes ratios	Maximum length	Correction factor γ	N_{eq} (μm^{-3})	Mean dist. between crystals (μm)
72	428	0.350	1/1.3/1.6	38.8	1.09	0.14e−03	19
	70	0.034	1/1.25/1.6	15.8	−	−	−
60	1316	0.30	1/1.9/2	31.1	1.31	0.49e−03	12.7
54	1543	0.32	1/1.25/1.3	15.2	1.15	2.04e−03	7.9
61	1539	0.277	1/1.9/1.9	23.8	1.33	1.36e−03	11.3
56	1408	0.301	1/1/2.2	15.3	1.44	6.43e−03	5.4
86	1200	0.310	1/1.2/3	15.6	1.32	38.72e−03	2.9
49	1939	0.318	1.0/1.3/2.7	15.6	1.15	10.6e−03	4.6
48	1020	0.305	1/1.2/3	17.6	1.41	8.72e−03	4.8
50	1905	0.268	1/1.50/5.00	13.8	1.36	24.7e−03	3.7

to the model distribution with the *CSD-Corrections* software. The shape was chosen using manual visual selection of the minimum misfit following instructions for the program. For run 56 more detailed work was performed. Based on these calculations the uncertainty of the shape factors is approximately 10–13% (1σ). To further use all the measured CSDs, the $lg(n(L))$ was approximated using polynomials on L of an order up to 4. Some distributions with a second part at small sizes were approximated using piece-wise continuous functions.

In the program *CSD-Corrections* we used a roundness parameter value of 0.5–0.7 as a value obtained by the averaging by the individual crystals from *ImageJ*. Therefore, on average, the volume fraction calculated for the ellipsoids while using our polynomial representations of density function ($\varepsilon_{s,csd}$) was underestimated in comparison to the observed value (ε_s) of ($\varepsilon_{s,csd} - \varepsilon_s$)/$\varepsilon_s = -0.32 \pm 0.17$. Theoretically, this parameter should have a value of $\Delta\varepsilon_s/\varepsilon_s = -(0.31-0.39)$ because *CSD-Corrections* distribution was scaled for crystals with a shape factor intermediate between block and ellipsoid in accordance with roundness parameter. Small deviations of $\Delta\varepsilon_s/\varepsilon_s$ from theoretical value can be explained by different representations of $n(L)$ at integration and the merging effects of different measurements sets (images) for a particular run. A scaling parameter γ was applied to each approximation of distributions to satisfy the equality of the volume fraction of the crystals and the measured relative cross-sectional area of the crystals (ε_{2D}) as follows

$$\gamma = \varepsilon_{2D} / \int_0^{L_{max}} n(l)V(l)dl, \quad V(L = L_3) = \frac{4}{3}\pi L_1 L_2 L_3$$

$$= \frac{4}{3}\pi L_3 \frac{L_1}{L_3} L_3 \frac{L_2}{L_3} L_3 = \alpha L^3 \quad (1)$$

where α is a shape factor equal to $\alpha = \frac{\pi}{6} \frac{L_1}{L_3} \frac{L_2}{L_3}$, and the axes of the approximating ellipsoid are ranked as follows $L_1 \leq L_2 \leq L_3$.

5.5 Experimental Results

5.5.1 Melting Diagram

We check the position of the pyroxene liquidus at $P_{total} = 2$ kbar and $C_{H2O} = 3.3$ wt%. Compositions Ab_{80}, Ab_{70}, Ab_{60} were held at $T = 1140\,°C$ for 5 h and pure quenching glasses were obtained (see Fig. 5.1) arguing for their location in the liquid field of the melting diagram. At $T = 1060\,°C$, the Ab_{80} composition was also quenched in the pure glass. Position of the Cpx liquidus was further ascertained with probing crystal technique. Would diopside crystal be placed in the Cpx undersaturated melt it dissolves with concentration on the boundary becoming close to the Cpx liquidus in Ab-Di system. At the start of crystal probing experiment we preheat Ab_{80} + Di and Ab_{70} + Di assemblages at 1250 °C for 30 min and then held at temperature of experiment $T = 1050$ and $T = 1150\,°C$ respectively for 48 h. During the preheating stage, crystals were partially dissolved, and a wide diffusion zone was formed. Upon cooling to the temperature of the experiment, an excess of diopside in the diffusion zone precipitated. Thus in the main stage of experiment diffusion between two-phase zone and pure melt had place. Concentration of Di on the boundary between two-phase zone and pure melt corresponds to the Di liquidus. The products of the quenching of the pure melt and the two-phase zone differed significantly which makes it possible to evaluate the equilibrium composition at their boundary. To estimate the composition of the pre-quenching melt, a sufficiently large rectangular area of crystals and quench glass up to $50 \times 5\ \mu m$ was used during the EDS. Two starting and equilibrium compositions were connected as shown Fig. 5.1 by the arrows. In the range of the compositions under study, we found a good agreement with the data published for the Ab-Di system at $P_{H2O} = 2$ kbar (see Fig. 5.1). The effect of excess 5–6 wt% of SiO_2 in the Ab-Di melt compensates effect of the water concentration of 3.3 wt% lower than $C_{H20} \approx 5.5$ wt% in the melt saturated at $P_{H2O} = 2$ kbar.

5.6 Crystallization Experiments

5.6.1 Sample Textures

For the selected composition of the $Ab_{55}Di_{45}$, the liquidus temperature is $T_l = 1150\,°C$. The temperature oscillations have a rectangular shape: during the first half-cycle, the temperature was set at $T = 1155\,°C$ or 5 °C above the liquidus, and during the second half-period it was reduced by 15 °C below the liquidus at $T = 1135\,°C$.

A wide variety of runs regimes led to a great variety of relatively spatially uniform textures, specified by the size and habitus of Cpx crystals. The maximal size of Cpx varied from c.a. 10–50 μm. The volume fraction of Cpx approached the full precipitation limit, which was estimated as 0.355 for the $Ab_{55}Di_{45}$ composition. The equilibrium volume fraction of Cpx reached during the main stage of experiment was much smaller. For the melt $Ab_{55}Di_{45}$ it was 0.046 for undercooling ΔT of 15 °C. This meant that the main volume of Cpx is precipitated during the quenching stage. Volume fraction of the equilibrium part of crystals is about 0.14, which translates into an equilibrium radius of approximately half of the total.

The habit of the crystals varies from more or less isometric (the ranked ratio of the axes of the approximating ellipsoid is $1/1.2/1.3$ ($L_1/L_2/L_3$) for run 54) to a highly elongated ($1/1.3/5$ for run 68). To correctly compare the distributions of crystals with the different shapes, we mutually plot these dependences using the corrected (equivalent) size parameter $L* = L\sqrt[3]{(L_1/L_3)(L_2/L_3)}$, where the largest crystal size is L_3. This relation can be reformulated as $L* = \sqrt[3]{L_1 L_2 L_3} \propto \sqrt[3]{volume}$.

During all experiments with melt, the final stage is quenching with different technically available rates. The crystals continue to nucleate and grow during quenching, thus modifying the CSD formed during the main stage of the experiment. The quenching effect can be eliminated, and the original CSD can be reconstructed. In this study, we assumed that CSD shape does not deform during growth, and only shifts to the larger sizes. This is true when the growth rate is crystal-size independent and the melt composition is homogeneous. In this case, it is clear how to restore CSD for the main stage of the experiment, removing the quenching effect. We calculated dL (L—is the largest crystal dimension), so that the volume fraction of the crystals for the measured density with the argument $n(l + dL)$ ($l + dL \leq Lmax$) becomes equal to the equilibrium value as follows:

$$\int_{0}^{Lmax - dL} \gamma a \cdot n(l + dL) \cdot (l/2)^3 dl = \varepsilon_{eq},\tag{2}$$

where ε_{eq} the equilibrium crystal volume fraction. This linear transformation implies that the smallest crystals ($l = 0$) at the end of the main stage of the experiment obtain the size dL after quenching. The value of dL splits the density function into parts related to the main and quenching stages of growth. The partition of acquired CSDs is shown in Figs. 5.2, 5.3 and 5.4.

After separation of the equilibrium part of the CSD, we calculated the number density of Cpx crystals N_{eq} with the eliminated quenching effect as follows:

$$N_{eq} = \int_{0}^{Lmax - dL} \gamma \cdot n(l + dL) dl \tag{3}$$

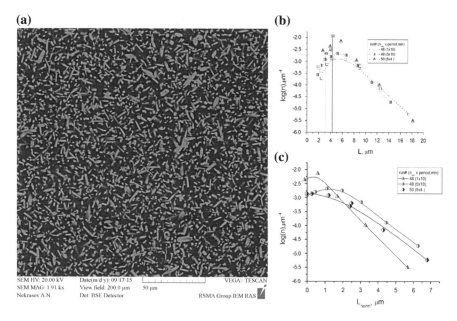

Fig. 5.2 Experiments in which the half-period of the temperature oscillations is less than the dissolution time of Cpx crystals. **a** BSE image of sample from run 50—5 oscillations with period 4 min; the texture demonstrates very weak ripening. **b** Full CSDs, vertical lines divide plots on the quenching and main parts. **c** The restored CSDs formed before quenching, densities plotted versus effective (proportional to crystal volume in order 1/3) maximum crystal dimension. The CSD of the Cpx crystals from run 49 is the most maturated in this series

The values of the calculated N_{eq} are shown in Table 5.3. It can be seen that the restored crystal density changes by three orders of magnitude, depending on the experimental conditions. Based on the measured CSDs, we calculated the average crystal size as the first moment of distribution and the mean inter-crystalline distance as $l \approx 2/(N_{eq})^{0.333}$; the values of these parameters are also listed in Table 5.3. The inter-crystalline distance characterizes the diffusion ability to provide a low supersaturation in the melt volume at rapid cooling. When this distance was sufficiently large, the flat crystal faces became morphologically unstable during quenching and the hopper overgrowth is formed.

5.7 Products Composition

The EMPA analyses demonstrated (see Table 5.4) that the experimentally grown diopside was a solid solution containing up to 7 wt% of enstatite. This content is near to the solvus composition of Cpx ($Di_{90}En_{10}$) in the En-Di system at T = 1000 °C (Boyd and Schairer 1964). The alumina content in Cpx is in the range 0.3–2 wt%. To characterize the composition of the residual glass, we used SEM-EDS analysis

(a)

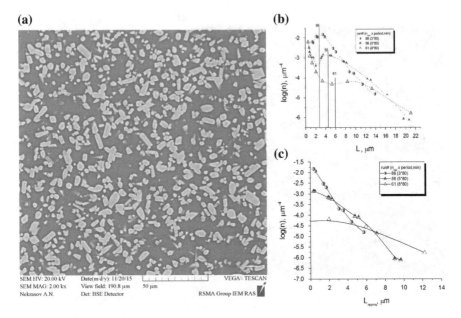

Fig. 5.3 Experiments in which the half-period of the temperature oscillations is near the dissolution time of the Cpx crystals, additional crystals nucleated in the cold half-cycle. The original log-linear CSD evolves towards the right-skewed form with a maximum. **a** BSE image of sample from run 56—5 oscillations with period of 80 min. **b** Full CSDs, vertical lines indicate dL and divide plots on the quenching and main parts, for run 56, the interval (level 1σ) for dL is outlined by a rectangle. **c** The restored CSDs formed before quenching

averaged over the surface of polygons selected between crystals with an area of approximately 20–100 μm². The glasses were enriched in Na and Al and depleted in Mg and Ca in comparison to the starting composition. The proportions of the increments in the concentration of elements can approximately be described as the diffusive mixing of albite and diopside in a melt instead of individual oxides. During experiments with a higher number density of crystals (48, 49, 50), the averaged melt composition was more enriched in albite. For these runs, the volume fraction of Cpx calculated on the basis of the glass composition was only slightly less than the volume fraction observed in the BSE images (relative difference 2–5%). For the other experiments, the residual glass contained less albite, and the difference between the measured and calculated solid fraction was greater (up to 29% in run 72). From this comparison, it follows that during the last stage (at a temperature approaching T_g) of quenching, the residual melt composition of the later runs, on the average, was not on pace with the albite flux from the growing Cpx crystals when forming sharp diffusive boundary layers. This effect is the most pronounced at the longest inter-crystalline distance corresponding to the smallest number density of the crystals during the run 72 that had unstable rims with hopper morphology.

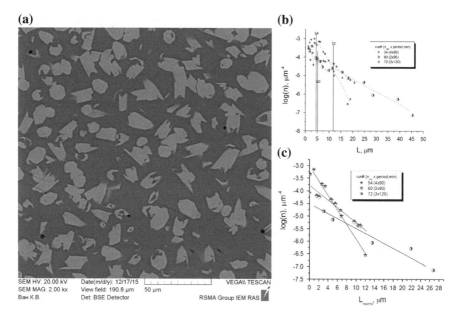

Fig. 5.4 The CSDs of Cpx in experiments in which the half-period is longer than the dissolution time. All the crystals were completely dissolved in the hot half-cycles and nucleated in the cold ones. **a** BSE image of sample from run 60—2 oscillations with period 1 h 30 min. **b** Full CSDs, vertical lines indicate dL and divide plots on the quenching and main parts. **c** The restored CSDs formed before quenching are near the log-linear in shape

5.8 Effect of the Variable Period of the Oscillations

In experiments with an oscillating temperature during the cold half-cycle, the nucleation and crystal growth of Cpx occurred at a moderate undercooling of a $\Delta T = 15\ °C$, and during the hot half-cycle crystal dissolution occurred at a temperature exceeding the liquidus at 5 °C. Superheating above the liquidus means that for a sufficiently large oscillation period, all the crystals dissolve. The time required for the dissolution of crystals of different sizes was estimated on the basis of experimental data on diffusion in the magma and will be discussed below. According to the oscillation period experiments can be divided into four groups.

(1) At the largest thermal oscillations period of three hours all crystals dissolved in the hot stage. At the same time, the concentration of clusters with large dimensions, but less critical for nucleation at the cold stage, decreased. With such a long superheating treatment, during run 51 after the second cold half-cycle, pure quenching glass with a few dendrites heterogeneously nucleated from the capsule wall was formed. This was possible only if the delay time of the nucleation after the hot half-cycle was longer than 1.5 h.

(2) During experiments with the shortest period of 4–10 min (runs 48–49–50), the crystals dissolved only slightly. Cpx crystals in these runs showed the largest

elongation (for the run 50 shape coefficients are 1/1.5/5, see Fig. 5.2a). In this group, the density functions, restored for the equilibrium stage, increased in maturation in the sequence of runs 50–48–49 (see Fig. 5.2b, c). With an increase in the degree of ripening an explicit maximum appears in the series of the dissolution-precipitation events and shifts to larger dimensions, as the model predicted (Simakin and Bindeman 2008). The maximum is recognized better on the initial plot not subjected to the quenching effect elimination (Fig. 5.2b). Because of intrinsically stochastic nature of the nucleation process, the initial stage of experiments 48, 49 and 50 with respect to N_{eq} can differ in order of magnitude there is no single dependence $N_{eq}(t)$.

(3) In the third group of experiments with a period of 40 min (runs 86, 56, 61), the crystals dissolved significantly during the hot half-cycle; therefore, new crystals nucleated during the cold half-cycle. The crystals in this group have more isometric than in the previous group habitus (Fig. 5.3a, Table 5.3). Relatively large number density and small inter-crystalline distance provide low supersaturation and smooth overgrowth at quenching. The restored CSDs after 3 and 5 oscillations had a log-linear form (Fig. 5.3b, c). The average size increased and the number density of the crystals appreciably decreased. After 8 oscillations, the CSD reached a weakly expressed maximum which is better seen in the initial CSD (Fig. 5.3b).

(4) Finally, in the forth group of experiments, the period of oscillation was 90–120 min. Maturation in this set is prominent, and the largest (up to 45 μ) nearly isometric crystals were formed (Fig. 5.4a). Two and five oscillations with a period of 90 min (runs 60 and 54) provided the CSDs with a shape that was near log-linear (Fig. 5.4b, c), as in the experiment 72 (3 oscillations with a period of 120 min). At quenching the crystals were overgrown with morphologically unstable (hopper-like) rim due to low crystal number density and large inter-crystalline distance (Table 5.3). We assumed that all the crystals dissolve during the hot stage and nucleated back during the cold stage. It is noteworthy that N_{eq} for run 60 (2 cycles) is less than that for run 54 (4 cycles) (see Table 5.3). Such irregularity may reflect complete dissolution and homogeneous nucleation with each oscillation with a stochastic result. Composition of the sufficiently large Cpx crystals with the largest dimension up to 45 μm from run 72 (Fig. 5.5a) was studied with the EDS.

The core region with an increased Al_2O_3 content was surrounded by an inner rim with a low Al_2O_3 content; the outer quenching rims are returned back to a higher Al_2O_3 content (see Fig. 5.5b). A minimal Al_2O_3 content of 0.35 wt% was observed in the inner rim surrounding the core. For the series of measurements (n = 12), the minimal concentration of aluminum oxide was 0.34 ± 0.12 wt%. The distribution of MgO (and opposite CaO) in these crystals also demonstrated a clear compositional zoning (see Fig. 5.5c): the core with a high CaO content is surrounded by an inner rim with the highest MgO content, and the outer quenching rims return back to a higher CaO content. Correlation coefficient of Ca and Mg concentrations in the Cpx from run 72 is -0.89 in accordance with Di-En mixing.

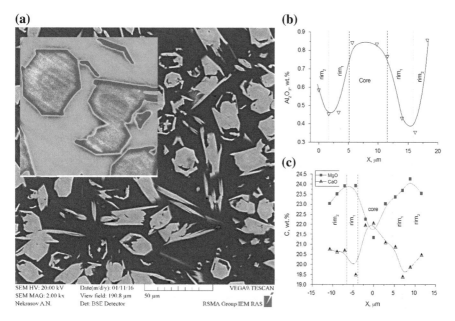

Fig. 5.5 Compositional zoning of clinopyroxene crystals from run 72. **a** BSE image of a Cpx crystal, on the inset an enlarged view of several crystals in false colors; three distinct zones can be seen, the outer quenching rim is morphologically unstable. **b** Profile of alumina concentration across a Cpx crystal; the core, the equilibrium and quenching rims can be distinguished. **c** Profiles of CaO and MgO concentrations across a Cpx crystal, there are the same zones as in the plot (**b**)

Detailed mapping of the composition of the largest crystals during the run 72 allowed us to more thoroughly check the correctness of our assumptions regarding the preservation of the CSD shape during quenching. The BSE images reflect the compositional zoning of the crystals and clearly demonstrate their cores. The CSD of these cores (surrounded by a zone with a low Al_2O_3 content), outlined manually on the BSE images, was acquired on the basis of the 70 counts. The procedure previously described (Eq. 2) and applied to the full CSD of run 72 led to an estimate of the quenching rim of the equilibrium crystals (dL) of 11 μm. The shift of the measured density function to a smaller size by 11 μm provided an estimate of the CSD at the equilibrium stage. The observed difference in the measured CSD of the cores with the transformed total CSD is not excessively large (see Fig. 5.6) in view of the low accuracy of the cores measurement data.

5.9 Discussion

It was found long ago that the thermal history of the melt in terms of the amount of the superheating above liquidus and its duration greatly affects the texture of the crystallization products (e.g., Donaldson 1979; Corrigan 1982). This effect took place

Table 5.4 The composition of experimental phases

Run #	Phase n/1σ	SiO$_2$	Al$_2$O$_3$	MgO	CaO	Na$_2$O	K$_2$O	Total	Ab in glass (wt%)
48	Glass	66.43	13.89	1.86	1.74	8.30	0.16	92.52	89.43
	(6)	*1.45*	*1.52*	*2.63*	*2.39*	*0.93*	*0.03*		
	Di	55.87	1.14	20.34	21.40	0.55	0.03	99.58	
	(4)	*0.43*	*0.16*	*0.28*	*0.27*	*0.09*	*0.04*		
52	Glass	62.97	11.34	7.09	7.33	5.86	0.08	94.67	61.83
	(3)	*0.49*	*0.08*	*0.07*	*0.47*	*0.16*	*0.01*		
	Di	55.74	1.45	20.73	20.42	0.57	0.01	98.92	
	(3)	*0.54*	*0.57*	*1.36*	*0.84*	*0.04*	*0.03*		
60	Glass	68.90	14.79	1.05	0.07	9.26		94.06	96.30
	(5)	*0.67*	*0.24*	*0.21*	*0.16*	*0.25*			
	Di core	56.61	0.21	21.51	22.72	0.00		101.06	
	(3)	*0.48*	*0.37*	*0.85*	*0.67*	*0.00*			
	Di rim	57.25	0.69	21.18	22.05	0.00		101.17	
	(3)	*0.14*	*0.75*	*1.42*	*0.45*	*0.00*			
51	Glass	63.91	9.93	9.12	8.71	5.52	0.10	97.30	54.11
	(5)	*0.93*	*0.28*	*0.55*	*0.39*	*0.32*	*0.07*		
86	Glass	66.87	14.08	2.40	1.76	8.99	0.22	94.32	88.26
	(3)	*0.76*	*0.30*	*0.35*	*0.54*	*0.10*	*0.11*		
	Di	55.69	2.03	19.30	21.11	1.01	0.06	99.19	
	(3)	*0.28*	*0.49*	*0.60*	*1.62*	*0.25*	*0.07*		
72	Glass	68.29	14.65	0.77	0.51	7.72		92.16	95.82

(continued)

Table 5.4 (continued)

Run #	Phase n/1σ	SiO$_2$	Al$_2$O$_3$	MgO	CaO	Na$_2$O	K$_2$O	Total	Ab in glass (wt%)
	(3)	*0.36*	*0.65*	*0.20*	*0.07*	*0.11*			
	Di core	55.16	0.78	20.49	21.78	0.32		98.61	
	(5)	*0.01*	*0.40*	*1.12*	*0.49*	*0.20*			
	Di rim	54.74	0.76	22.13	20.58	0.31		98.62	
	(5)	*1.20*	*0.07*	*0.16*	*0.20*	*0.06*			
	Di core	55.15	1.06	19.70	21.43	0.46		97.80	
	Di rim 1	54.61	0.46	21.44	21.55	0.26		98.32	
	Di rim 2	55.58	0.71	22.24	20.44	0.35		99.32	
54	Glass	68.74	14.77	1.22	0.75	8.44	0.14	94.18	93.97
	(5)	*0.50*	*0.27*	*0.22*	*0.21*	*0.32*	*0.05*		
	Di core	55.05	0.63	22.01	20.76	0.34	0.01	98.99	
	(5)	*0.36*	*0.13*	*0.21*	*0.65*	*0.08*	*0.02*		
	Di rim	55.47	0.76	22.24	20.32	0.38	0.06	99.46	
	(2)	*0.25*	*0.06*	*0.57*	*0.06*	*0.06*	*0.08*		
50	Glass	66.91	13.86	1.93	2.28	8.09	0.15	93.32	87.91
	(5)	*0.65*	*0.31*	*0.44*	*0.93*	*0.16*	*0.06*		
	Di*	**61.83**	**5.56**	**16.60**	**15.07**	**2.18**	**0.06**	**101.39**	
	(3)	*0.43*	*0.52*	*0.18*	*0.23*	*0.15*	*0.11*		
49	Glass	68.84	14.96	1.37	0.87	9.29	0.18	95.67	93.52
	(5)	*1.04*	*0.37*	*0.28*	*0.47*	*0.30*	*0.08*		
	Di	57.69	2.51	19.06	20.26	1.16	0.06	100.95	

(continued)

Table 5.4 (continued)

Run #	Phase n/1σ	SiO₂	Al₂O₃	MgO	CaO	Na₂O	K₂O	Total	Ab in glass (wt%)
	(5)	*0.82*	*1.56*	*1.90*	*2.93*	*0.66*	*0.04*		
	Di core	55.73	1.18	20.00	22.41	0.68		100.00	
	Di core	56.8	1.29	19.85	22.69	0.58		101.21	
	Di rim	56.86	1.74	20.08	20.47	0.85		100.00	
56	Glass	68.71	15.11	2.31	2.46	8.43	0.14	97.37	87.13
	(5)	*0.73*	*0.29*	*0.19*	*0.35*	*0.16*	*0.04*		
	Di	*55.96*	*1.03*	*21.42*	*21.71*	*0.45*	*0.03*	*100.78*	
	(5)	*0.41*	*0.32*	*0.45*	*0.25*	*0.10*	*0.03*		
61	Glass	70.04	13.48	1.80	0.78	8.05	0.12	94.31	91.52
	(4)	*0.33*	*0.96*	*0.96*	*0.27*	*0.40*	*0.05*		
	Di core	55.98	0.69	21.12	22.25	0.34	0.02	100.41	
	(5)	*0.28*	*0.28*	*0.39*	*0.35*	*0.17*	*0.03*		
	Di rim	56.44	0.87	22.12	20.94	0.48	0.06	100.91	
	(6)	*0.71*	*0.57*	*1.24*	*0.70*	*0.33*	*0.05*		

The number of averaged analyses is in the parentheses under the phase name; line with single standard deviations is under the line with the corresponding oxide content. Di* corresponds to the mineral composition with some glass contribution due to the tiny crystal size, empty cells in K₂O column correspond to contents b.d.l.

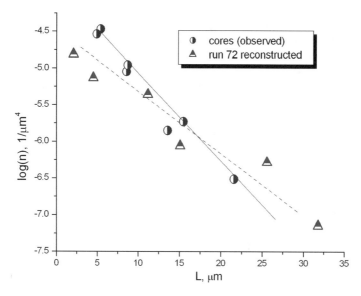

Fig. 5.6 The size distributions of the largest Cpx crystals from run 72; semi-filled stars depict the estimated CSD formed before quenching ($\varepsilon_s = 0.046$), semi-filled circles display the CSD of the cores of large crystals approximately identified on the BSE images ($\varepsilon_s = 0.036$). A fairly good agreement between the observed and theoretically reconstructed CSDs can be seen

in our experiments in which the melt overheated during the half-period of oscillations. The crystals were dissolved and the melt overheated with a sufficiently large period of oscillation. The speed and time of dissolution of the crystal is controlled by the diffusion rate of the slowest component, which is usually SiO_2, unless cations with a high ionic potential are involved, such as Zr^{4+} or P^{5+} (Zhang et al. 2010).

Modeling dissolution of a crystal assemblage with a variable size stochastically distributed in space is a complex problem. The attribution of a certain volume of melt to each crystal (Bindeman and Melnik 2016) cannot be a realistic approximation in our case, because the number density sharply decreases in the experiments with a period of oscillation approaching the dissolution time. The number density of the residual crystals decreased during the hot half-cycle 25–30 times, which corresponds to an increase in the intercrystalline distance from 17 to 50 μm.

The simplest solution for dissolution time (t_{dissol}) of a spherical crystal in a diluted suspension, which is commonly used for scaling estimates, is as follows:

$$t_{dissol} = \frac{\rho_s \Delta C^* (1 - C_b) R_0^2 D_{SiO_2}}{2\rho_l}, \tag{4}$$

here ΔC^* is the difference between the solubility of Di at a given $T = 1155\,°C$ (C_b) and initial content of the diopside in the melt during the equilibrium stage at T = $1135\,°C$ (C_{eq}) is corrected tentatively by a factor of 1/2 to account for the decreasing of the melt undersaturation over time. The smallest silica diffusivity D_{SiO2} controls the dissolution rate, and unequal crystal (ρ_s) and melt (ρ_l) densities are taken into account. The main uncertainty in the dissolution time is introduced by an estimate of the diffusion coefficient. As the first proxy we used the empirical model proposed in Mungal (2002), based on the composition of the melt and its viscosity calculated in accordance with Hui and Zhang (2007), and the diffusion coefficient of silica was evaluated by the value $D_{SiO2} = 0.68 \times 10^{-8}$ cm^2/s. The concentration difference was read from the melting diagram (Fig. 5.1) $\Delta C = 0.042$. The dissolution time according to Eq. 4 is estimated to be 15–20 min at the maximal radius of the crystals $R_0 = 8$–9 μm, respectively. A more accurate semi-analytical integration of the equation for the dissolution of a spherical crystal provides an approximately twice as long dissolution time of 30–40 min for the same R_0 range. This difference arises from the effective concentration difference on the surface of a dissolving crystal of actually less than one-half of the ΔC, when the effects of decreasing the volume of the melt and the composition of the melt approaching C_0 with time are included. Thus, the theoretical estimate of 30–40 min corresponds to the threshold value between incomplete and complete dissolution of crystals in the hot stage (groups three with half-period 20 min and four with half period 45–60 min).

Our experiments show that the almost complete dissolution of the crystals during the hot cycle gives the best result in terms of the rate of maturation (Fig. 5.7). During short times, both Mills and Glazner (2013) and Cabane et al. (2005) reported the average volume number content of crystals (N_v—zero-moment of CSD), which decrease with a rate proportional to approximately $t^{-1/5}$. In our experiments with an almost complete dissolution (group four), the volume content drops from 6.7e−03 μm^{-3} (initial N_v) to 0.49e−03 μm^{-3} in only two full cycles (run60) and to 2.04e−03 μm^{-3} for four full cycles (run54). In these series, the CSDs shapes are nearly log-linear, with the slope rotating as the number of oscillations increases. In our experiments, the rate of nucleation in the cold half-period determines the number density and average crystal size. In general, the nucleation process in silicate melts is non-stationary (e.g., Tsuchiyama 1983) with a certain lag time in which stochastic processes take place in atomic-scale microclusters, with reaching a critical size and the formation of equilibrium microcrystals. Superheating greatly influences the distribution of clusters, so that with a sufficiently high or prolonged (or both) superheating, the delay time of nucleation increases and the rate of nucleation decreases. Davis and Ihinger (2002) are to our knowledge the only researchers that quantitatively modeled the impact of the size distribution of subcritical clusters on the kinetics of magma crystallization. Using their approach, it is possible to model the effect of superheating in the oscillating temperature regime on the nucleation rate and resultant CSD and quantitatively interpret our data.

Our experimental data justify an approximation of the growth rate independent of the crystal size, at a sufficiently small distance between the crystals. Whether crystal growth is controlled by diffusion or kinetics is important not only in the reconstruction the pre-quenching CSD (in the experiment, for minerals from volcanic rocks), but also for the evolution of CSD with repeated dissolution-precipitation events. This evolution can be modelled with a PDE solution for the distribution function (Simakin and Bindeman 2008). Van Westen and Groot (2018) used a simpler

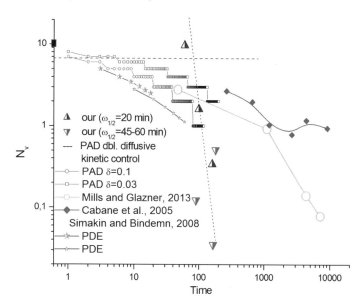

Fig. 5.7 Dependences of the volume crystal number content versus time during maturation in a crystal-melt mixture, as determined in various experimental studies and calculated with different methods. Our experimental data are shown with triangles, $N_v/4$ in μm^{-3} is plotted, time for all experiments in minutes. Results of the modeling with Particles Assemblage Dynamics (PAD) method (van Westen and Groot 2018) with open squares, squares and dashed line, δ is a volume fraction of crystals dissolved; N_v is total number of crystals, time is number of cycles; stars depict results of PDE solution for CSD from (Simakin and Bindeman 2008). In bi-logarithmic coordinates dimensionalities of N_v and time are inessential since slopes are invariants of scaling

method, manipulating a set of a finite number of crystals. The diffusion control is approximated by the square root dependence of each individual radius on time, which leads to the equation for the radii of the crystals R_i (crystals are ranked in order of increasing size) at the dissolution stage:

$$\sum_{i=n1}^{N} (\sqrt{R_{i,k}^2 - \lambda_{1,k}})^3 = v_0(1 - \delta), \tag{5}$$

where δ is the volume fraction of dissolved crystals, n_1 is the number of the first crystal that survived the dissolution, and λ_1 is an unknown constant. For simplicity, we consider a small number of crystals $N = 10$; the crystals have an initial size $R_i = a\,i^2$ $(i = 1…N)$. The minimum number $(n1)$ satisfying Eq. (5) should be found numerically in each cycle. After finding λ_1^k new radii of the survived crystals with number $\geq n1$ are calculated

$$R_{i,k+1/2} = \sqrt{R_{i,k}^2 - \lambda_{1,k}}, \; i = n1..N \tag{6}$$

For a constant (independent of size) growth rate, the assumption of complete precipitation of the dissolved material in a cold half cycle requires:

$$\sum_{i=n1}^{N} (R_{i,k+1/2} + \lambda_{2,k})^3 = v_0, \tag{7}$$

With diffusion growth control, we obtain another equation similar to Eq. (5):

$$\sum_{i=n1}^{N} (\sqrt{R_{i,k+1/2}^2 + \lambda_{3,k}})^3 = v_0, \tag{8}$$

Final crystals sizes after k + 1 cycle of temperature oscillations are:

$$R_{i,k+1} = R_{i,k+1/2} + \lambda_{2,k} \text{ and}$$
$$R_{i,k+1} = \sqrt{R_{i,k+1/2}^2 + \lambda_{3,k}}, \ i = n_1..N, \tag{9}$$

respectively. Numerical solution of the systems of Eqs. (5, 6) and (5, 7) gives completely different results. With kinetic control, the number of crystals reduces with time only to one. The calculated dependence $N(t) \approx a/t^{1/5}$ over a short time (number of steps) demonstrated a close resemblance to the experimental data of Cabane et al. (2005) and Mills and Glazner (2013) (see Fig. 5.7). Moreover, the theoretically obtained increase in the logarithmic slope of $N(t)$ at the larger time was observed by Mills and Glazner (2013). The Ostwald ripening corresponds to a larger logarithmic slope n = $a/t^{1/3}$ (Lifshitz and Slyozov 1961). Reducing the degree of dissolution [setting a smaller δ value in Eq. (5)] leads to the slower ripening with a close initial logarithmic slope just as shown in Cabane et al. (2005), where small stochastic temperature oscillations were caused by temperature regulation. When both dissolution and growth are controlled by diffusion, only the first cycle in our calculations reduces the number of crystals without changing the number and size of crystals in subsequent cycles. With double diffusion control, similar results are obtained when using a larger number of crystals for modeling, which allows to calculate the density function of CSD (Van Westen and Groot 2018). In the first cycles, the distribution was changed, and then modified only by Ostwald ripening, due to the size dependence of the solubility of the crystals, whereas the temperature cycle did not affect the shape of the CSD.

5.10 Conclusions

The crystallization of clinopyroxene from a melt in the Ab-Di-H$_2$O system at P = 200 MPa with temperature oscillations around the liquidus was investigated. A procedure is proposed for the recovery of CSD of clinopyroxene at the stage before quenching, which correctly removes the quenching portion of the distribution. This procedure is based on the assumption that the growth rate does not depend on the crystal size. The recovery of CSD was successfully validated for the run with sufficiently large Cpx crystals, for which cores formed at the main stage of the experiment can be distinguished in the BSE images. Our experiments showed that the texture of crystallization products, characterized by CSD, strongly depends on the period of temperature oscillations. The maximum temperature during the oscillations was above the liquidus of diopside, which completely dissolved when the duration of the hot half period was above 30–40 min. Then the nucleation and growth of diopside in each cold half-period results in a log-linear CSD. With a period of oscillation of 4-10 min, Cpx crystals were only partially dissolved in the melt at the hot stage and re-precipitated at the cold one. In this case, CSD of clinopyroxene gradually developed to a log-normal like shape, as in previous studies. By simple calculations, it was demonstrated that the crystal number density decreases and mean radius increases only when the crystals grow in a kinetic (size independent) mode and dissolve at the diffusion control.

Acknowledgements Study was done within theme AAAA-A18-118020590141-4 of the State task of IEM RAS for 2019–2021.

References

Avrami M (1939) Kinetics of phase change. I. General theory. J Chem Phys 7(12):1103–1112

Bindeman IN, Melnik OE (2016) Zircon Survival, Rebirth and recycling during crustal melting, magma crystallization, and mixing based on numerical modelling. J Petrol 57(3):437–460

Boyd FR, Schairer JF (1964) The system $MgSiO_3$-$CaMgSi_2O_6$. J Petrol 5:275–309

Brandeis G, Jaupart C, Allegre CJ (1984) Nucleation, crystal growth, and the thermal regime of cooling magmas. J Geophys Res 89:10161–10177

Cabane H, Laporte D, Provost A (2005) An experimental study of Ostwald ripening of olivine and plagioclase in silicate melts: implications for the growth and size of crystals in magmas. Contrib Mineral Petr 150(1):37–53. https://doi.org/10.1007/s00410-005-0002-2

Cashman KV, Marsh BD (1988) Crystal size distribution (CSD) in rocks and the kinetics and dynamics of crystallization, II, Makaopuhi lava lake. Contrib Mineral Petrol 99:292–305

Chekhmir AS, Simakin AG, Epel'baum MB (1991) Dynamic phenomena in fluid-magma systems. Nauka, Moscow, 141 p. (in Russian)

Corrigan GM (1982) Supercooling and the crystallization of plagioclase, olivine, and clinopyroxene from basaltic magmas. Mineral Mag 46:31–42

Davis MJ, Ihinger PD (2002) Effects of thermal history on crystal nucleation in silicate melt: numerical simulations. J Geophys Res 107(B11):2284. https://doi.org/10.1029/2001jb000392

Donaldson CH (1979) An experimental investigation of the delay in nucleation of Olivine in Mafic magmas. Contrib Mineral Petr 69:21–32

Epel'baum MB (1980) Silicate melts with volatile components. Nauka, Moscow, 255 p. (in Russian)

Higgins MD (2000) Measurement of crystal size distributions. Am Mineral 85:1105–1116

Higgins MD (2002) A crystal size-distribution study of the Kiglapait layered mafic intrusion, Labrador, Canada: evidence for textural coarsening. Contrib Mineral Petr 144:314–330

Hort M (1998) Abrupt change in magma liquidus temperature because of volatile loss or magma mixing: effects on nucleation, crystal growth and thermal history of the magma. J Petrol 39(5):1063–1076

Hui H, Zhang Y (2007) Toward a general viscosity equation for natural anhydrous and hydrous silicate melts. Geochim Cosmochim Acta 71:403–416

Lifshitz IM, Slyozov VV (1961) The kinetics of precipitation from supersaturated solid solution. J Phys Chem Solids 19:35–50

Marsh BD (1988) Crystal size distribution (CSD) in rocks and the kinetics and dynamics of crystallization. I Theory Contrib Mineral Petr 99(3):277–291

McMillan PF, Holloway JR (1987) Water solubility in aluminosilicate melts. Contrib Mineral Petr 97:320–332

Mills RD, Glazner AF (2013) Experimental study on the effects of temperature cycling on coarsening of plagioclase and olivine in an alkali basalt. Contrib Mineral Petr 166:97–111

Mungall JE (2002) Empirical models relating viscosity and tracer diffusion in magmatic silicate melts. Geochim Cosmochim Acta 66(1):125–143

Ni H, Keppler H, Walte N, Schiavi F, Chen Y et al (2014) In situ observation of crystal growth in a basalt melt and the development of crystal size distribution in igneous rocks. Contrib Mineral Petr 167:1003

Pati JK, Arima M, Gupta AK (2000) Experimental study of the system diopside-albite-nepheline at P(H$_2$O) = P(total) = 2 and 10 kbar and at P(total) = 28 kbar. Can Mineral 38:1177–1191

Simakin AG, Bindeman IN (2008) Evolution of crystal sizes in the series of dissolution and precipitation events in open magma systems. J Volcanol Geoth Res 177:997–1010

Simakin AG, Chevychelov VY (1995) Experimental studies of feldspar crystallization from the granitic melt with various water content. Geochem Intl 38:523–534

Simakin AG, Salova TP (2004) Plagioclase crystallization from a hawaiitic melt in experiments and in a volcanic conduit. Petrology 12(1):82–92

Simakin AG, Trubitsyn VP, Kharybin EV (1998) The size and depth distribution of crystals settling in a solidifying magma chamber. Izv-Phys Solid Eart 34(8):639–646

Simakin A, Armienti P, Epel'baum M (1999) Coupled degassing and crystallization: experimental study at continuous pressure drop, with application to volcanic bombs. Bull Volcanol 61:275. https://doi.org/10.1007/s004450050297

Simakin AG, Salova TP, Armienti P (2003) Kinetics of clinopyroxene growth from a hydrous hawaiite melt. Geochem Int 41(12):1165–1175

Toramaru A (1991) Model of nucleation and growth of crystals in cooling magmas. Contrib Mineral and Petrol 108:106. https://doi.org/10.1007/bf00307330

Tsuchiyama A (1983) Crystallization kinetics in the system CaMgSi$_2$O$_6$-CaAl$_2$Si$_2$O$_8$: the delay in nucleation of diopside and anorthite. Am Mineral 68:687–698

Van Westen T, Groot RD (2018) Effect of temperature cycling on ostwald ripening. Cryst Growth Des 18(9):4952–4962. https://doi.org/10.1021/acs.cgd.8b00267

Vona A, Romano C (2013) The effects of undercooling and deformation rates on the crystallization kinetics of Stromboli and Etna basalts. Contrib Mineral Petr 166(2):491–509

Vona A, Romano C, Dingwell DB, Giordano D (2011) The rheology of crystal-bearing basaltic magmas from Stromboli and Etna. Geochim Cosmochim Ac 75:3214–3236

Yoder HS (1965–1966) Carnegie Institution, Washington, Year book, 65 274 1966. Carnegie Institution of Washington, Washington DC

Zhang Y, Ni H, Chen Y (2010) Diffusion data in silicate melts. Rev Mineral Geochem 72:311–408

Chapter 6
Solubility and Volatility of MoO_3 in High-Temperature Aqueous Solutions

A. V. Plyasunov, T. P. Dadze, G. A. Kashirtseva, and M. P. Novikov

Abstract The thermodynamic properties of the neutral molybdic acid H_2MoO_4 are evaluated at 273–623 K and the saturated water vapor pressure from our own solubility data at 563–623 K and literature results at lower temperatures. Combining the Gibbs energies of H_2MoO_4 in the state of the aqueous solution with those in the ideal gas state, we calculated Henry's constants and the vapor–liquid distribution constants of H_2MoO_4 at 273–623 K, and with the use of the relevant asymptotic relations, extrapolated values of Henry's constants, k_H, and vapor–liquid distribution constants, K_D, toward the critical point of pure water. Our results show that over the whole temperature range of the existence of the vapor–liquid equilibrium of water, the neutral molybdic acid H_2MoO_4 is somewhat less volatile compared with $Si(OH)_4$, and the difference in volatility of these species decreases with the temperature.

Keywords Molybdenum trioxide · Molybdic acid · Aqueous solubility · Vapor–liquid distribution · Henry's constant · Krichevskii parameter

6.1 Introduction

Since the opening of the Institute of Experimental Mineralogy, one of directions of research here was the study of the solubility of minerals in hydrothermal solutions, with the aim to quantify the transport and deposition of ore minerals. The long list of studied systems and relevant publications is beyond the scope of this contribution, and we will only briefly mention several solubility investigations conducted by the authors of the current research team. These studies include the solubility of amorphous SiO_2 in water and in aqueous solutions of acids at 373–673 K and pressure of 101.3 MPa (Sorokin and Dadze 1980), of SnO_2 in water and aqueous solutions of acids and salts at 473–673 K and pressure of up to 150 MPa (Dadze et al. 1981; Dadze and Sorokin 1986); the solubility of zinc oxide ZnO in aqueous solutions of alkalis (Plyasunov et al. 1988; Plyasunov and Plyasunova 1993a), salts (Plyasunov and Ivanov 1991)

A. V. Plyasunov (✉) · T. P. Dadze · G. A. Kashirtseva · M. P. Novikov
D.S. Korzhinskii Institute of Experimental Mineralogy, Russian Academy of Sciences, Academician Osipyan Street 4, Chernogolovka, Moscow Region, Russia 142432
e-mail: andrey.plyasunov@gmail.com

© The Editor(s) (if applicable) and The Author(s), under exclusive license to Springer Nature Switzerland AG 2020
Y. Litvin and O. Safonov (eds.), *Advances in Experimental and Genetic Mineralogy*, Springer Mineralogy, https://doi.org/10.1007/978-3-030-42859-4_6

and alkali + salt mixtures (Plyasunov and Plyasunova 1993b) at 473–873 K and pressures up to 200 MPa; investigations of solubility of gold in aqueous H_2S-bearing solutions (Dadze et al. 1999, 2000, 2001; Dadze and Kashirtseva 2004; Gorbachev et al. 2010), including solubility in low-density hydrothermal fluids (Zakirov et al. 2008, 2009).

The current contribution is concerned with the thermodynamic properties of the neutral molybdic acid H_2MoO_4 (its real stoichiometry is likely $MoO_2(OH)_2$ (Akinfiev and Plyasunov 2013)) in aqueous solutions at elevated temperatures. The primary motivation for this study were the unusual results by Rempel et al. (2009), who studied the vapor–liquid distribution of Mo from $(NH_4)_2MoO_4$ solutions at 573–643 K, and reported that $H_2MoO_4(aq)$ is more volatile (in terms of Henry's constants) than, for example, Ar. As the interactions between water and nonpolar Ar are much weaker than those for H_2MoO_4, which form hydrogen bonds with H_2O molecules, such finding appears unlikely. However, an earlier publication by Khitarov et al. (1967) on the partition of Mo(VI) from dilute solutions of Na_2MoO_4 over a temperature range 523–623 K also found unexpectedly high concentrations of molybdenum in the vapor phase. Such was the situation before the start of our investigation: Mo(VI) compounds appeared to be highly anomalous in terms of their partition from aqueous solutions, challenging our knowledge of regularities in water-solute interactions. Nevertheless, one may speculate that trapping a portion of vapor phase in a two-phase system, as was done in studies (Khitarov et al. 1967; Rempel et al. 2009), is an error-prone method, especially when the incomplete separation of liquid and vapor phases is recognized as one of the major issues in the experimental studies of the vapor–liquid partitioning of aqueous solutes (Palmer et al. 2004). Indeed, in 2016, Kokh et al. (2016) published results of the study, where a direct sampling of the coexisting vapor and liquid phases at 623 K was employed. The liquid phase was ~1 m salt solution, and contained small amounts of Mo and other metals. In the S-free, CO_2-free series of experiments concentrations of Mo(VI) in steam were 3–3.5 orders of magnitude lower than in earlier experiments (Khitarov et al. 1967; Rempel et al. 2009).

Existing discrepancies in the Mo vapor–liquid partitioning from aqueous solutions were the main reasons to begin a new study of the distribution of H_2MoO_4 between the vapor and liquid phases of water. We have chosen an indirect way to obtain the values of the vapor–liquid distribution constants of the neutral form H_2MoO_4. First, we determined the thermodynamic properties of this species from the solubility of molybdite, MoO_3, in liquid acid solutions at 563–623 K. Thermodynamic properties of H_2MoO_4 in the ideal gas state are already known (Akinfiev and Plyasunov 2013). The difference of the Gibbs energies of this form in the standard aqueous solution and in the ideal gas state gives Henry's constant of H_2MoO_4. Henry's constant and the vapor–liquid constant are connected through a relation, which is simple, but requires the knowledge of the fugacity coefficient of the neutral molybdic acid, which can be evaluated, as will be discussed in Sect. 6.3.2.

Over the course of this work, it was found necessary to perform a series of additional experimental investigations for a proper interpretation of the obtained results. First, our studies of the solubility of MoO_3 in HCl and $HClO_4$ solutions at an acid molality above 0.03 m showed an increase in the solubility in HCl, but not in $HClO_4$

solutions, indicating the formation of chloride complex(es) of Mo(VI), in agreement with earlier works (Kudrin 1985; Borg et al. 2012). In order to address this issue, we (Dadze et al. 2018b) performed at 573 K and 10 MPa the study of solubility of MoO$_3$ in aqueous solutions of the mixture HCl–HClO$_4$–NaCl (up to 1 m of chloride ion). These data can be explained by the formation of the species MoO$_2$(OH)$_2$Cl$^-$, with $\log_{10} K^\circ = -(0.87 \pm 0.20)$ for the solubility reaction MoO$_3$(cr) + H$_2$O(l) + Cl$^-$ = MoO$_2$(OH)$_2$Cl$^-$. Second, the thermodynamic interpretation of the solubility values in Na$^+$-bearing solutions may be affected by the formation of the complex species NaHMoO$_4$(aq), as suggested by Kudrin (1989). Therefore, we (Dadze et al. 2017b) studied, at 573 K and 10 MPa, the solubility of MoO$_3$ in aqueous solutions of NaClO$_4$ up to the salt concentration of 2.21 m, and found no need to invoke a complex formation between Na$^+$ and HMoO$_4^-$, as solubility data could be fully explained by the variations of the activity coefficients of H$^+$ and HMoO$_4^-$ in sodium perchlorate solutions. Finally, we note that at room temperatures the Mo(VI) speciation at the metal concentrations above 0.001 m and pH < 6 is dominated by highly charged polymeric species, containing 6, 7, 8, ... up to 36 Mo atoms per polymer (Cruywagen 2000). However, we expect that at high temperatures simple monomeric species will predominate in aqueous Mo-bearing solutions, and such an assumption was explicitly used at interpretation of our MoO$_3$ solubility data. The reason is that highly charged polymers experience an electrostatic repulsion, which has to be overcome when polynuclear species are formed from mononuclear constituents. The high value of the dielectric constant ($\varepsilon \sim 78$ at 298 K) decreases the electrostatic forces at room temperatures, however, the dielectric constant of water falls at high temperatures (for example, at 573 K $\varepsilon \sim 20$), forcing highly charged polynuclear forms to dissociate into monomeric species (Plyasunov and Grenthe 1994) with the increase of temperature. Nevertheless, in order to verify the assumption that only monomers HMoO$_4^-$ and H$_2$MoO$_4$ dominate speciation of Mo(VI) in weakly acid aqueous solutions at elevated temperatures, we also studied (Dadze et al. 2018a) the solubility of calcium molybdate CaMoO$_4$ (the mineral powellite) at 573 K and 10 MPa in acid (HCl, HClO$_4$) and salt (NaCl, NaClO$_4$) solutions, because the Mo(VI) concentrations at the CaMoO$_4$ solubility are about 10–200 times lower that those at the MoO$_3$ solubility. Modeling the CaMoO$_4$ solubility data using the thermodynamic properties of HMoO$_4^-$ and H$_2$MoO$_4$(aq) evaluated from data on dissolution of MoO$_3$, i.e., at much higher contents of dissolved molybdenum, showed a satisfactory agreement of experimental and calculated CaMoO$_4$ solubility values, thus providing an indirect confirmation of our assumption that at 573 K polynuclear forms contribute insignificantly to the material balance of dissolved Mo(VI).

All these auxiliary results are described in corresponding publications (Dadze et al. 2017a, b, 2018a, b) and will not be further addressed here. The current contribution is concerned with the thermodynamic properties of the neutral molybdic acid H$_2$MoO$_4$ and its volatility from aqueous solutions.

6.2 Experimental Methods

6.2.1 Materials

Crystals of MoO_3 were prepared by the three-stage calcining of ammonium molybdate as described earlier (Dadze et al. 2017a). Calcium molybdate $CaMoO_4$ (the mineral powellite) was synthesized from a mixture of MoO_3 and $CaCO_3$ taken in equivalent proportions and heated up to $T = 1473$ K. Both the synthesized molybdite and powellite were hydrothermally treated at 573 K for several days in pure water to remove small particles, edges, defects, etc. MoO_3 formed elongated crystals with sizes of 1–50 μm, see SEM photo in Supplemental Information to Dadze et al. (2017a). The typical size of the isometric $CaMoO_4$ crystals used for solubility runs was 10–30 μm, see SEM photo in Dadze et al. (2018a).

X-ray powder diffraction patterns were obtained using copper or cobalt Kα radiation on a Bruker D8 Discovery diffractometer, both before and after the experiments. Scanning electron microscopy (SEM) images using secondary electrons at an accelerating voltage of 20 kV were obtained using Tescan VEGA TS 5130MM equipped with the INCAEnergy EDXS (energy dispersive X-ray spectroscopy) microanalysis system. According to the averaged results of 4 electron microprobe analyses, the Ca:Mo atomic ratio in the synthesized calcium molybdate is equal to (0.981 ± 0.014):1.0.

Aqueous solutions of HCl, $HClO_4$, NaCl, and $NaClO_4$ were prepared from commercial reagents.

6.2.2 Experimental Procedure

Experiments were carried out in autoclaves made of the titanium alloy VT-8 of ~22 cm in the length and an internal volume of ~20 cm³. Prior to the experiments, autoclaves were loaded with 20 wt% nitric acid solution with the degree of filling of 0.72 and kept for a day at 573 K to produce a protective TiO_2 layer (Dadze et al. 2017a). The autoclaves were placed in a vertical cylindrical furnace during the experiments. The temperature gradients were less than 2 K along the autoclave in the case of a typical loading of 10 autoclaves per furnace. A temperature controller Miniterm-300 was used to monitor the temperature to an accuracy of ±1–2 K. Temperature was measured with a type K (chromel-alumel) thermocouple using a multilogger thermometer HH506RA (OMEGA Engineering). Temperature variations during the experiments were typically within ±2 K. The pressure in the autoclaves was not directly measured, but determined by the degree of the filling of the autoclave with a solution. Preliminary experiments showed poor reproducibility at the two-phase (liquid + vapor) conditions. Therefore, in order to avoid the vapor phase in the experimental vessel, the experiments were performed at pressures slightly exceeding the vapor pressure of water at a given temperature. The degree of filling was typically chosen to slightly exceed the density of liquid water at saturation at a given temperature. The necessary values of the densities of water were calculated using the Wagner and Pruß (2002) equation of state for water. Recommendations for NaCl solutions by Archer (1992) and for NaClO$_4$ solutions by Abdulagatov and Azizov (2003) were applied. Densities of dilute (0.2 m and less) solutions of HCl and HClO$_4$ were assumed to be equal to densities of NaCl solutions of the same molality.

A weighed quantity of crystalline MoO$_3$ (~100 mg) was placed in a titanium container, which was suspended in the upper part of the autoclave in such a way that it did not contact the liquid solution unless the latter expanded up to 95% of the volume of the autoclave. Due to this design, the solid phase was never in contact with the quenched solution. The volume of the added solution was within 12–16 cm^3. After the run, the autoclaves were cooled down to room temperature under running cold water for 5–7 min. The time to reach the steady-state for studies involving MoO$_3$ was estimated to be 10–15 h, and for runs with CaMoO$_4$—4 to 7 days, based on a series of kinetic runs at 573 K and 10 MPa, see Dadze et al. (2017a, 2018a).

6.2.2.1 Stability of Dilute Aqueous Perchlorate Solutions

Experiments were performed in dilute (not more than 0.22 m) aqueous solutions of perchloric acid, HClO$_4$. It is well known that concentrated aqueous solutions of perchloric acid decompose when heated above 523 K according to the reaction H$^+$ + ClO$_4^-$ = 0.5 Cl$_2$ + 0.5 H$_2$O + 1.75 O$_2$ (Henderson et al. 1971). Experiments of Henderson et al. (1971) in a titanium reactor showed that rates of decomposition depend strongly on initial HClO$_4$ concentrations, decreasing by more than 2 orders of magnitude with the decreasing molality of perchloric acid from ~4.6 to ~1.0 m. For every 10 K decrease in temperature, the rate of decomposition decreases approximately three-fold. However, the decomposition is negligibly small for 0.2 m acid at 573 K for experiments of a one-day duration (Henderson et al. 1971). In our experiments, the 0.011 m solution of perchloric acid (no solid phase present) with the initial pH = 1.98 ± 0.03 did not show any change of pH values, within errors of measurements, after being held at 573 K and 10 MPa (no vapor phase present) for 1–3 days. In contrast, a decomposition of aqueous HClO$_4$ was detected at 623 K. In blank runs (no solid phase present) of a one-day duration, a strong odor of chlorine from the quenched solution containing initially 0.10 m HClO$_4$ was detected. The pH value of the quenched solution increased from the initial value of 1.04 to 1.43, indicating significant decomposition of HClO$_4$. At the same temperature (623 K) in the quenched solution initially containing 0.01 m perchloric acid, no chlorine odor was detected, and the pH value of the quenched solution corresponded to that of the initial solution, pH = 2.03. In contrast, dilute aqueous HCl solutions are stable up to very high (magmatic) temperatures.

6.2.3   Chemical Analyses

The molybdenum concentration in quenched solutions after the experiment was determined using the spectrophotometer Spekol-11 at $\lambda = 453$ nm with the thiocyanate method, based on the formation of yellow color thiocyanate complexes of Mo(V) (Marchenko 1971), and by using thiourea as a reducing agent. The detection limit for Mo is 5 μg in a sample, i.e. at the largest aliquot, 10 ml, the minimal measurable molality of Mo is ~5 · 10^{-6} m. At 10 μg of Mo in a sample, the analytical uncertainty is about 20%, and at 20 μg and above—10%. Analytical results at relatively high solubility values (in excess of 10^{-3} m) in water and dilute acid solutions were duplicated by the weight loss (WL) method, using the WA-33 (Texma-Robot, Poland) laboratory balance with an accuracy of 0.2 mg. We found a satisfactory agreement of the analytical and WL results on the concentration of molybdenum in the solution (Dadze et al. 2017a, 2018a).

For a few runs with $CaMoO_4$ (Dadze et al. 2018a), analytical concentrations of Ca and Mo have been determined using ICP-MS and/or inductively coupled plasma atomic emission spectrometry (ICP-AES) using a PlasmaQuad II mass spectrometer with a quadrupole mass analyzer (VG Elemental, GB). Independent measurements of Mo concentrations in quenched solutions by either photometry/colorimetry or ICP-MS show a good agreement (within first per cent).

6.2.4   Treatment of Experimental Data

The data analysis was carried out using the program OptimA (Shvarov 2015), which processes experimental solubility data to derive the standard Gibbs free energies of specified aqueous complexes. Details of calculations, including corrections for activity coefficients, are described in our publications (Dadze et al. 2017a, b, 2018a, b).

6.3   Results and Discussion

6.3.1   Solubility Results for MoO₃ in Water and Acid Solutions

Experimental results of the solubility of MoO_3 in acid solutions, when chloride complexes do not form in appreciable amounts (i.e. at $m(HCl) < 0.03$), are shown in Fig. 6.1 for the most studied isotherm 573 K. As seen, solubility falls with the increase of the acid concentration and can be explained by the formation of two hydroxo complexes of Mo(VI) according to the reactions:

Fig. 6.1 Solubility of MoO₃ in HCl and HClO₄ solutions at 573 K, 10 MPa. Symbols designate experimental data, the solid line represents calculated solubility values, while dashed lines shows contributions of $HMoO_4^-$ and H_2MoO_4(aq)

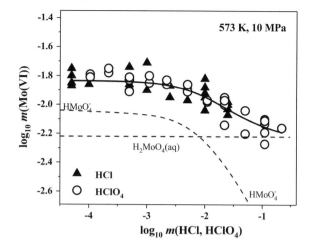

$$MoO_3(cr) + H_2O(l) = HMoO_4^- + H^+ \tag{6.1}$$

and

$$MoO_3(cr) + H_2O(l) = H_2MoO_4(aq) \tag{6.2}$$

The most extensive set of data was obtained at 573 K and 10 MPa and it is shown in Fig. 6.1. The following values of $\log_{10}K^\circ$ have been evaluated from experimental data: $\log_{10}K^\circ(1)$ −4.25, −4.31, −44.20, −4.55; $\log_{10}K^\circ(1)$ −2.40, −2.22, −2.41, −1.95 at temperatures 563, 573, 593, 623 K. The expected uncertainty of $\log_{10}K^\circ$ values is ±0.2 at 563–593 K and may be somewhat higher at 623 K.

In a later publication (Dadze et al. 2018a), we used both our results and the literature values to determine the temperature dependence of equilibrium constants for a number of reactions involving Mo(VI) species (the first and second ionization constants of H₂MoO₄, solubility product of powellite CaMoO₄, etc.), and to calculate the thermodynamic properties of H₂MoO₄(aq), $HMoO_4^-$, and MoO_4^{2-} at 273.15–623 K at the saturated water vapor pressure. The rest of this publication will be concerned with thermodynamic values of the neutral molybdic acid H₂MoO₄ in the vapor and liquid phases of water, including recommendations of Henry's constants and vapor–liquid distribution constants over the whole temperature range of existence of the vapor–liquid equilibrium of water, from 273.15 to 647.096 K (the critical temperature of water, T_c).

6.3.2 Volatility of MoO₃ from Aqueous Solutions

It is known (Alvarez et al. 1994; Palmer et al. 2004) that overwhelmingly neutral (i.e. non-charged) species partition from aqueous solutions into a vapor phase, as it is energetically very expensive to transfer an ion from a liquid phase into a gas phase that has a much lower value of a dielectric constant. Because of this, the rest of this study will be concerned with the thermodynamic properties of the neutral molybdic acid H_2MoO_4. The partition between the liquid and gaseous phases is described by two quantities: the vapor–liquid distribution constant, K_D, and Henry's constant, k_H. The vapor–liquid distribution constant, K_D, is defined as

$$K_D = \lim_{x \to 0} y/x, \qquad (6.3)$$

where y and x stand for the mole fractions of a solute in coexisting vapor and liquid phases, respectively. In our case, the simplest way to evaluate K_D is to use the relation between the vapor–liquid distribution constant and Henry's constant, k_H, defined as

$$k_H = \lim_{x \to 0} f_2/x, \qquad (6.4)$$

where f_2 stands for the solute's fugacity. There is a simple relation between two properties (Japas and Levelt Sengers 1989; Alvarez et al. 1994):

$$k_H = \phi_2^\infty \cdot P_1^* \cdot K_D \qquad (6.5)$$

where ϕ_2^∞ stands for solute's fugacity coefficient at infinite dilution, and P_1^* is the pure solvent pressure at vapor–liquid saturation.

Results of Dadze et al. (2017a) and Akinfiev and Plyasunov (2013), reporting the Gibbs energies of the species H_2MoO_4 in the state of an aqueous solution and in the ideal gas state, allow calculating Henry's constants as described below at temperatures between 273 and 623 K at P_1^*, see Table 6.1.

The logarithm of Henry's constant is proportional to the difference of the chemical potential of a species in the ideal gas state, $G^o(g)$, and in the state of the standard aqueous solution, G_2^∞:

$$\ln k_H(\text{bar}) = -[G^o(g) - G_2^\infty - RT \ln N_w]/RT \qquad (6.6)$$

and

$$\ln k_H(\text{MPa}) = \ln k_H(\text{bar}) - \ln 10 \qquad (6.7)$$

where $N_w \approx 55.508$ is the number of moles of water in 1 kg of water, the last term in Eq. (6.6) is needed because G_2^∞ is in the molality concentration scale, while Henry's constant is in the mole fraction scale.

Table 6.1 Thermodynamic properties of liquid water (the vapor pressure P_1^*, the fugacity f_1^* at P_1^*, the density of the liquid water $\rho_1^*(L)$ at P_1^*, and of molybdic acid in water, G_2^∞ (H₂MoO₄(aq)), and in the ideal gas state G^o (H₂MoO₄(g)), Henry's constant, k_H, the fugacity coefficient at infinite dilution, ϕ_2^∞, the vapor–liquid distribution constant, K_D, of molybdic acid in water)

T (K)	$\ln P_1^*$ (MPa)[a]	$\ln f_1^*$ (MPa)[a]	$\rho_1^*(L)$ (mol cm⁻³)[a]	G_2^∞ (H₂MoO₄(aq)) (kJ mol⁻¹)[b]	G^o (H₂MoO₄(g)) (kJ mol⁻¹)[c]	$\ln k_H$ (MPa)[d]	$\ln \phi_2^{\infty}$[e]	$\ln K_D$[f]
273.15	−7.400	−7.401	0.055497	−881.09	−783.04	−41.46	0.00	−34.05
283.15	−6.702	−6.703	0.055489	−881.69	−786.39	−38.77	−0.01	−32.06
293.15	−6.058	−6.059	0.055406	−882.40	−789.79	−36.28	−0.01	−30.21
298.15	−5.754	−5.756	0.055342	−882.78	−791.50	−35.11	−0.01	−29.34
303.15	−5.462	−5.463	0.055265	−883.19	−793.22	−33.98	−0.01	−28.50
313.15	−4.908	−4.911	0.055074	−884.06	−796.70	−31.84	−0.02	−26.91
323.15	−4.394	−4.398	0.054842	−885.03	−800.22	−29.85	−0.03	−25.43
333.15	−3.915	−3.920	0.054574	−886.08	−803.77	−28.00	−0.04	−24.05
343.15	−3.467	−3.474	0.054273	−887.21	−807.36	−26.27	−0.05	−22.76
348.15	−3.255	−3.263	0.054110	−887.81	−809.17	−25.45	−0.06	−22.14
353.15	−3.049	−3.058	0.053941	−888.43	−810.99	−24.66	−0.06	−21.55
363.15	−2.657	−2.669	0.053582	−889.74	−814.65	−23.15	−0.08	−20.41
373.15	−2.289	−2.304	0.053196	−891.12	−818.35	−21.74	−0.11	−19.35
398.15	−1.460	−1.486	0.052124	−894.92	−827.74	−18.58	−0.18	−16.94
423.15	−0.742	−0.783	0.050902	−899.21	−837.32	−15.88	−0.29	−14.85
448.15	−0.114	−0.174	0.049529	−903.97	−847.10	−13.55	−0.43	−13.01
473.15	0.441	0.355	0.047996	−909.18	−857.05	−11.54	−0.60	−11.38
498.15	0.936	0.819	0.046280	−914.84	−867.17	−9.79	−0.82	−9.91

(continued)

Table 6.1 (continued)

T (K)	$\ln P_1^*$ (MPa)[a]	$\ln f_1^*$ (MPa)[a]	$\rho_1^*(L)$ (mol cm^{-3})[a]	G_2^∞ (H_2MoO_4(aq)) (kJ mol^{-1})[b]	G^o (H_2MoO_4(g)) (kJ mol^{-1})[c]	$\ln k_H$ (MPa)[d]	$\ln \phi_2^{\infty,e}$	$\ln K_D^f$
523.15	1.380	1.227	0.044345	−920.93	−877.45	−8.28	−1.07	−8.59
548.15	1.783	1.588	0.042131	−927.44	−887.89	−6.96	−1.36	−7.38
573.15	2.150	1.908	0.039530	−934.35	−898.48	−5.81	−1.70	−6.27
598.15	2.489	2.193	0.036321	−941.67	−909.21	−4.81	−2.07	−5.23
623.15	2.805	2.448	0.031901	−949.36	−920.08	−3.94	−2.50	−4.24
633.15	2.927	2.542	0.029286	(−950.23)	−924.47	(−3.18)	(−3.02)	(−3.09)
643.15	3.047	2.632	0.025058	(−951.09)	−928.88	(−2.44)	(−3.57)	(−1.92)
647.096	3.094	2.667	0.017874	(−947.49)	−930.62	(−1.42)	(−4.51)	0

[a]Wagner and Pruß (2002); [b]Dadze et al. (2018a) at 273.15–623.15 K at P_1^*, at 633.15–647.096 K—see text; [c]Akinfiev and Plyasunov (2013); [d]Eqs. (6.6)–(6.7) at 273.15–623.15 K; at 633.15–647.096 K—Eq. (6.10); [e]Eq. (6.8) at 273.15–623.15 K at P_1^*, at 633.15–647.096 K—see text; [f]Eq. (6.5) at 273.15–623.15 K, at 633.15–647.096 K—Eq. (6.9)

The relation (6.5) can be used to calculate K_D provided that the values of ϕ_2^∞ can be estimated. Elsewhere (Akinfiev and Plyasunov 2013; Dadze et al. 2017a) we recommended for H_2MoO_4 the approximation

$$\ln \phi_2^\infty \approx k \cdot \ln \phi_1^* = 7 \cdot \ln \phi_1^*, \tag{6.8}$$

where ϕ_1^* is the fugacity coefficient of pure water (Wagner and Pruß 2002), and the rules to estimate k depending on the composition of a compound were discussed in Akinfiev and Plyasunov (2013). It was shown that Eq. (6.8) allows a satisfactory prediction of solubility of MoO₃ in steam at 573–873 K at water densities up to ~250 kg m⁻³, see Dadze et al. (2017a) for detail. Values of K_D evaluated with relations (6.5) and (6.8) at 273.15–623.15 K are given in Table 6.1.

6.3.2.1 Extrapolation of K_D and k_H Toward the Critical Point of Water

The value of the vapor–liquid distribution constant at the critical point of water is equal to 1, as the liquid and the vapor phase became indistinguishable here. The question is, what is the law governing the course of K_D toward the critical point of a solvent? This question was solved theoretically by Japas and Levelt Sengers (1989) and confirmed by the reliable data (Fernández-Prini et al. 2003): at temperatures above ~500 K and up to the critical temperature of water, variations of K_D can be quantitatively described by the relation:

$$RT \ln K_D = A_{Kr} \cdot \frac{2\left(\rho_1^*(L) - \rho_c\right)}{\rho_c^2}, \tag{6.9}$$

where A_{Kr} is the Krichevskii parameter, the main thermodynamic characteristic of a solute close to the critical point of a solvent, see Levelt Sengers (1991) for the corresponding discussion; ρ_c stands for the critical density of the solvent (322 kg m⁻³ for water), $\rho_1^*(L)$ is the pure water density along the liquid side of the saturation vapor–liquid curve (note that in order to obtain values of A_{Kr} in MPa, density values should be in cm⁻³ mol). Equation (6.9) was employed by us earlier (Dadze et al. 2017a) to evaluate $A_{Kr} = -228 \pm 33$ MPa for H_2MoO_4 in water. The relation (6.9) was used to estimate K_D at 633.15, 643.15 and T_c, see values in parentheses in the corresponding column of Table 6.1.

The theoretically-based asymptotic (i.e. valid in the neighborhood of the critical point of a solvent) relation was derived (Japas and Levelt Sengers 1989) for Henry's constant as well

$$RT \ln \frac{k_H}{f_1^*} = C_o + A_{Kr} \cdot \frac{\left(\rho_1^*(L) - \rho_c\right)}{\rho_c^2}, \tag{6.10}$$

The evaluation of the constant C_o for H_2MoO_4 can be made in the same way as for $Si(OH)_4$ (Plyasunov 2012), i.e. by plotting the values of the auxiliary function Z

Fig. 6.2 Evaluation of C_o for H_2MoO_4 from k_H data

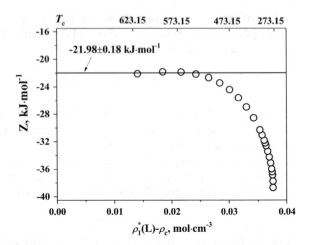

$$Z = RT \ln \frac{k_H}{f_1^*} - A_{Kr} \cdot \frac{\left(\rho_1^*(L) - \rho_c\right)}{\rho_c^2} \qquad (6.11)$$

versus $\rho_1^*(L) - \rho_c$, see Fig. 6.2. According to Eq. (6.10), the auxiliary function Z close to the critical point of water should approach the constant value equal to C_o, and data for H_2MoO_4 confirm that at temperatures above 550 K with $C_o = -21.98 \pm 0.18$ (2σ) kJ mol^{-1}. With relation (6.10) we estimated k_H at 633.15, 643.15 and T_c, see values in parentheses in the corresponding column of Table 6.1.

6.3.2.2 Temperature Dependence of K_D and k_H from 273.15 K to T_c

The asymptotic relations, given by Eqs. (6.9)–(6.10), should serve as foundations for the equations for k_H and K_D covering the whole temperature range of the vapor–liquid equilibrium for water, from 273.15 to the critical temperature of water, $T_c = 647.096$ K (Wagner and Pruß 2002). After some trials, the following forms were selected for the regression of K_D and k_H data from Table 6.1:

$$RT \ln K_D = A_{Kr} \cdot \frac{2\left(\rho_1^*(L) - \rho_c\right)}{\rho_c^2} \cdot \left\{ 1 + a_1\left(1 - T/T_c\right) + a_2\left(1 - T/T_c\right)^2 \right\}, \qquad (6.12)$$

and

$$RT \ln \frac{k_H}{f_1^*} = C_o + A_{Kr} \cdot \frac{\left(\rho_1^*(L) - \rho_c\right)}{\rho_c^2} \cdot \left\{ 1 + b_1\left(1 - T/T_c\right) + b_2\left(1 - T/T_c\right)^2 \right\}, \qquad (6.13)$$

Fig. 6.3 Temperature dependence of K_D: the solid line—Eq. (6.12) for H₂MoO₄, solid circles—values from Table 6.1 at 273–623 K, empty circles—values evaluated from the asymptotic relation (6.9) data at 633.15, 643.15 and T_c. The dashed and dotted lines show the corresponding dependences for Si(OH)₄ (Plyasunov 2012) and Ar (Fernández-Prini et al. 2003)

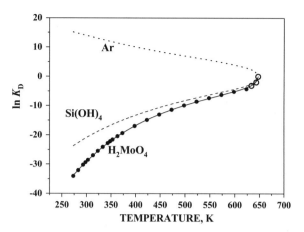

Fig. 6.4 Temperature dependence of k_H. The legend is the same in Fig. 6.3

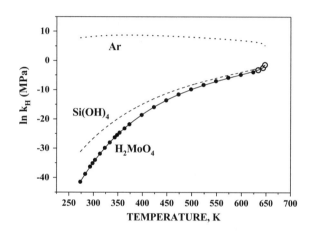

where the Krichevskii parameter $A_{Kr} = -228$ MPa; $a_1 = -0.373912$, $a_2 = 1.93796$; the constant $C_0 = -21{,}980$ J mol^{-1}, $b_1 = -0.516457$, $b_2 = 4.06929$. The fitted values of K_D and k_H are shown in Figs. 6.3 and 6.4. The combination of $G^o(g)$ and k_H allowed evaluating G_2^∞ for H₂MoO₄ at 633.15, 643.15 and T_c, see the corresponding column of Table 6.1.

6.3.2.3   Comparison with Literature Data for Mo and Some Other Solutes

A comparison of K_D values recommended in this study, and those reported in the literature (Khitarov et al. 1967; Rempel et al. 2009; Kokh et al. 2016) is not straightforward. Indeed, the literature partition values are reported as stoichiometric distribution constants, i.e. as ratios of analytical concentrations of elements in coexisting phases. However, as briefly discussed in Sect. 6.1, the partitioning solute is a non-charged species, H_2MoO_4 in the case of Mo(VI) solutions, and its concentration is a strong function of pH of the liquid solution. Nonetheless, a semi-quantitative estimation of K_D values in terms of H_2MoO_4 concentrations in coexisting vapor and liquid phases at 623 K based on data of Khitarov et al. (1967), Rempel et al. (2009), Kokh et al. (2016) was undertaken in Dadze et al. (2017a). It was found that the difference with the experimental results (Khitarov et al. 1967; Rempel et al. 2009) exceeded 8 and ~3 orders of magnitude, respectively. By contrast, a relatively close agreement (0.5–1 $\log_{10}$ units) was found with vapor phase concentrations measured by Kokh et al. (2016). Considering a complicated composition of the liquid phase in experiments of Kokh et al. (2016)—0.5 m NaCl, 0.5 m KCl, 0.1 m HCl, 0.01 m $ZnCl_2$, 0.01 m CuCl, 0.01 m Na_2MoO_4, plus solid phases SnO_2, SiO_2, Fe_2O_3, Fe_3O_4, Pt, Au and simplifications in calculations, this level of agreement appears to be as good as it gets. We remind that Kokh et al. (2016) applied a direct sampling of the coexisting vapor and liquid phases, while Khitarov et al. (1967) and Rempel et al. (2009) attempted to trap a portion of a vapor phase in a two-phase system, which appears to be an error-prone method.

As was discussed in Sect. 6.1, Henry's constants of H_2MoO_4 reported in Rempel et al. (2009) exceed those for Ar. According to Fig. 6.4, this result appears grossly in error. Previously we (Dadze et al. 2017a) suggested that $Si(OH)_4$ would be a good proxy for H_2MoO_4. As Figs. 6.3 and 6.4 show this is a better, but still not a perfect approximation for H_2MoO_4.

6.4   Conclusions

Experimental results of the solubility of MoO_3 in acid aqueous solutions at 563–623 K were used to determine the thermodynamic properties of the neutral molybdic acid H_2MoO_4 at these temperatures (Dadze et al. 2017a). The joint treatment of our own and literature data resulted in a recommendation of the thermodynamic properties of this species at 273–623 K and the saturated water vapor pressure (Dadze et al. 2018a). Combining the Gibbs energies of H_2MoO_4 in the state of the aqueous solution with those in the ideal gas state (Akinfiev and Plyasunov 2013), we calculated Henry's constants and the vapor–liquid distribution constants of H_2MoO_4 at 273–623 K, and with the use of the relevant asymptotic relations (Japas and Levelt Sengers 1989) extrapolated values of K_D and k_H toward the critical point of pure water. Our results show that over the whole temperature range of existence of the vapor–liquid equilibrium of water, the neutral molybdic acid H_2MoO_4 is somewhat less volatile compared with $Si(OH)_4$, and that the difference in volatility of these species decreases with the temperature, see Figs. 6.3 and 6.4.

Finally, we note that many vapor–liquid partitioning results in the geochemical literature are reported as ratios of analytical concentrations of elements in coexisting phases, often having multicomponent compositions. The rigorous thermodynamic treatment of such data is usually not possible, leaving such results suitable for semi-quantitative modeling at best (or littering the literature with improbable thermodynamic values at worst). If the quantitative modeling of geochemical processes is a goal, even a remote one, than the experimental investigation of the simplest systems, which are easiest for rigorous thermodynamic interpretation, will be of value.

Acknowledgements This research was partially supported by the Russian Foundation for Basic Research (Grant # 15-05-2255). The authors thank A. N. Nekrasov and T. N. Dokina (IEM RAS) for SEM and XRD measurements and Dr. V. K. Karandashev (IPTM RAS) for ICP-MS analyses.

References

Abdulagatov IM, Azizov ND (2003) High-temperature and high pressure densities of aqueous NaClO₄ solutions. High Temp High Press 35–36:477–498

Akinfiev AN, Plyasunov AV (2013) Steam solubilities of solid MoO₃, ZnO and Cu₂O, calculated on a basis of a thermodynamic model. Fluid Phase Equilib 338:232–244

Alvarez J, Corti HR, Fernández-Prini R, Japas ML (1994) Distribution of solutes between coexisting steam and water. Geochim Cosmochim Acta 58(13):2789–2798

Archer DG (1992) Thermodynamic properties of the NaCl + H₂O system. II. Thermodynamic properties of NaCl(aq), NaCl·2H₂O(cr), and phase equilibria. J Phys Chem Ref Data 21(4):793–829

Borg S, Liu W, Etschmann B, Tian Y, Brugger J (2012) An XAS study of molybdenum speciation in hydrothermal chloride solutions from 25–385°C and 600 bar. Geochim Cosmochim Acta 92:292–307

Cruywagen JJ (2000) Protonation, oligomerization, and condensation reactions of vanadate(V), molybdate(VI), and tungstate(VI). Adv Inorg Chem 49:127–182

Dadze TP, Kashirtseva GA (2004) Solubility and occurrence mode of gold in acid sulfide solutions. Dokl Earth Sci 395(2):235–237

Dadze TP, Sorokin VI (1986) Solubility of SnO₂ in water at 200–400°C and 1.6–150 MPa. Dokl Akad Nauk SSSR 286(2):426–428 (in Russian)

Dadze TP, Sorokin VI, Nekrasov IYa (1981) Solubility of SnO₂ in water and in aqueous solutions of HCl, HCl + KCl, and HNO₃ at 200–400°C and 1013 bar. Geochem Int 18(5):142–152

Dadze TP, Akhmedzhanova GM, Kashirtseva GA, Orlov PYu (1999) The Au solubility in H₂S-bearing aqueous solutions at 300°C. Dokl Earth Sci 369A(9):1275–1276

Dadze TP, Kashirtseva GA, Ryzhenko BN (2000) Gold solubility and species in aqueous sulfide solutions at T = 300°C. Geochem Int 38(7):708–712

Dadze TP, Akhmedzhanova GM, Kashirtseva GA, Orlov PYu (2001) Solubility of gold in sulfide-containing aqueous solutions at T = 300°C. J Mol Liquids 91(1):99–102

Dadze TP, Kashirtseva GA, Novikov MP, Plyasunov AV (2017a) Solubility of MoO₃ in acid solutions and vapor-liquid distribution of molybdic acid. Fluid Phase Equilib 440:64–76

Dadze TP, Kashirtseva GA, Novikov MP, Plyasunov AV (2017b) Solubility of MoO₃ in NaClO₄ solutions at 573 K. J Chem Eng Data 62(11):3848–3853

Dadze TP, Kashirtseva GA, Novikov MP, Plyasunov AV (2018a) Solubility of calcium molybdate in aqueous solutions at 573 K and thermodynamics of monomer hydrolysis of Mo(VI) at elevated temperatures. Monatsh Chem/Chem Month 149(2):261–282

Dadze TP, Kashirtseva GA, Novikov MP, Plyasunov AV (2018b) Solubility of MoO₃ in aqueous acid chloride-bearing solutions at 573 K. J Chem Eng Data 63(5):1827–1832

Fernández-Prini R, Alvarez JL, Harvey AH (2003) Henry's constants and vapor–liquid distribution constants for gaseous solutes in H₂O and D₂O at high temperatures. J Phys Chem Ref Data 32(3):903–916

Gorbachev NS, Dadze TP, Kashirtseva GA, Kunts AF (2010) Fluid transfer of gold, palladium, and rare earth elements and genesis of ore occurrences in the Subpolar Urals. Geol Ore Deposit 52(3):215–233

Henderson MP, Miasek VI, Swaddle TW (1971) Kinetics of thermal decomposition of aqueous perchloric acid. Can J Chem 49(2):317–324

Japas ML, Levelt Sengers JMH (1989) Gas solubility and Henry's law near the solvent's critical point. AIChE J 35(5):705–713

Khitarov NI, Arutyunyan LA, Malinin SD (1967) On the possibilities of molybdenum migration in the vapor phase above molybdate solutions at elevated temperatures. Geokhimiya 2:155–159 (in Russian)

Kokh MA, Lopez M, Gisquet P, Lanzanova A, Candaudap F, Besson P, Pokrovski GS (2016) Combined effect of carbon dioxide and sulfur on vapor-liquid partitioning of metals in hydrothermal systems. Geochim Cosmochim Acta 187:311–333

Kudrin AV (1985) The solubility of tugarinovite MoO_2 in aqueous solutions at 300–450°C. Geochem Int 22(9):126–138

Kudrin AV (1989) Behavior of Mo in aqueous NaCl and KCl solutions at 300–450°C. Geochem Int 26(8):87–99

Levelt Sengers JMH (1991) Solubility near the solvent's critical point. J Supercrit Fluids 4(4):215–222

Marchenko Z (1971) Photometric determination of elements. Mir, Moscow, 501 p (in Russian)

Palmer DA, Simonson JM, Jensen JP (2004) Partitioning of electrolytes to steam and their solubilities in steam. In: Palmer DA, Fernàndez-Prini R, Harvey AH (eds) Aqueous systems at elevated temperatures and pressures. Elsevier, New York, pp 409–439

Plyasunov AV (2012) Thermodynamics of $Si(OH)_4$ in the vapor phase of water: Henry's and vapor–liquid distribution constants, fugacity and cross virial coefficients. Geochim Cosmochim Acta 77:215–231

Plyasunov AV, Grenthe I (1994) The temperature dependence of stability constant for the formation of polynuclear cationic complexes. Geochim Cosmochim Acta 58(17):3561–3582

Plyasunov AV, Ivanov IP (1991) The solubility of zinc oxide in sodium chloride solutions up to 600°C and 1000 bar. Geochem Int 28(6):77–90

Plyasunov AV, Plyasunova NV (1993a) Study of solubility of zinc oxide in KOH solutions at 400–500°C and pressure 0.5–2.0 kbar. Dokl Akad Nauk SSSR 328(5):605–608 (in Russian)

Plyasunov AV, Plyasunova NV (1993b) Study of solubility of zinc oxide in KOH + NaCl solutions at 400–500°C and pressure 0.5–2.0 kbar. Dokl Akad Nauk SSSR 329(2):228–231 (in Russian)

Plyasunov AV, Belonozhko AB, Ivanov IP, Khodakovsky IL (1988) Solubility of zinc oxide in alkaline solutions at 200–350°C under saturated steam pressure. Geochem Int 25(10):77–85

Rempel KU, Williams-Jones AE, Migdisov AA (2009) The partitioning of molybdenum(VI) between aqueous liquid and vapour at temperatures up to 370°C. Geochim Cosmochim Acta 73(11):3381–3392

Shvarov YuV (2015) A suite of programs, OptimA, OptimB, OptimC, and OptimS compatible with the Unitherm database, for deriving the thermodynamic properties of aqueous species from solubility, potentiometry and spectroscopy measurements. Appl Geochem 55:17–27

Sorokin VI, Dadze TP (1980) Solubility of amorphous SiO_2 in water and in aqueous solutions of HCl and HNO_3 at temperatures 100–400°C and pressure of 101.3 MPa. Dokl Akad Nauk SSSR 254(3):735–739 (in Russian)

Wagner W, Pruß A (2002) The IAPWS formulation for the thermodynamic properties of ordinary water substance for general and scientific use. J Phys Chem Ref Data 31(2):387–535

Zakirov IV, Dadze TP, Sretenskaya NG, Kashirtseva GA (2008) Experimental data on gold solubility in low-density hydrothermal fluids. Dokl Earth Sci 423(2):1492–1494

Zakirov IV, Dadze TP, Sretenskaya NG, Kashirtseva GA, Volchenkova VL (2009) Gold solubility in low-density fluids in the $Au–H_2O–H_2S–Cl$ system: experimental data. Geochem Int 47(3):311–314

Chapter 7
Experimental Determination of Ferberite Solubility in the KCl–HCl–H$_2$O System at 400–500 °C and 20–100 MPa

Alexander F. Redkin and Gary L. Cygan

Abstract The solubility of ferberite, FeWO$_4$ was studied at 400–500 °C, pressures of 20, 25, 40, 50 and 100 MPa, oxygen fugacity corresponding to the Ni–NiO, Fe$_3$O$_4$–Fe$_2$O$_3$ buffers, in $0.7 \div 8.9$ mKCl solutions and acidity controlled by quartz-microcline-muscovite buffer assemblage. The parameters of the experiments cover both the field of homogeneous solutions and the region of immiscibility in the KCl–H$_2$O system. The total W concentration depends upon mCl, T and fO$_2$ in the system and range from $1 \cdot 10^{-4}$–0.05 mol kg^{-1} in 0.7 mKCl to 0.01–0.15 in 8.9 mKCl. The results suggest a large bulk solubility in the dense, salt-rich phase of the two-phase fluid. Ferberite dissolution in KCl solutions under pH and fO$_2$ buffered conditions at 400–500 °C proceeds congruently as well as incongruently with accompanying potassium tungsten bronzes formation, K$_x$WO$_3$, (x = 0.2–0.3). Thermodynamic calculations performed for a homogeneous solution at $P = 100$ MPa indicate that the predominant aqueous species of tungsten in KCl–HCl solutions at fO$_2 = f$O$_2$(Ni–NiO) may be W(V, VI) species: WO$_3^-$, HWO$_4^-$, H$_2$W$_2$O$_7^-$ at 500 °C and HWO$_4^-$, W$_5$O$_{16}^{3-}$ at 400 °C. Application of the extended FeWO$_4$ solubility model to natural systems suggest that deposition of tungsten from ore-bearing solutions is due to interaction with wall rocks containing feldspars, and iron oxides together with decreasing temperatures. In the magnetite bearing system, the equilibrium tungsten concentration does not exceed $2 \cdot 10^{-5}$ mol kg^{-1} at temperatures of 400 °C.

Keywords Ferberite · Potassium tungstate bronze · Solubility · Buffered system · Homogenous solution · Immiscibility region · Tungsten species

7.1 Introduction

Wolframite, (Fe, Mn)$^{+2}$WO$_4$, as well as its end members ferberite, FeWO$_4$, and hüebnerite, MnWO$_4$, are the main ore minerals associated with hydrothermal deposits of

A. F. Redkin (✉)
Institute of Experimental Mineralogy, Russian Academy of Sciences, Moscow, Russia 142432
e-mail: redkin@iem.ac.ru

G. L. Cygan
U.S. Environmental Protection Agency, Chicago, IL 60604, USA

© The Editor(s) (if applicable) and The Author(s), under exclusive license to Springer Nature Switzerland AG 2020
Y. Litvin and O. Safonov (eds.), *Advances in Experimental and Genetic Mineralogy*, Springer Mineralogy, https://doi.org/10.1007/978-3-030-42859-4_7

137

tungsten related to the metasomatic processes of granite greisenization and albitization. Tungsten ore deposition usually occurs late in the paragenetic sequence and is typically located in the apical segments of massifs. Mineralization typically occurs at depths equated with the range of 100–150 m. Temperature conditions of ore formation are estimated to be in the interval 300–540 °C (Smirnov et al. 1981). According to mineral-geochemical investigations ore deposition took place in several stages, typically as part of the post-greisen stage of development. The greisen metasomatic process created conditions appropriate for leaching W and other ore elements (e.g., Fe, U, REE) from the surrounding rocks interacting with the hydrothermal solutions. The analysis of gas-liquid inclusions in quartz from greisen of different generations (Ivanova et al. 1986) indicate wolframite deposits were formed under the influence of solutions containing high concentrations of salts (dominantly NaCl, KCl), carbon dioxide and rich in fluorides. Hydrothermal solutions containing tungsten leached iron from the host rock, effectively bleaching the host rock. It appears that the calcium content in these hydrothermal solutions is not critical since we observe sheelite ($CaWO_4$) mineralization principally associated with skarns. Tungsten ore deposition in these, high-temperature, high-concentration salt solutions occurred during the greisen stage of the hydrothermal process. The hydrothermal solution had the same acidity as that of the solutions forming quartz-topaz and quartz-mica (muscovite) greisens. For these reasons, a study of ferberite solubility in highly concentrated salt solutions, particularly the two-phase region of aqueous-salt fluids, under buffered conditions is a principal problem in tungsten geochemistry. In addition, investigations into the effect of pressure on the solubility of ore minerals is also of considerable interest since it has been demonstrated by Hemley et al. (1986, 1992), Malinin and Kurovskaya (1996) that pressure decreases can increase metal solubility in acid-buffered systems.

7.2   Experimental Procedures

Experiments were performed at 400–500 °C and pressures of 20, 25, 40, 50 and 100 MPa using large volume hydrothermal pressure vessels (30 mm internal diameter × 200 mm long): cold-seal pressure vessels (Tuttle bombs) and a pinch-off vessel (Ivanov et al. 1994). A drawing of the Pinch-off device in thin-wall capsule of gems is represented in Fig. 7.1a, b. The experimental charge in the welded capsule is positioned in the clamping mechanism and placed between the two halves of the prop above the connector. The clamp location on the capsule may be moved so that the ratio of the separated, or "pinched-off" capsule may be varied (Fig. 7.1a). This pinch-off mechanism is operated by way of a hydraulic push of the rod that moves the pinching tool upwards causing the pincering effect to occur. The pincer consists of two identical halves with a conical external surface, specially designed internal longitudinal hollows and transverse connectors. Crushing and collapsing of the capsule does not occur because of these longitudinal grooves and connectors. The walls of the capsule are pressed tightly enough that a weld occurs at the pinch point

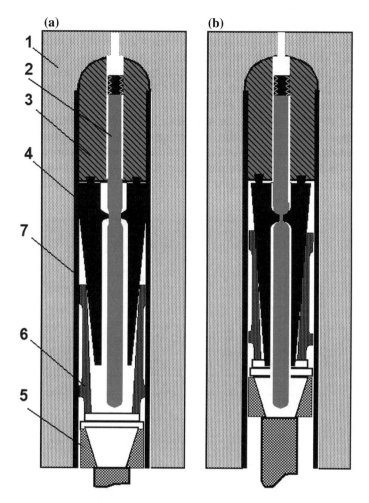

Fig. 7.1 Schematic showing the pinch-off vessel used for separation of media under hydrothermal conditions: **a** initial state; **b** after the separator with the one connector has been activated. Details of the device: 1—Ni-alloy reactor, 2—capsule, 3—stop holder (two details), 4—separator or capsule clamp (two details), 5—movable socket rod, 6—movable sleeve, 7—tube

(Fig. 7.1b). The hot spot of the furnace was measured and found to have a thermal gradient of 2–3 °C over 12 cm. The capsule pinching was done after the necessary time required for establishing equilibrium. The pinch-off vessel was used to refine KCl concentration in brine at 500 °C and pressure of 40 and 50 MPa.

Using large volume vessel (cold seals Tuttle bomb), four experiments on ferberite solubility were run simultaneously. Temperature was regulated using a high-precision controller using chromel-alumel thermocouples between the experimental capsules. The thermocouple uncertainty is ±2–3 °C. Experiments at pressures of 20–50 MPa were set using a Burdon manometer for pressures of up to 60 MPa and an uncertainty of ±0.2 MPa. Experiments conducted at 100 MPa used a 2500-style manometer with an accuracy of ±15 MPa. Experimental run times were from 14 to 30 days, depending upon P and T, though previous studies in similar systems containing the ore minerals SnO_2, UO_2, $CaWO_4$, indicate equilibrium with the solutions is approached in several hours to less than 7 days (Dadze et al. 1981; Kovalenko et al. 1986; Redkin et al. 1989; Malinin and Kurovskaya 1996).

Study of ferberite solubility in aqueous-salt KCl–HCl solutions was made in platinum capsules, 60–90 mm long × 7–8 mm diameter. The starting solutions for the experiments were made from commercial reagents of high-purity KCl, HCl and doubly-distilled water. Pressures and starting KCl compositions were chosen such that the experiments would be in the two-phase region (Fig. 7.2a, b). At 500 °C and

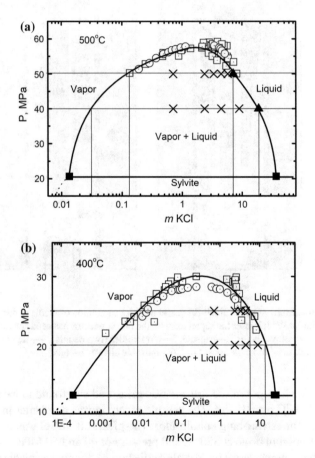

Fig. 7.2 Vapor and liquid compositions along the 500 (**a**) and 400 °C (**b**) isothermal coexistence curves for the KCl–H$_2$O system. Solid square symbols for three-phase (sylvite-liquid-vapor) by Hovey et al. (1990). For two-phase region: dotted square by Cygan et al. (1994), dotted round by Shmulovich et al. (1995), solid triangles represent our data obtained using the pinch-off device. The initial mKCl values used in the runs on ferberite solubility are shown by "X"

a pressure of 40 and 50 MPa, 5, 15, 25 and 20 or 30 wt% solutions of KCl were used (Fig. 7.2a), and at 400 °C 20, 25 and 100 MPa, 5, 15, 25 and 40 wt% of KCl were used (Fig. 7.2b). The initial solutions also contained HCl of 0.01 and 0.1 or 0.001 mol/kg H_2O. Varying starting mKCl/mHCl ratios made it possible to approach equilibrium in the QMM buffer system from both sides as was done by Hemley (1959).

Ferberite starting material was synthesized through recrystallization of a mixture of FeO and WO_3 in a solution of 30 wt% LiCl in gold capsules. The ratio of the aqueous solution to the solid mixture was about 2:1 by weight. This mixture was then reacted at 500 °C and 100 MPa (Klevtsov et al. 1970) for 14 days. This produced grain sizes of 5–10 μm in diameter, and subsequent recrystallization in the same solution and T at $P = 40$ MPa for 14 days yielded grains diameters of 15–25 μm. X-ray diffraction scans of synthetic ferberite correspond to the standard sample 27–256 (PDF 1980), while microprobe analyses indicate the composition corresponds to the stoichiometric formula of ferberite, $FeWO_4$. Fourier transform infrared (FTIR) spectrum of the synthetic ferberite, depicted in Fig. 7.3, also indicates that the synthesized product is pure and complete conversion of FeO and WO_3 has occurred.

The initial solid charge consisted of a mixture of 20 mg of ferberite with 60 mg of the alumina-silicate buffer, quartz (SiO_2)-microcline ($KAlSi_3O_8$)-muscovite ($KAl_3Si_3O_{10}(OH)_2$)-(QMM). The initial Na_2O concentration of a natural pegmatitic

Fig. 7.3 Infra-red spectra of synthetic ferberite and the starting compounds FeO and WO_3 used in the ferberite synthesis

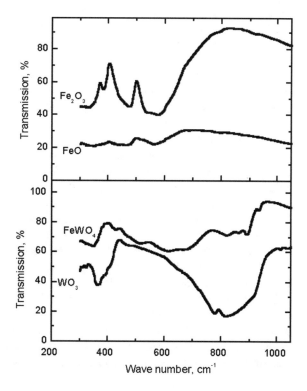

microcline (as Mic phase) was decreased through exposure to molten KCl. The exchange yielded a microcline composition of (in wt%) SiO_2-65.59, Al_2O_3-17.72, K_2O-16.44, FeO < 0.12, TiO_2 < 0.06, Na_2O, MnO, MgO, Cr_2O_3, CaO < 0.02. Pegmatic $2M_1$ muscovite was processed by rubbing plates of the pure natural mineral (composition (in wt%) SiO_2-47.79, TiO_2-0.64, Al_2O_3-32.46, Fe_2O_3-3.17, MnO-0.13, MgO-1.317, CaO-0.070, Na_2O-0.61, K_2O-10.77, H_2O + 2.52, F_2 < 0.04). The muscovite was sieved to 20–300 μm and then purified in dilute HCl. Before crushing in an agate mortar, synthetic quartz was heated briefly in a furnace to 700 °C and quenched in pure water to form fractures. The QMM buffer is required in this study to provide a constant KCl/HCl ratio in the aqueous phase to control the system's acidity through the following reaction:

$$1.5Mic + HCl_{(aq)} = 3Qtz + 0.5Ms + KCl_{(aq)} \qquad (7.1)$$

In the two-phase region of the $KCl–H_2O$ system, the QMM buffer will provide constant concentrations of KCl and HCl in each phase at a given T and P. A change in the initial KCl concentration alters the relative masses or volumes of the L and V phases, but not the KCl and HCl concentrations (Cygan et al. 1994).

Oxygen fugacity in the experiments was controlled using the nickel-bunsenite (NNO, $fO_2 = 10^{-17.70}$ at 500 °C and $10^{-22.42}$ Pa at 400 °C) and hematite-magnetite oxygen buffers (HM, $fO_2 = 10^{-13.63}$ Pa at 500 °C). The buffers consisted of approximately a 1:1 ratio of fine-crystalline metallic nickel and bunsenite or synthetic magnetite and commercial hematite, a total amount of 3–4 g was placed into the Ni-alloy container. The presence of Ni and Fe_3O_4 in the reaction mixture after the runs was checked by magnet.

Platinum capsules with the charge of solid phases (80 mg) and solution (0.4 ml) were assembled. The air above the charge was removed by blowing pure argon through the capsule before welding. We feel this contributed to a more rapid attainment of equilibrium fO_2.

The charges were quenched to room temperature using an air-water mixture for 7–10 min. The capsules were recovered, dried and weighed to check for the integrity of the capsule seal. Capsules showing less than ±5 mg change in weight were considered valid runs. The capsules were opened and the solutions and the charge were washed into plastic test-tubes using approximately a 20-fold volume of doubly-distilled water. The dilution of the recovered aqueous solution was determined by weighing the washed material to an accuracy of ±0.03 mg. These solutions were then centrifuged at a velocity of 5000–6000 rpm for 2–3 min. An aliquot of pure solution for chemical analysis was removed. The recovered solid materials were washed with doubly-distilled water and then centrifuged and dried in ceramic cups at 100–120 °C.

The transparent, colorless solutions were subjected to spectrophotometric analysis for tungsten using a Spekol-11 (Germany) and the bromopyrogallol red technique (Andreeva et al. 1992). For iron we used the ortho-phenantroline and α, α-dipyridill technique (Marczenko 1976), and for SiO_2, mainly the molybdate yellow method (Teleshova 1973). Aqueous Al analyses were performed using an atomic absorption spectrometer (Perkin-Elmer model 403). All our attempts to use spectrophotometric analysis for Al using the antrazochrom technique (Basargin et al. 1976) were unsuccessful due to the strong interfering effect that high W concentrations had on measurements. The analytical error for Fe is estimated to be <5%, for W and SiO_2 < 10%, and for Al < 30%. Solid charge products were subjected to a quantitative microprobe analysis using a Camebax MBX at $E_o = 15$ keV, the size of beam of electrons $\varnothing = 1$–2 μm and X-ray diffraction investigations using a DRON-2 (made in Russia) with Cu-Kα.

7.3 Solubility-Salt Concentration Correlation

The solubility of a congruently-dissolving mineral in the two-phase region at constant pressure and acidity of the V and L phases can be described by the equation

$$S_{total} = S^V \cdot X_V + S^L \cdot X_L = S^V + X_L \cdot (S^L - S^V), \qquad (7.2)$$

where S_{total} is a total mineral solubility in the two-phase region (wt%), S^L and S^V are the mineral solubility in L (high-density) and V (low-density) phases, and X_L and X_V are the mass fractions of L and V. Equation (7.2) represents a general expression of the lever rule for phases in equilibrium, and describes the distribution of KCl as well

$$C_{salt}^T = C_{salt}^L \cdot X_L + C_{salt}^V \cdot X_V,$$

where C_{salt}^T, C_{salt}^L, and C_{salt}^V are the total concentration of salt and concentrations of salt in the coexisting L and V phases in wt%, respectively. It follows,

$$X_L = \left(C_{salt}^T - C_{salt}^V\right)/\left(C_{salt}^L - C_{salt}^V\right). \qquad (7.3)$$

Replacing X_L in (7.2) by (7.3) we get the expression where mineral solubility depends linearly on total salt concentration, C_{salt}^T, whereas S^V, S^L, C_{salt}^L, and C_{salt}^V are constant:

$$S_{total} = S^V + (S^L - S^V) \cdot \left(C_{salt}^T - C_{salt}^V\right)/\left(C_{salt}^L - C_{salt}^V\right), \qquad (7.4)$$

and $S_{total} = S^V$ at $C_{salt}^T = C_{salt}^V$ and $S_{total} = S^L$ at $C_{salt}^T = C_{salt}^L$. It is clear that in the two-phase region, we can speak about solubility of a compound either in a liquid or in the vapor phase. However, the value S_{total} itself represents an apparent solubility, that is, a total solubility obtained after fluid homogenization after quench. Nevertheless, these apparent mineral solubility data may be described by using a linear function of the total salt concentration in a fluid.

7.4 Results and Discussions

7.4.1 Solutions After the Runs

Experimental results are represented in Table 7.1 and Fig. 7.4a, b. Quench pH values of parallel experiments, that is, experiments made at the same initial concentration of KCl but different mHCl (0.1 and 0.01), are quite similar indicating the acidic-alkaline equilibrium between the QMM buffer and KCl–HCl–H_2O solution is operating. The divergence in mHCl values of some runs, especially those conducted in 40 wt% KCl starting solutions (Fig. 7.5), may be related to the effect of high concentrations of Fe and W on the quenched pH value (mHCl). Total W and Fe concentrations are a function of the mKCl and the initial mHCl. These varied from $1 \cdot 10^{-4}$–0.05 in 0.7 mKCl starting solution experiments to 0.01–0.15 m in 8.9 mKCl (see Fig. 7.4a). Incongruent dissolution of ferberite is indicated by the fact that mW was consistently not equal to mFe in the analyzed solutions. Despite the significant concentrations of W and Fe (0.02–0.15 mol kg^{-1}) the quenched solutions were acidic, transparent, and colorless (or light blue (Busev et al. 1976)) suggesting that iron in the solutions was present as species of Fe(II). During preparation for analysis, mixing the sample with NaOH caused the formation of very small needle-like crystals of Na, K blue oxichlorwolframites(V) which then easily redissolved into the solution.

Ferberite dissolution took place in solutions saturated with QMM buffer components. Concentration of Al in these solutions was in the range of 0.0003–0.003 mol kg^{-1}, i.e., considerably lower than the W and Fe molality. In the homogeneous, single-phase region of the KCl–HCl system, quartz solubility did not change in experiments with mKCl ranging from 0.038 to 2.367. However, quartz solubility decreased reaching 0.022 mSiO$_2$ at 500 °C (Fig. 7.6a) and 0.02 mSiO$_2$ at 400 °C in 8.942 mKCl (Fig. 7.6b) experiments. Taking into account such an inadequate character of behavior of SiO$_2$ and W one can assume that in weakly-acidic concentrated aqueous-salt solutions at 500 °C and NNO buffer a formation of the marked numbers of predominant Si–W heteropolycomplexes did not take place. However, these complexes could be formed when we analyzed quenched solutions for SiO$_2$ using the blue method (Marczenko 1976) affecting the results to lower mSiO$_2$.

7.4.2 Solid Phases After the Runs

X-ray diffraction and microprobe analysis of the recovered solid phases after the experiments showed that the initial components of the QMM buffer, i.e., Mic, Ms and Qtz, did not undergo phase changes and all three minerals were present in the run products. No new alumina-silicate phases were found, and the compositions of Mic and Ms remained unchanged.

Table 7.1 Experimental results for ferberite FeWO$_4$ solubility in KCl–HCl solutions at QMM buffers

P (MPa)	Time (days)	Initial solution (mol kg^{-1})		Final solution (log m)				
		KCl	HCl	W	Fe	Al	SiO$_2$	HCl[*]
1	2	3	4	5	6	7	8	9
500 °C, NNO								
100	18	0.706	0.10	−2.234	−1.970	<−3.5	−1.468	−1.9
100	22	0.706	0.01	−1.941	−2.654	<−3.4	−1.457	−2.1
100	18	2.367	0.10	−1.787	−1.272	<−3.5	−1.414	−1.9
100	22	2.367	0.01	−1.691	−1.487	<−3.4	−1.430	−1.9
100	18	4.471	0.10	−1.680	−0.980	<−3.5	−1.455	−1.9
100	22	4.471	0.01	−1.380	−1.121	<−3.1	−1.493	−2.2
100	18	8.942	0.10	−1.683	−0.812	<−3.5	−1.704	−3.7
100	22	8.942	0.01	−1.260	−1.045	<−3.4	−1.613	−3.6
50	25	0.706	0.10	−2.712	−1.355	−3.0	−1.782	−1.2
50	27	0.706	0.01	−2.800	−1.642	−3.1	−1.835	−1.2
50	25	2.367	0.10	−2.146	−1.022	−2.9	−1.599	−1.0
50	27	2.367	0.01	−1.655	−1.250	−3.2	−1.661	−1.4
50	25	3.353	0.10	−1.995	−0.914	−3.2	−1.557	−1.3
50	27	4.471	0.01	−1.181	−1.073	−3.2	−1.556	−1.7
50	25	4.471	0.10	−1.778	−0.882	−3.0	−1.523	−1.5
50	27	5.749	0.01	−1.173	−1.101	−3.2	−1.512	−2.2
40	40	0.706	0.10	−2.536	−1.224	−2.5	−1.724	−1.0
40	28	0.706	0.01	−2.459	−1.352	−2.5	−1.745	−0.9
40	40	2.367	0.10	−2.152	−0.991	−3.1	−1.852	−1.3
40	28	2.367	0.01	−1.840	−1.147	−3.0	−1.816	−1.4
40	40	4.471	0.10	−1.692	−0.887	−3.2	−1.883	−1.4
40	28	4.471	0.01	−1.419	−1.061	−3.0	−1.914	−1.7
40	40	8.942	0.10	−1.177	−0.888	−3.3	−2.009	−2.4
40	28	8.942	0.01	−0.988	−0.983	−3.3	−1.786	−3.3
500 °C, HM								
100	18	0.706	0.01	−2.911	−2.017	−1.574	n.a.	−1.9
100	22	0.706	0.001	−2.505	−1.934	−1.603	n.a.	−1.9
100	18	2.367	0.01	−2.287	−1.540	−1.688	n.a.	−1.7
100	22	2.367	0.001	−1.965	−1.456	−1.496	n.a.	−1.9
100	18	4.471	0.01	−2.284	−1.437	−1.697	n.a.	−2.7
100	22	4.471	0.001	−1.744	−1.355	−1.692	n.a.	−3.0
100	18	8.942	0.01	−2.202	−1.465	−1.986	n.a.	−3.7

(continued)

Table 7.1 (continued)

P (MPa)	Time (days)	Initial solution (mol kg^{-1})		Final solution (log m)				
		KCl	HCl	W	Fe	Al	SiO$_2$	HCl*
1	2	3	4	5	6	7	8	9
100	22	8.942	0.001	−1.915	−1.383	−2.062	n.a.	−3.9
50	25	0.706	0.01	<−4.1	−1.646	−1.750	n.a.	−1.8
50	27	0.706	0.001	−2.584	−1.792	–	n.a.	−1.5
50	25	2.367	0.01	−2.067	−1.455	−1.688	n.a.	−1.9
50	27	2.367	0.001	−1.882	−1.692	–	n.a.	−1.6
50	25	4.471	0.01	−1.742	−1.306	−1.556	n.a.	−2.7
50	27	4.471	0.001	−1.713	−1.675	–	n.a.	−2.3
50	25	5.749	0.01	−1.564	−1.261	−1.609	n.a.	−3.1
50	27	5.749	0.001	−1.567	−1.645	–	n.a.	−3.0
40	40	0.706	0.01	−2.997	−1.663	−1.971	n.a.	−1.7
40	28	0.706	0.001	<−4.1	−1.731	–	n.a.	−2.2
40	40	2.367	0.01	−1.926	−1.642	–	n.a.	−2.0
40	28	2.367	0.001	−1.750	−1.702	–	n.a.	−2.5
40	40	4.471	0.01	−2.181	−1.495	−1.737	n.a.	−2.2
40	28	4.471	0.001	−2.143	−1.675	–	n.a.	−1.9
40	40	8.942	0.01	−1.892	−1.526	−1.847	n.a.	−2.7
40	28	8.942	0.001	−1.830	−1.752	–	n.a.	−3.7
400 °C, NNO								
100	24	0.706	0.01	−3.604	−3.768	n.a.	−1.605	−3.0
100	36	0.706	0.001	−3.433	<−4.2	n.a.	−1.640	−3.2
100	24	2.367	0.01	−2.551	−2.467	n.a.	−1.546	−3.5
100	36	2.367	0.001	−2.685	−2.230	n.a.	−1.607	−2.8
100	24	4.471	0.01	−1.831	−1.827	n.a.	−1.594	−2.9
100	36	4.471	0.001	−1.997	−1.817	n.a.	−1.713	2.8
100	24	8.942	0.01	−1.366	−1.359	n.a.	−1.704	−2.6
100	36	8.942	0.001	−1.473	−1.306	n.a.	−1.812	−2.5
25	24	0.706	0.01	−3.566	−2.856	n.a.	−1.853	−3.0
25	36	0.706	0.001	<−4.1	−1.743	n.a.	−2.151	−2.0
25	24	2.367	0.01	−2.833	−2.279	n.a.	−1.727	−3.0
25	36	2.367	0.001	−2.300	−1.810	n.a.	−1.845	−2.2
25	24	3.353	0.01	−2.398	−2.104	n.a.	−1.685	−3.0
25	36	3.353	0.001	−2.097	−1.716	n.a.	−1.867	−2.5
25	24	4.471	0.01	−2.000	−1.867	n.a.	−1.682	−3.0

(continued)

Table 7.1 (continued)

P (MPa)	Time (days)	Initial solution (mol kg^{-1})		Final solution (log m)				
		KCl	HCl	W	Fe	Al	SiO$_2$	HCl*
1	2	3	4	5	6	7	8	9
25	36	4.471	0.001	−1.696	−1.589	n.a.	−1.891	−2.6
20	24	0.706	0.01	−3.490	−2.070	n.a.	−2.392	−1.8
20	36	0.706	0.001	−3.577	−1.389	n.a.	−2.061	−1.2
20	36	2.367	0.01	−2.132	−2.132	n.a.	−1.886	−3.8
20	36	2.367	0.001	−2.131	−1.197	n.a.	−2.013	−1.7
20	24	4.471	0.01	−1.629	−1.422	n.a.	−1.827	−1.9
20	24	4.471	0.01	−1.645	−1.447	n.a.	−1.859	−1.7
20	24	8.942	0.01	−1.409	−1.340	n.a.	−1.820	−2.5
20	36	8.942	0.001	−1.234	−1.088	n.a.	−1.973	−2.6

* The molality of HCl was estimated from pH measurement of the quenched diluted solution

Among W-bearing minerals, indent in addition to the initial ferberite, a new phase was found represented by elongated black crystals. SEM analysis indicated the composition corresponded to a chemical composition similar to the K–W bronzes (PTB) K$_x$WO$_3$. The small quantities of PTB (<<5 wt%) found did not allow for positive identification using X-ray diffraction. The largest crystals (10–50 μm thick) were chosen for microprobe analysis. The analytical results indicate their composition at 500 °C, 100 MPa corresponded to the formula K$_{0.24\pm0.01}$WO$_3$ (19 tests, runs in 25 and 40 wt% KCl and 0.1 mHCl), at 50 MPa K$_{0.23\pm0.01}$WO$_3$ (11 tests, run in 25 wt% KCl and 0.1 mHCl), and at 40 MPa K$_{0.21\pm0.02}$WO$_3$ (15 tests, run in 40 wt% KCl and 0.1 mHCl). We did not find additional system components present in PTB. A detailed study of ferberite crystals showed however, that in some crystals there occurred an increase in the W/Fe molar ratio during the experiment and an accompanying accumulation of potassium was observed. One can assume that such crystals represented thin growths of ferberite and PTB.

7.5 Thermodynamic Simulations

In solving the problem of the behavior and conditions of ore deposition in hydrothermal deposits, an important role is played by the solubility of ore minerals and the forms of the existence of ore components in their equilibrium solutions. The main component of hydrothermal fluid in tungsten deposits in greisenes is alkali metal chlorides, which concentrations vary widely from 0 to 60 wt% (Wood and Vlassopoulos 1989; Wood and Samson 2000; Gao et al. 2014). In these solutions, according to various researchers, tungsten could exist in the form of tungsten acid H$_2$WO$_4^0$, the

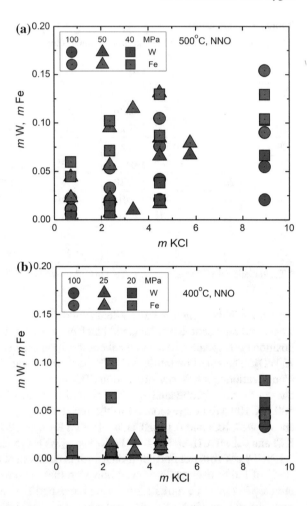

Fig. 7.4 W and Fe concentrations (mol kg^{-1}) in equilibrium with ferberite in the KCl–HCl–H$_2$O system buffering by QMM and NNO at 500 (**a**) and 400 °C (**b**)

products of its dissociation HWO_4^-, WO_4^{2-} (Ivanova and Khodakovskii 1972) and polymerization $W_4O_{13}^{2-}$ (Bryzgalin 1976), $H_7(WO_4)_6^{5-}$, $H_{10}(WO_4)_6^{2-}$ (Wesolowski et al. 1984), associations with alkaline metal ions $NaHWO_4^0$ (Wood and Vlassopoulos 1989; Wood and Samson 2000), chlorine $WO_2Cl_2^0$ (Veispäls 1979) and mixed form such as HWO_3Cl^0, $NaWO_3Cl^0$ (Khodorevskaya et al. 1990; Kolonin and Shironosova 1991). It is generally not possible to identify species, as the existing experimental data are not sufficient for this purpose. Therefore, the relevance of studies of W complexation in hydrothermal systems in connection with the problem of transport and deposition of tungsten minerals (Eugster 1986; Kolonin and Shironosova 1991; Volina and Barabanov 1995) is still unaddressed.

To describe the solubility of W-containing minerals at high *T-P* parameters, there are mutually agreed thermodynamic data (equilibrium constants and Gibbs free energies) for only 5 species W(VI): $H_2WO_4^0$, HWO_4^-, WO_4^{2-}, $NaHWO_4^0$ and $NaWO_4^-$ (Wood and Samson 2000). For the species $KHWO_4^0$ and KWO_4^- approximation of

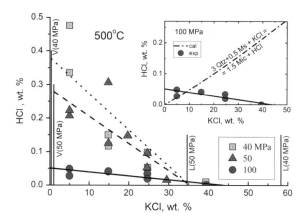

Fig. 7.5 Total quench HCl results as a function of KCl concentration in the experiments on ferberite solubility in Qtz-Mic-Ms-Ni-NiO buffered system at 500 °C and 40, 50, 100 MPa and their correlation with calculated values of HCl and KCl of the alumina-silicate buffer assemblage at 100 MPa. The dotted, dashed and solid lines are results of linear fit in addition to the L, V lines

the association constants has been evaluated. The data on particles and complexes of W(V) are completely absent, although in experiments on the solubility of WO_3 and (Fe, Mn)WO_4 in NaCl solutions, the formation of tungsten blue with a variable content of W(VI)/W(V) occurred (Bryzgalin 1976; Wesolowski et al. 1984; Manning and Henderson 1984; Wood and Vlassopoulos 1989; Zaraisky 1995; Redkin et al. 1999). These compounds, according to the composition and X-ray data, corresponded to sodium tungsten bronzes, Na_xWO_3, where $x = 0.1$–0.4, and were formed already at fO_2 with the oxygen buffer corresponding to Fe_2O_3–Fe_3O_4 (Red'kin 2000). The effect of fO_2 on the solubility of wolframite and its extreme members of ferberite and hubnerite (huebnerite) (MnWO$_4$) has not been considered.

We used numerical simulation of the Qtz + Mic + Ms + Ferb with KCl–HCl–H_2O solution at 400–500 °C and 100 MPa in an open system with fO_2 corresponding to the Ni–NiO (NNO: $fO_2 = 10^{-22.423}$ Pa at 400 °C–$10^{-17.706}$ Pa at 500 °C) and Fe_2O_3–Fe_3O_4 (HM: $10^{-13.63}$ Pa at 500 °C) buffers to determine the composition of the W complexes (Robie and Hemingway 1995). Determination and refinement of thermodynamic properties of aqueous solution species is solved using the software package HCh (Shvarov and Bastrakov 1999) and the program supporting technology OLE Automation (Shvarov 2008), realized in the program OptimA (Shvarov 2015). Thermodynamic properties of most simple particles and complexes K^+, KCl^0, KOH^0, Na^+, $NaCl^0$, $NaOH^0$, SiO_2aq^0, $HSiO_3^-$, Fe^{2+}, WO_4^{2-}, Cl^-, H_2O, OH^-, H^+, were taken from SUPCRT'92 (Johnson et al. 1992; Shock et al. 1997), aluminum species Al^{3+}, $AlOH^{2+}$, $Al(OH)_2^+$, $HAlO_2^0$ $Al(OH)_3^0$, AlO_2^- $Al(OH)_4^-$ by Pokrovskii and Helgeson (1995), Gibbs Free energy of HCl^0, by Hemley et al. (1992), $FeCl^+$, $FeCl_2^0$ by Wood and Samson (2000). Free energy and HKF parameters of hydroxo and chloride complexes Fe(II): $Fe(OH)^-$, $FeOaq^0$, $HFeO_2^-$, $FeCl^+$ and $FeCl_2^0$ are calculated from the equilibrium constants of the hydrolysis reactions presented in Wood and Samson

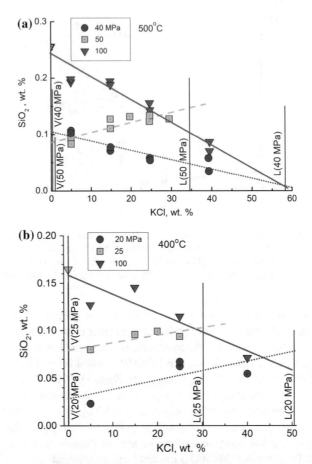

Fig. 7.6 The dependence of dissolved silica (in wt%) from Qtz-Mic-Ms buffers on the initial KCl concentration in experiments at 500 (**a**) and 400 °C (**b**), their approximation by linear fit (solid and dashed lines) and calculated quartz solubility in relation to Si species composition at 1000 bar. Vertical lines show KCl concentration in vapor 'V' and liquid 'L' phases at the corresponding pressure in MPa

(2000). For Fe(III): $FeOH^{2+}$, $Fe(OH)_2^+$, $Fe(OH)_3^0$, $Fe(OH)_4^+$ hydroxocomplexes the best convergence of the calculated and data free energies was obtained using the approximation of the stability constants of the complexes (Wood and Samson 2000) modified Ryzhenko-Bryzgalin equation (Ryzhenko 1981; Bryzgalin and Ryzhenko 1981; Bryzgalin and Rafalsky 1982). The contribution of Fe(III) species to the total iron concentration was negligible (Wood and Samson 2000).

We did not consider the species of the $KFeCl_3^0$ composition, although according to Tagirov (1998), the contribution of these particles to the total solubility of the iron-containing phases in highly concentrated water-salt solutions may be dominant.

The activity coefficients of aqueous species with were calculated using an extended Debye-Hückel equation

$$\log \gamma_i = -\frac{A z_i^2 \sqrt{I}}{1 + B \mathring{a} \sqrt{I}}, \tag{7.5}$$

where I (mol kg^{-1}) stands for ionic strength of the solution, A and B represent the Debye Hückel coefficients for pure solvent dependent on P, T; z_i is the charge and $\mathring{a}$ is the ion size parameter that was accepted after Rafalsky (1973) to be equal 4.5 Å for all species.

The free energies of formation of quartz (Qtz) and microcline (Mic) are taken as the basis for solid mineral phases (Robie et al. 1978). The free energies of Ab (albite), And (andalusite), Ms, Prg (paragonite), Pf (pyrophillite) were corrected according to experimental constants of hydrolysis reactions (Hemley and Jones 1964) and taken as stable phases in the system $K_2O–Na_2O–Al_2O_3–SiO_2–H_2O–HCl$ at 500–400 °C and 100 MPa.

For reaction (7.1), which served in our experiments as an acid-base buffer, the accepted value for the equilibrium ratio log $mKCl/mHCl = 2.08$ (at 500 °C, 100 MPa) and 2.70 (at 400 °C, 100 MPa) (Hemley 1959; Redkin 1983). This also led to an aluminum saturation concentration corresponding to our experimental data (approximately $10^{-4}–10^{-5}$ mol kg^{-1} H_2O).

When calculating the free energies of ferberite, we used the equilibrium constants of the reaction of dissolving ferberite calculated at 400 and 500 °C and 100 MPa according to Wood and Samson (2000):

$$FeWO_{4(s)} + 2HCl^0 = H_2WO_4^0 + FeCl_2^0, \tag{7.6}$$

pK(6) = 3.2 and 3.0, respectively.

When determining the composition of aqueous species and the valence state of tungsten in solutions (see Table 7.1) with the parameters of the experiments, the following assumptions were taken into account;

1. Most aqueous tungsten was assumed to be present as complex of W(V) as the cooled solutions were W, colorless and transparent despite high concentrations ($mW_{total} > 0.002$ mol kg^{-1}). When performing chemical analyzes on tungsten, using the colorimetric method with bromopyrogallol red in the presence of cetylpyridinium bromide as a surfactant (Mamedova et al. 2004), it was noted that the solutions after the experiments when mixed with a strong alkali, with the addition of hydroxylamine, actively interacted, producing small ammonia bubbles (alkaline reaction of the wet indicator paper in the atmosphere of bubbles). These gas bubbles dissolved in the liquid phase. It is known (Brikun et al. 1967) that strong reducing agents (HI, H_2S, $SnCl_2$, zinc dust, etc.) reduce hydroxylamine to NH_3. Since there were no other reducing agents in the solutions, it is reasonable to assume that the species W(V) or even W(IV) behaved as the reducing agent in the analyzed solutions.

An additional but indirect argument for the presence of W(V) can be potassium tungsten bronzes K_xWO_3 (PTB), where $x = 0.20$–0.24, which is formed in the experiments and represents compounds of W(V, VI).

2. W(V) in solution at high temperature is represented by monomeric and polymeric species, which may include K^+, Cl^-, OH^-, H_2O. Silica-12-tungsten acid and its salts $K_{4-x}H_x[SiW_{12}O_{40}] \cdot nH_2O$ ($x = 0$–4, $n = 5$, 14, 24 or 30), as well as other silica heteropolytungstate compounds were not formed in appreciable amounts. This follows from the analysis of solubility data of quartz (Table 7.1 and Fig. 7.6a, b) at 500 and 400 °C and 100 MPa.

3. Changes in the acidity of the solutions may be mainly due to hydrolysis or dissociation of tungsten compounds upon cooling, since ferrous chloride species, $FeCl^+$ and $FeCl_2^0$, which are stable in the test mode, do not hydrolyze when cooled to room temperature, but completely dissociate into Fe^{2+} and Cl^- without changing $mHCl_{total}$ ($mHCl^0 + mH^+$).

For test the calculated and experimental data of congruent and incongruent (with the formation of PTB-02, $K_{0.2}WO_3$) solubility of ferberite, presented in Table 7.1, we examined more than 15 W(V) species containing from 1 to 6 tungsten atoms and having a charge of $+1$ to -3, as well as 5 anionic species with a charge of -1 to -3, containing both W(V) and 20–66 mol% W(VI). The $KHWO_4^0$ and KWO_4^- species proposed in Wood and Samson (2000), line WS in Figs. 7.7 and 7.8 were the dominant contributors to the solubility of ferberite in strong KCl solutions, but did not reflect the effect of fO_2, were omitted from consideration.

We assumed that W(V) compounds as well as compounds W(VI) compounds in a hydrothermal solution will have similar structure. This was taken into consideration when selecting the composition of possible aqueous species W(V).

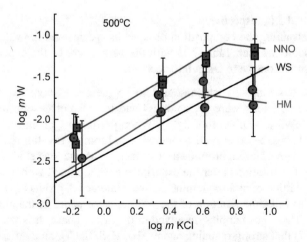

Fig. 7.7 Influence of $mKCl$ on mW in a solution equilibrated with ferberite in the QMM buffered system at 500 °C and 100 MPa by experimental (symbols) and calculated (lines) data. The WS line is from calculations by Wood and Samson (2000)

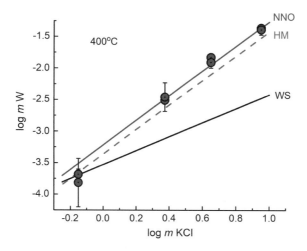

Fig. 7.8 Influence of mKCl on mW in a solution equilibrated with ferberite in the QMM buffered system at 400 °C and 100 MPa by experimental (symbols) and calculated (lines) data. The WS line is from calculations by Wood and Samson (2000)

According to the data presented in Table 7.1, the total concentrations of iron and tungsten in the solutions after the experiment were differed. This could be due to the initial purity of synthetic ferberite (a slight excess of iron oxide), an insufficient amount of quartz, and the formation of a containing phase in the course of the experiment, i.e. potassium tungsten bronzes (PTB). The congruent solubility of ferberite in potassium chloride solution, with an acidity given by Qtz-Mic-Ms buffer, can be calculated by the equation

$$m\text{FeWO}_4 = \sqrt[\alpha+1]{(m\text{W}_{\text{total}})^\alpha \times m\text{Fe}_{\text{total}}}, \tag{7.7}$$

where $\alpha = \sum_i \frac{\alpha_i}{i}$, $\alpha_i = \frac{i \cdot m\text{W}_i}{m\text{W}_{\text{total}}}$ mol fractions of fluid species containing the same number i of atoms W, taking into account the fractions of species of different degrees of polymerization in calculations using the obtained equilibrium constants. Considering that the main contribution to W_{total} is made by W(V, VI) species in the form of monomers, Eq. (7.7) can be simplified to the form

$$m\text{FeWO}_4 = \sqrt{(m\text{W}_{\text{total}} \times m\text{Fe}_{\text{total}})}, \tag{7.8}$$

representing the geometric mean of the concentrations of tungsten and iron. In Fig. 7.7, square and round symbols show the solubility of ferberite calculated using Eq. (7.8) for NNO and HM buffers, respectively. Extreme concentrations of W_{total} and Fe_{total} are accepted for the magnitude of the $m\text{FeWO}_4$ definition error.

The best agreement of experimental and calculated data (see Fig. 7.7) at 500 °C and 100 MPa was achieved for two newly introduced species WO_3^- (or $\text{H}_2\text{W}^{\text{V}}\text{O}_4^-$)

and $H_2W_2O_7^-$ ($H_2W^VW^{VI}O_4^-$). According to calculations, $FeCl_2^0$ are the predominant iron species in solution at both NNO and HM oxygen buffers over the entire KCl concentration range studied. The structure of the predominant tungsten species, conversely, depended on both fO_2, and $mKCl$. At HM buffer conditions the W(VI) content ranged from 32 to 43.5% of mW_{total} in the $mKCl$ range from 0.3 to 7.7. At low concentrations of KCl, tungsten-bearing species WO_3^-, HWO_4^-, $H_2W_2O_7^-$ (in order of importance) prevailed, while in concentrated solutions the $H_2W_2O_7^-$ species dominated.

At NNO buffer conditions, calculations indicate the WO_3^- and $H_2W_2O_7^-$ species dominated in the $mKCl$ range investigated. W(VI) concentrations increased in solution from 9 to 34.8% with increasing $mKCl$ from 0.3 to 5.3. Incongruent dissolution of Ferb with concurrent formation of PTB and a slight decrease in both mW_{total} and fraction of W(VI) in solution (Fig. 7.7) shows good agreement between calculated and experimental data.

The influence of potassium chloride concentration on solubility of the ferberite in the solutions at 400 °C, 100 MPa, NNO buffer and at the acidity given by Mic-Ms-Qtz buffer is shown in Fig. 7.8. The slope of mW with mK at 400 °C is steeper than at 500 °C. This indicates that there are aqueous tungsten species present that contain more than 2 W atoms. The best agreement between the experimental and calculated data (see Fig. 7.8) was obtained by introducing the species $W_5O_{16}^{3-}$ (or $W^VW_4^{VI}O_{16}^{3-}$) previously proposed in the work (Redkin and Kostromin 2010). According to calculations, the fraction of the $W_5O_{16}^{3-}$ species increases dramatically at $mKCl > 0.05$ reaching 99% in the solution containing 0.7 $mKCl$. At the same time, the share of W(V) is 20% of the particles concentration of $W_5O_{16}^{3-}$. In Fig. 7.8 (WS line) also shows the results of calculations based on data (Wood and Samson 2000) using $KHWO_4^0$ and KWO_4^- species.

Concentrations of sodium salts in natural hydrothermal solutions are typically greater than that of potassium, therefore we made calculations for ferberite solubility in solutions of NaCl–KCl–HCl. Since the composition of ore-bearing solutions can only be assumed, but since it is known the hydrothermal solutions produce quartz-muscovite alteration in the host rocks, it is possible to speculate on geochemical boundary conditions. The Qtz + Ms association field at 500 and 400 °C, 100 MPa and constant mCl is limited by the following non-variant points: Qtz-Ab-Mic-Ms (1), Qtz-Ab-Prg-Ms (2) and Qtz-And-Prg-Ms (3 at 500 °C) or Qtz-Pf-Prg-Ms (3 at 400 °C) after Hemley and Jones (1964). If the concentration of chlorides in a system changes, then non-variant points turn into the hydrolysis equilibria, which determine the $mKCl/mHCl$ and $mNaCl/mHCl$ ratios. Thus, the solubility of ferberite on the lines of hydrolysis reactions of aluminosilicates will correspond to extreme values. The systems in which iron oxides are present are also of interest. Calculations show that at low chloride concentrations, magnetite accompanies ferberite.

Figures 7.9 and 7.10 show the results of the calculations at 400 and 500 °C and pressure of 100 MPa. As expected, the solubility of Ferb increases with the increasing temperature, mCl ($mNaCl + mKCl$), $f(H_2)$, $mKCl/mHCl$ (or when changing associations $1 \rightarrow 3$) and decreases in the presence of Mgt. The maximum solubility of ferberite is found in KCl solutions equilibrated with Qtz-And-Ms at 500 °C or

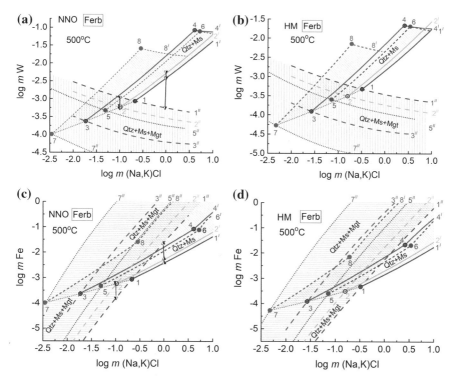

Fig. 7.9 The projections of 3D surfaces of ferberite solubility on the m(Na, K)Cl–mW and m(Na, K)Cl–mFe plane at 500 °C, 100 MPa, NNO (**a, c**) and HM (**b, d**) buffers in chloride solutions controlled by various alumina-silicate assemblages. *Legend* 1—nonvariant point (NP) Qtz + Ab + Mic + Ms + Ferb + Mgt; 2—Qtz + Ab + Prg + Ms + Ferb + Mgt; 3—Qtz + And + Ms + Prg + Ferb + Mgt; 4—Qtz + And + Ms + Prg + Ferb + PTB-02; 5—Qtz + Mic + Ms + Ferb + Mgt (NaCl free solutions); 6—Qtz + Mic + Ms + Ferb + PTB-02 (no NaCl); 7—Qtz + And + Ms + Ferb + Mgt (no NaCl); 8—Qtz + And + Ms + Ferb + PTB-02 (no NaCl); 1′, 2′, 4, 6—lines of saturation of solutions with Ferb (congruent dissolution); 4′ and 6′—Ferb + PTB, 1″, 2″, 3″, 4″, 5″, 7″, 8″—Ferb + Mgt, correspondingly; the shaded areas show the limits of variation in the concentrations of tungsten (**a, c**) and iron (**b, d**) for the Qtz + Ms + Ferb and Qtz + Ms + Ferb + Mgt, respectively

Qtz-Pf-Ms at 400 °C. Further change in the composition of solution towards reduction of mKCl/mHCl will cause the formation of andalusite (or pyrophyllite), and in solutions containing fluorine—topaz. Therefore, solutions saturated with ferberite, marked by hatching in these figures, are of practical interest for modeling tungsten ore formation. From the data obtained, it follows that at 500 °C the change in $f(O_2)$ from NNO to HM buffer reduces the tungsten concentrations in solution by 1.5–2.7 times, and at 400 °C by 1.4 times.

The influence of aluminosilicate assemblage and the presence of magnetite on the content of tungsten and iron in solution saturated with ferrite is of interest. Figure 7.9a, c and 7.10a, b triangular symbols show trends of the changing mW and mFe in solution initially equilibrated with the Qtz-Ms-Ferb assemblage, then interacting with wall

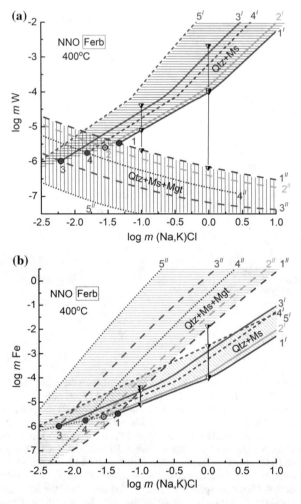

Fig. 7.10 The projections of 3D surfaces of ferberite solubility on the m(Na, K)Cl–mW (**a**) and m(Na, K)Cl–mFe (**b**) plane at 400 °C, 100 MPa, NNO buffer in chloride solutions controlled by various alumina-silicate assemblages. *Legend* 1—nonvariant point Qtz + Ab + Mic + Ms + Ferb + Mgt; 2—ND Qtz + Ab + Prg + Ms + Ferb + Mgt; 3—Qtz + Pf + Ms + Prg + Ferb + Mgt; 4—Qtz + Mic + Ms + Ferb + Mgt (no NaCl added); 5—Qtz + Pf + Ms + Ferb + Mgt + PTB-02 (no NaCl); 6—Qtz + Mic + Ms + Ferb + PTB-02 (no NaCl); 1′, 2′, 3′, 4′, 5′—congruent dissolution of Ferb; 1″, 2″, 3″, 4″, 5″,—Ferb + Mgt; the shaded areas show the limits of variation in the concentrations of tungsten (**a**) and iron (**b**) for the Qtz + Ms + Ferb and Qtz + Ms + Ferb + Mgt, respectively

rocks containing Mic, Ab, Mgt. In solutions containing 0.76 mNaCl + 0.24 mKCl + 0.01 mHCl, equilibrated at 500 °C with Qtz-Ms-Ferb assemblage, the smallest change (1.4 times) is due to the appearance of Mic and Ab, while saturation of the solution with magnetite reduces the initial mW by 10 times. In 0.1 mCl solution of 0.076 mNaCl + 0.024 mKCl + 0.001 mHCl, the mW variations are less obvious and have the opposite tendency, i.e. the appearance of magnetite in association with Qtz-Ms-Ferb leads to a slight decrease in mW, while saturation of the solution with albite, microcline and magnetite from side rocks increases the solubility of ferberite (mW) (see Fig. 7.9a).

The solution containing 0.76 mNaCl + 0.24 mKCl + 0.01 mHCl, equilibrated at 400 °C and 100 MPa with Qtz-Ms-Ferb assemblage, contains 0.002 mW. The interaction of such a solution with side rocks containing Mic or Mic + Ab, leads to the deposition of 94.5–95.2% tungsten in the form of ferberite from the solution, while the presence of Mgt in wall rocks reduces the concentration of tungsten in the affecting solution by 99.97% (Fig. 7.10). Tungsten content in the 0.076 mNaCl + 0.024 mKCl + 0.001 mHCl equilibrium solution with the Qtz-Pf-Ms-Ferb assemblage is 50 times lower, than in the 1.0 mCl solution. Although tungsten deposition from this solution due to geochemical barriers associated with the appearance of Mic and Mgt in the host or wall rocks is very significant (from 80 to 95%), solutions containing 0.1 mCl cannot be considered promising for the formation large deposits of tungsten ores associated with granites.

Particular attention should be paid to the effect of temperature and wall rock mineralogy on the solubility of ferberite. Figure 7.11 shows a case when the chloride solutions containing 1.0 and 0.1 mCl are first equilibrated at 500 °C and 100 MPa with Qtz-Ms-Ferb assemblage are than isobarically cooled to 400 °C, imposing the solutions to interact with wall rocks containing feldspars and magnetite. Calculations indicate that the decrease in temperature of hydrothermal ore-bearing solution and its interaction with both feldspars and magnetite leads to the deposition of 99% of

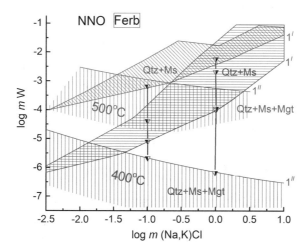

Fig. 7.11 Influence of temperature, mKCl and wall rock minerals (Mic, Ab, Mgt) on mW change in solutions initially equilibrated with the quartz-muscovite-ferberite assemblage at 500 °C, 100 MPa, NNO (the projections of 3D surfaces of ferberite solubility on the m(Na, K)Cl–mW plane). *Legend* 1′—nonvariant equilibria Qtz + Ab + Mic + Ms + Ferb; 1″—Qtz + Ab + Mic + Ms + Ferb + Mgt

tungsten. Moreover, taking into account the high solubility of ferberite at 500 °C and 100 MPa in 0.1 mCl solution (6.5×10^{-4} mW), such solutions can be considered as ore-bearing ones.

The presented results of thermodynamic modeling were performed for a homogeneous region of NaCl–KCl–HCl–H_2O fluid. In the region of immiscibility, the water-salt fluid splits into two equilibrium fluids of different densities, containing different amounts of HCl and (Na, K)Cl, which have a major influence on the peculiarity of dissolving ferberite in coexisting fluid phases.

The thermodynamic calculations used for a homogeneous solution cannot be used for the region of the fluid immiscibility (Redkin et al. 2015). As soon as apparent solubility of ferberite on concentration of KCl in two-phase region corresponds to Eq. (7.4) we can calculate the solubility of ferberite in the V- and L-phases of KCl fluid. The results of calculations in wt% are shown in Table 7.2. The correlations between apparent solubility of ferberite and concentration of KCl in the two-phase region are given in the right-hand column. The values of S_{FeWO4} obtained clearly demonstrate its dependence on fO_2. Otherwise the ferberite dissolution has to be considered as a redox reaction in range fO_2 from $10^{-17.7}$ (NNO) to $10^{-13.63}$ Pa (HM) at 500 °C. Taking in account that the predominant valence of iron relates to Fe(II) as in ferberite as well as in solute species, because high dissolution of iron in the system studied, W can change valance only from W(VI) in ferberite to W(V or IV) in solution.

Table 7.2 Estimated values of ferberite solubility in vapor and liquid phases of the fluid KCl–HCl–H_2O

T (°C)	P (MPa)	V-phase (wt%)		L-phase (wt%)		fO_2	Correlations used
		KCl	FeWO$_4$	KCl	FeWO$_4$	Buffer	Ferberite solubility (S) (wt%)
500	50	0.96	0.21	34.59	2.03	NNO	S = 0.154 + 0.0543 · C$_{KCl}$ (± 0.15)
500	40	0.32	0.50	58.25	2.57	NNO	S = 0.493 + 0.0356 · C$_{KCl}$ (± 0.11)
500	50	0.96	0.19	34.59	0.77	HM	S = 0.178 + 0.017 · C$_{KCl}$ (± 0.15)
500	40	0.32	0.34	58.25	0.36	HM	S = 0.342 + 0.00032 · C$_{KCl}$ (± 0.15)
400	25	0.067	0.001	30.00	0.41	NNO	S = 5 · 10^{-5} + 0.0136 · C$_{KCl}$ (± 0.15)
400	20	0.011	0.1	51.16	1.33	NNO	S = 0.101 + 0.024 · C$_{KCl}$ (± 0.15)

7.6 Conclusions

1. Ferberite dissolution in KCl solutions under pH and fO_2 buffered conditions (Qtz-Mic-Ms, and NNO or HM) at 400–500 °C behaves both congruently as well as incongruently with accompanying potassium tungsten bronzes formation (PTB), K_xWO_3, ($x = 0.2$–0.3).

2. Using bulk composition data, the congruent and apparent solubility of ferberite at 400 and 500 °C in homogeneous KCl–HCl solutions at $P = 100$ MPa and in the region of fluids immiscibility at $P = 20, 25, 40$, and 50 MPa has been estimated. Tungsten concentration values do not show strong pressure dependence, however, there is some solubility increase accompanying pressure decreases in mKCl $<$ 3 mol kg^{-1} solutions.

3. The data at $P = 100$ MPa and mKCl from 0.7 to 3 mol kg^{-1} suggest the general contribution of total tungsten solubility may favor the mononuclear species which, in turn, causes increased acidity of the quenched solution in the homogeneous region. Conversely, in 4.0–8.9 mKCl solutions and in the region of fluids immiscibility at 500 °C, pressure of 40, 50 MPa and $fO_2 = 10^{-17.7}$ Pa (NNO), we identified polynuclear tungsten(V, VI) species contributing to increased alkalinity during quench.

4. As has been shown in thermodynamic calculations at $P = 100$ MPa, the predominant aqueous species of tungsten in KCl–HCl solutions at $fO_2 = fO_2$(NNO) may be W(V, VI) species: WO_3^-, HWO_4^-, $H_2W_2O_7^-$ at 500 °C and HWO_4^-, $W_5O_{16}^{3-}$ at 400 °C.

5. The total mW in equilibrium with ferberite in the aluminosilicate buffered systems depends upon mCl, T and fO_2. Deposition of tungsten from ore-bearing solutions is due to interaction with wall rocks containing feldspars, and iron oxides and decrease of temperature. In the magnetite bearing system, the equilibrium tungsten concentration does not exceed $2 \cdot 10^{-5}$ mol kg^{-1} at 400 °C. Therefore, aqueous tungsten mobility should decrease sharply in iron-enriched rocks.

Acknowledgements This study was suggested through discussions with Dr. J. J. Hemley (USGS), Prof. G. P. Zaraisky (IEM RAS), and Corresponding Member of the RAS V. I. Velichkin (IGEM RAS) about the major role of immiscibility in water-salt systems and its impact on the processes responsible for formation of giant ore deposit systems. The authors are grateful also to T. K. Chevichelova, G. V. Bondarenko, A. N. Nekrasov, O. A. Mozgovaya all from IEM RAS, G. E. Kalenchuk from IGEM RAS, Olga Smetanina for analytical assistance and for their help in manuscript preparation, Prof. A. F. Koster van Groos and David M. Petrovski (USEPA) and the anonymous referees for critical comments on an earlier draft of this paper. Some recalculations and revisions of this manuscript were done after discussion with Prof. S. A. Wood. This research was supported from the Grants of Russian Foundation for the Basic Research and project No AAAA-A18-118020590151-3 of the IEM RAS.

References

Andreeva IYu, Lebedeva LI, Kavelina GL (1992) Determination of small amounts of molybdenum and tungsten as complexes of the metals with bromopyrogallol red and some superactants. Russ J Anal Chem 12:2202–2206 (in Russian)

Basargin NN, Bikova VS, Polupanova LI (1976) Photometric analyses of aluminum in the silicate rocks with the aid of antrazochrom. In: Theoretical and practical questions of organic reagents utilization in the analyses of the mineral objects. Nedra Press, Moscow, pp 119–124 (in Russian)

Brikun IK, Kozlovsky MT, Nikitina LV (1967) Hydrazine and hydroxylamine and their application in analytical chemistry. Nauka, Alma-Ata, 175 p (in Russian)

Bryzgalin OV (1976) On the solubility of tungstic acid in an aqueous salt solution at high temperatures. Geochem Int 13(3):155–159

Bryzgalin OV, Rafalsky RP (1982) Approximate estimation of constants of instability of complexes of ore elements under high temperatures. Geokhimia 6:839–849 (in Russian)

Bryzgalin OV, Ryzhenko BN (1981) Prediction of temperature and baric dependence of the dissociation constants of electrolytes based on the elementary electrostatic model. Geokhimia 12:1886–1890 (in Russian)

Busev AI, Ivanov VM, Sokolova TM (1976) Analytical chemistry of tungsten. Nauka Press, Moscow, 240 p (in Russian)

Cygan GL, Hemley JJ, Doughten MW (1994) Fe, Pb, Zn, Cu, Au, and HCl partitioning between vapor and brine in hydrothermal fluids—implications for porphyry copper deposits. In: USGS research on mineral deposits. Part A, 9th V.E. McKelvey forum, USGS circular 1103-A, pp 26–27

Dadze TP, Sorokin VI, Nekrasov IYa (1981) Solubility of SnO_2 in water and in aqueous solutions of HCl, HCl + KCl, and HNO_3 at 200–400°C and 1013 bar. Geochem Int 18(5):142–152

Eugster HP (1986) Minerals in hot water. Am Mineral 71:655–673

Gao YW, Li Z, Wang J, Hattori K, Zhang Z, Jianzhen GJ (2014) Geology, geochemistry and genesis of tungsten-tin deposits in the Baiganhu District in the Northern Kunlun Belt, Northwestern China. Econ Geol 109:1787–1799

Hemley JJ (1959) Some mineralogical equilibria in the system $K_2O–Al_2O_3–SiO_2–H_2O$. Am J Sci 257:241–270

Hemley JJ, Jones WR (1964) Chemical aspects of hydrothermal alteration with emphasis on hydrogen metasomatism. Econ Geol 59(4):538–569

Hemley JJ, Cygan GL, d'Angelo WM (1986) Effect of pressure on ore mineral solubilities under hydrothermal conditions. Geology 14:377–379

Hemley JJ, Cygan GL, Fein JB, Robinson GR, d'Angelo WM (1992) Hydrothermal ore-forming processes in the light of studies in rock-buffered system. 1. Iron-copper-zinc-lead sulfide solubility relations. Econ Geol 87:1–22

Hovey JK, Pitzer KS, Tanger JC IV, Bischoff JL, Rosenbauer RJ (1990) Vapor-liquid phase equilibria of potassium chloride-water mixtures: equation-of-state representation for $KCl–H_2O$ and $NaCl–H_2O$. J Phys Chem 94:1175–1179

Ivanov IP, Chernysheva GN, Dmitrenko LT, Korzhinskaya VS (1994) New hydrothermal facilities to study mineral equilibria and mineral solubilities. In: Experimental problems of geology. Nauka Press, Moscow, pp 706–720 (in Russian)

Ivanova GF, Khodakovskii IL (1972) About the state of tungsten in hydrothermal solutions. Geokhimia 11:1426–1433 (in Russian)

Ivanova GF, Naumov VB, Kopneva LA (1986) Physic-chemical parameters of formation of scheelite in the deposits of various genetic types. Geokhimia 10:1431–1442 (in Russian)

Johnson JW, Oelkers EH, Helgeson HC (1992) SUPCRT92: a software package for calculating the standard molal thermodynamic properties of minerals, gases, aqueous species, and reactions from 1 to 5000 bar and 0° to 1000°C. Comput Geosci 18(7):899–947

Khodorevskaya LI, Tikhomirova VI, Postnova LE (1990) Study of WO_3 solubility in HCl solutions at 450°C. Dokl Akad Nauk SSSR 113(3):720–722 (in Russian)

Klevtsov PV, Novgorodtseva NA, Kharchenko LYu (1970) Hydrothermal synthesis of the $FeWO_4$ crystals. Crystallographia 15(3):609–610 (in Russian)

Kolonin GR, Shironosova GP (1991) The state of the art in the field of experimental studies on tungsten forms in hydrothermal solutions. In: 12th all-union meeting on experimental mineralogy, Miass, USSR, 24–26 Sept 1991, Abstracts, p 58

Kovalenko NI, Ryzenko BN, Barsukov VL (1986) The solubility of cassiterite in HCl and HCl + NaCl (KCl) solutions at 500°C and 1000 atm under fixed redox conditions. Geochem Int 23(7):1–16

Malinin SD, Kurovskaya NA (1996) The effect of pressure on mineral solubility in aqueous chloride solutions under supercritical conditions. Geokhim Int 1(1):45–52

Mamedova AM, Ivanov VM, Akhmedov SA (2004) Interaction of tungsten(VI) and vanadium(V) with pyrogallol red and bromopyrogallol red in the presence of surfactants. Vestn Mosk Univ Ser Chem 45(2):117–123 (in Russian)

Manning DAC, Henderson P (1984) The behaviour of tungsten in granitic melt-vapour systems. Contrib Mineral Petrol 86(3):286–293

Marczenko Z (1976) Spectrophotometric determination of elements. Wiley, New York, 355 p

PDF (1980) Powder diffraction file 1980. Joint Committee on Powder Diffraction Standards, International Centre for Diffraction Data, Swarthmore, PA, USA

Pokrovskii VA, Helgeson HC (1995) Thermodynamic properties of aqueous species and the solubilities of minerals at high pressures and temperatures; the system Al_2O_3–H_2O–NaCl. Am J Sci 295(10):1255–1342

Rafalsky RP (1973) Hydrothermal equilibria and processes of minerals formation. Nauka, Moscow, 288 p. Free book site: http://www.geokniga.org/books/7820 (in Russian)

Red'kin AF (2000) Experimental study of the behavior of ore-forming compounds in the system WO_3–SnO_2–UO_2–NaCl–H_2O at 400–500 °C, 200–1000 bar and the hematite-magnetite buffer. Geochem Int 38(Suppl. 2):S227–S236

Redkin AF (1983) Experimental and thermodynamical investigation of frontier reactions controlling conditions of formation of wall-rock beresites. Thesis of Ph.D., Vernadsky Institute of RAS, Moscow, 27 p (in Russian)

Redkin AF, Kostromin NP (2010) On the problem of transport species of tungsten by hydrothermal solutions. Geochem Int 48(10):988–998

Redkin AF, Savelyeva NI, Sergeyeva EI, Omelyanenko BI, Ivanov IP, Khodakovsky IL (1989) Investigation of uraninite (UO_2) solubility under hydrothermal conditions. Strasbourg Sci Geol Bull 42(4):329–334

Redkin AF, Zaraisky GP, Velichkin VI (1999) An influence of the phase conversions in water-salt systems and rock-buffered action on solubility and partitioning of some ore elements (W, Sn, U). In: Abstracts: international symposium "physico-chemical aspects of endogenic geological processes" devoted to the 100-anniversary of D.S. Korzhinskii. Moscow, Russia, pp 182–183

Redkin AF, Kotova NP, Shapovalov YB (2015) Liquid immiscibility in the system NaF–H_2O at 800 °C and 200–230 MPa and its effect on the microlite solubility. J Sol Chem 44(10):2008–2026

Robie RA, Hemingway BS (1995) Thermodynamic properties of minerals and related substances at 298.15 K and 1 bar (105 pascals) pressure and at higher temperatures. US Geol Surv Bull 2131:461 p

Robie RA, Hemingway BS, Fisher JR (1978) Thermodynamic properties of minerals and related substances at 298.15 K and 1 bar (10^5 pascals) pressure and at higher temperatures. US Geol Surv Bull 1452:456 p

Ryzhenko BN (1981) Thermodynamics of equilibria in hydrothermal conditions. Nauka, Moscow, 191 p

Shmulovich KI, Tkachenko SI, Plyasunova NV (1995) Phase equilibria in fluid systems at high pressures and temperatures. In: Shmulovich KI, Yardley BWD, Gonchar GG (eds) Fluids in the crust: equilibrium and transport properties. Chapman and Hall, London, pp 193–214

Shock EL, Sassani DC, Willis M, Sverjensky DA (1997) Inorganic species in geologic fluids: correlations among standard molal thermodynamic properties of aqueous ions and hydroxide complexes. Geochim Cosmochim Acta 61:907–950

Shvarov YuV (2008) HCh: new potentialities for the thermodynamic simulation of geochemical systems offered by Windows. Geochem Int 46(8):834–839

Shvarov Yu (2015) A suite of programs, OptimA, OptimB, OptimC, and OptimS compatible with the Unitherm database, for deriving the thermodynamic properties of aqueous species from solubility, potentiometry and spectroscopy measurements. Appl Geochem 55:17–27

Shvarov YuV, Bastrakov E (1999) HCh: a software package for geochemical equilibrium modeling. User's guide 3.3. Australian Geological Survey Organization, 25 p

Smirnov VI, Ginsburg AI, Grigoriev VM, Yakovlev GF (1981) Course of ore deposits. High school manual. Nedra Press, Moscow, pp 161–174 (348 p) (in Russian)

Tagirov BR (1998) Experimental and computational study of the form iron transport in chloride hydrothermal solutions. Ph.D. thesis, IGEM RAN, Moscow, 22 p (in Russian)

Teleshova RL (1973) Differential spectrophotometric micromethod for silica determination in silicate minerals and rocks. Nauka Press, Moscow, pp 26–29 (in Russian)

Veispäls A (1979) Thermodynamic investigations of the chemical transport of tungsten trioxide. Izv Latv Acad Sci Seriya Phys Tech Sci 1:60–65 (in Russian)

Volina OV, Barabanov VF (1995) To the concern of tungsten existence forms in hydrothermal solutions. Proc Russ Mineral Soc 4:1–11 (in Russian)

Wesolowski D, Drummond SE, Mesmer RE, Ohmoto H (1984) Hydrolysis equilibria of tungsten(VI) in aqueous sodium chloride solutions to 300 °C. Inorg Chem 23:1120–1132

Wood SA, Samson IM (2000) The hydrothermal geochemistry of tungsten in granitoid environments: I. Relative solubility's of ferberite and scheelite as a function of T, P, pH, and mNaCl. Econ Geol 95:143–182

Wood SA, Vlassopoulos D (1989) Experimental determination of the hydrothermal solubility and speciation of tungsten at 500 °C and 1 kbar. Geochim Cosmochim Acta 53:303–312

Zaraisky GP (1995) The influence of acidic fluoride and chloride solutions on the geochemical behavior of Al, Si and W. In: Shmulovich KI, Yardley BWD, Gonchar GG (eds) Fluids in the crust: equilibrium and transport properties. Chapman and Hall, London, pp 139–162

Part II
Genesis of Minerals and Rocks

Chapter 8
Evolution of Mantle Magmatism and Formation of the Ultrabasic-Basic Rock Series: Importance of Peritectic Reactions of the Rock-Forming Minerals

Yu. A. Litvin, A. V. Kuzyura, and A. V. Spivak

Abstract Peritectic mechanisms for fractional evolution of the upper mantle, transition zone and lower mantle magmatism and petrogenesis of the ultrabasic-basic rock series are justified in theory and experiments. Compositions of the ultrabasic upper mantle peridotite and pyroxenite rocks belong to the multicomponent olivine–orthopyroxene–(jadeite–poor clinopyroxene)–garnet system. On experimental evidence at 4 GPa, evolution of the primary olivine-normative magma, generated at the system, must be accompanied by a disappearance of the orthopyroxene through the invariant peritectic reaction with melt and formation of the univariant cotectic assembly olivine + clinopyroxene + garnet + melt. The problem of a termination of the olivine formation arises in view of the fact that compositions of the basic eclogite and grospydite rocks belong to the silica-oversaturated (jadeite-rich omphacite)–garnet–corundum–coesite system. A further ultrabasic-basic evolution of the system has been made possible with fractional crystallization of the evolved melts which compositions become progressively enriched with jadeitic component. Experimental study of melting relations on the olivine–diopside–jadeite–garnet system at 6 GPa demonstrates a disappearance of the olivine through the invariant peritectic reaction with jadeite-rich melt and formation of the basic univariant cotectic omphacite + garnet + melt. Hence the ultrabasic-basic evolution of the primary upper mantle ultrabasic magma is effected principally by the peritectic reactions of orthopyroxene and olivine when coupled with the regime of fractional crystallization. In the case of the transition zone, compositions of the ultrabasic ringwoodite-bearing and basic stishovite-bearing rocks are rated to the ringwoodite–(majoritic garnet)–magnesiowustite–stishovite system. It has been found experimentally at 20 GPa that the ultrabasic-basic evolution of the primary ringwoodite-normative magma in the polythermal section ringwoodite—$(2FeO + SiO_2)$ of the boundary $MgO–FeO–SiO_2$ system is attended with a disappearance of ringwoodite in consequence of its invariant peritectic reaction with melt and formation of the univariant cotectic assembly stishovite + magnesiowustite + melt. For the lower mantle conditions, compositions of the ultrabasic bridgmanite-bearing and basic stishovite-bearing rocks are

Y. A. Litvin (✉) · A. V. Kuzyura · A. V. Spivak
D.S. Korzhinskii Institute of Experimental Mineralogy, Russian Academy of Sciences,
Academician Osipyan Street 4, Chernogolovka, Moscow Region 142432, Russia
e-mail: litvin@iem.ac.ru

© The Editor(s) (if applicable) and The Author(s), under exclusive license to Springer
Nature Switzerland AG 2020
Y. Litvin and O. Safonov (eds.), *Advances in Experimental and Genetic Mineralogy*,
Springer Mineralogy, https://doi.org/10.1007/978-3-030-42859-4_8

165

concerned with the bridgmanite–Ca–perovskite–(periclase ↔ wustite)$_{ss}$–stishovite system. Experimentally at 24 GPa we find that ultrabasic-basic evolution of the primary bridgmanite-normative magma in the polythermal section bridgmanite—(FeO + SiO$_2$) of the boundary MgO–FeO–SiO$_2$ system has been marked by a disappearance of bridgmanite through its invariant peritectic reaction with melt and formation of the univariant cotectic assembly stishovite + magnesiowustite + melt (effect of "stishovite paradox"). The ultrabasic-basic evolution of the primary transition zone and lower mantle ultrabasic magmas, effected by the peritectic reactions of ringwoodite and bridgmanite, respectively, can result exclusively in the regime of fractional crystallization. The peritectic reactions of the upper mantle orthopyroxene and olivine, transition zone ringwoodite and lower mantle bridgmanite are also operable for genesis of the diamonds and associated minerals in the relevant silicate–(±oxide)–carbonate–carbon systems subjected to the ultrabasic-basic fractional evolution.

Keywords Upper mantle · Transition zone · Lower mantle · Mantle magmatism · Diamond genesis · Ultrabasic-basic magma evolution · Physicochemical experiments · Orthopyroxene peritectic · Olivine peritectic · Ringwoodite peritectic · Bridgmanite peritectic · Fractional crystallization

8.1   Introduction

Evidence for ultrabasic-basic evolution of the Earth's mantle magmas at 150–800 km depths (pressure P is changeable within 3.5–30 GPa) has been well documented by the upper mantle peridotite and eclogite xenoliths in kimberlites as well as paragenetic mineral inclusions in diamonds derived from the upper mantle, transition zone and lower mantle depths (Meyer and Boyd 1972; Kaminsky 2017). A multicomponent substance of the deeper-Earth horizons is prohibitive for its in situ studying, but that's fragments have been carried out to the Earth's surface by kimberlite magmas as primary inclusions in diamonds. According to isochemical concept, the mantle chemical composition remains unchanged with depth and corresponds to that of the model pyrolite with mineralogy of garnet lherzolite (Ringwood 1975). It has been found in high-pressure experiments that subsolidus phases of the pyrolite are similar to mineral inclusions in diamonds derived from the upper mantle, transition zone and lower mantle depths (Akaogi 2007). This allows to conclude that the transition zone and lower mantle diamond-producing silicate–(±oxide)–carbonate–carbon melts–solutions has been originated with dissolution of minerals of the native deeper-mantle rocks in the primary carbonate melts of metasomatic origin (Litvin 2017). At the stage of diamond genesis, the dissolved mineral components were recrystallized into phases, similar to the native minerals, and the newly formed minerals could be trapped by growing diamonds as paragenetic inclusions. The mineralogical and compositional resemblance of the native deeper-mantle minerals to the paragenetic phases included into diamonds provides a possibility of estimating the general compositions of the native mantle and diamond-producing systems from compositions of the diamond-hosted primary mineral inclusions.

The garnet-bearing rocks of the Earth's upper mantle are characteristic from depths of 80–100 km right up to the boundary with the transition zone at 410 km depth. The upper-mantle ultrabasic peridotites and pyroxenites and basic eclogites and grospy-dites are represented by the mantle derived xenoliths in kimberlites (MacGregor and Carter 1970; Ringwood 1975; Sobolev 1977; Dawson 1980). Mineralogical compo-sitions of the totality of garnet-bearing ultrabasic and basic rocks have been gener-alized as the olivine–(clinopyroxene/omphacite)–corundum–coesite system (Litvin et al. 2016). The interrelations between the rocks are profitable to present in a form of the complex diagram composed of the unit tetrahedral simplexes (Fig. 8.1). It is believed that the ultrabasic peridotite and pyroxenite xenoliths (simplex A) dominate essentially over the basic eclogite samples (simplexes C, D, E jointly), statistically as about 95–5% for African and Siberian kimberlites (MacGregor and Carter 1970). Pyroxenite xenoliths have occasionally revealed as relatively more numerous in the kimberlitic pipe Matsoku, Lesoto (Dawson 1968). It is also known that the mantle xenoliths may be represented completely or mainly by the basic eclogites, as for instance, in the pipes Roberts Victor, Bobbeian and Ritfontein (S. Africa), Orapa (Botswana), Garvet-Ridge and Mozes-Rock (USA) and Zagadochnaya (Yakutiya, Russia) (Dawson 1980). Therewith, the ultrabasic olivine-bearing eclogites (sim-plex B) have not been occurred in essence among the xenoliths (O'Hara and Yoder 1967; O'Hara 1968; Yoder 1976).

It is symptomatic that primary mineral inclusions in the upper mantle derived dia-monds belong also to both the ultrabasic and basic paragenesises (Sobolev 1977). The less common association of olivine together with eclogitic omphacite and ferriferous garnet has revealed in the diamond-hosted inclusions (Wang 1998). This paragenesis can be assigned rather to the ultrabasic olivine-bearing eclogite assemblage than to

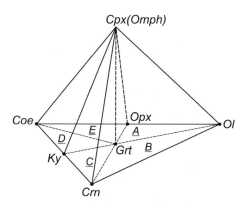

Fig. 8.1 Complex coordinate framework of peridotite–pyroxenite-eclogite compositions and its unit simplexes: (A) peridotite–pyroxenite Ol–Opx–Cpx–Grt, (B) olivine-corundum-eclogite Ol–Crn–Cpx–Grt, (C) corundum-kyanite-eclogite Crn–Ky–Omph–Grt, (D) kyanite–coesite–eclogite Ky–Coe–Omph–Grt, (E) coesite–orthopyroxene–eclogite Coe–Opx–Omph–Grt

the "mixed" ultrabasic-basic peridotite–eclogite association. Hence the mineralogical evidence for both the native upper mantle rocks and diamond-hosted mineral paragenesises indicates convincingly the feasibility of ultrabasic-basic evolution of the upper mantle silicate magmas as well as the parental silicate–($\pm$oxide)–carbonate–carbon melts which are common for diamonds and genetically associated phases (Litvin 2017). The necessity of considering the effect of ultrabasic-basic evolution is taken properly into account when constructing the generalized composition diagram of the upper mantle diamond-producing melts (Litvin et al. 2016).

A probability of the upper mantle ultrabasic-basic magmatic evolution is also marked by the petrochemical trends with uninterrupted compositional conversions for clinopyroxenes and garnets from the olivine-bearing peridotite to silica-enriched eclogite rocks (MacGregor and Carter 1970). The mineralogical indications have been pictorially upheld by continuous petrochemical trends being common for both the ultrabasic and basic upper mantle rocks (Fig. 8.2) (Marakushev 1984) as well as for minerals primarily trapped by diamonds as inclusions (Sobolev 1977). The problems of ultrabasic-basic evolution of the upper-mantle magmatic systems is of prime interest for the deep magma ocean activity in the early Earth as well as genesis and fractionation of the Archean komatiitic magma.

Previously the probability of evolution of the upper-mantle magma and petrogenesis of basic eclogites as derivatives from ultrabasic garnet lherzolite has being discussed by O'Hara and Yoder (1967). However the problem could not be solved for lack of necessary information on physicochemical behavior of the multicomponent petrological systems under the upper mantle conditions. The questions of Yoder (1976), how is crystallization of the partial melting products of garnet peridotite accompanied by the disappearance of two major phases (olivine and orthopyroxene)

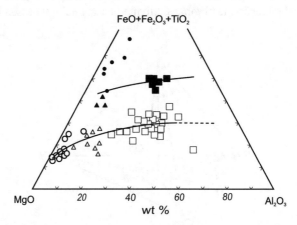

Fig. 8.2 Petrochemical diagram for ultrabasic and basic upper mantle rocks of xenoliths in kimberlites from data of Marakushev (1984). Mg–Al rocks (white signs): circles—garnet dunites and peridotites; triangles—garnet pyroxenites; squares—bimineral eclogites and Ky-eclogites. Fe–Ti-rocks (black signs): circles—phlogopite-ilmenite peridotites; triangles—phlogopite-ilmenite pyroxenites; squares—rutile eclogites

and formation of only garnet and clinopyroxene (i.e., eclogite), had not obtained the answers at that time.

The range of the Earth's transition zone extends from 410 km ($P = 14$ GPa) to the boundary with the lower mantle at 660 km depth ($P = 24$ GPa). In abundance of major silicate and oxide components in diamond-hosted inclusions, the native rocks of the transition zone belong to the system $MgO–FeO–Al_2O_3–CaO–Na_2O$. Therewith the general compositions of diamond-producing media can be represented by the system $MgO–FeO–Al_2O_3–CaO–Na_2O–MgCO_3–FeCO_3–CaCO_3–Na_2CO_3–(\pm K_2CO_3)–C$. The compositions of both magmatic and diamond-producing systems are very changeable as evidenced by ringwoodite $(Mg,Fe)_2SiO_4$ (the indicatory mineral of ultrabasic rocks) and stishovite SiO_2 (the indicatory mineral of basic rocks). Majorite garnet solid solutions with components of majorite $Mg_4Si_4O_{12}$ (Ringwood and Major 1971) and Na-majorite $Na_2MgSi_5O_{12}$ (Bobrov et al. 2009; Bobrov and Litvin 2011; Dymshits et al. 2015) are also the indicatory transition zone minerals among the primary inclusions in diamonds. Therewith the diamond-hosted inclusions of Ca-perovskite $CaSiO_3$, wustite $(Fe,Mg)O$ and carbonates of Mg, Fe, Ca, Na, K are reported (Harte 2010; Kaminsky 2012, 2017).

The depths of the Earth's lower mantle extends from 660 km ($P = 24$ GPa) to the boundary with the core at 2900 km depth ($P = {\sim}120$ GPa), but the lower mantle substance has been sampled as primary inclusions in diamonds from the depth around 800 km ($P = {\sim}28$ GPa). The possibility that diamond genesis can result at the deeper mantle horizons must not be ruled out. However, any mineralogical indications for this are presently absent. The native rocks of the lower mantle belong mainly to the system $MgO–FeO–Al_2O_3–CaO–Na_2O–SiO_2$ on evidence derived from the diamond-hosted primary inclusions. The compositions of diamond-producing media have been assigned to the system $MgO–FeO–Al_2O_3–CaO–Na_2O–SiO_2–MgCO_3–FeCO_3–CaCO_3–Na_2CO_3–(\pm K_2CO_3)–C$. The compositions are changeable from the assembly of ferropericlase $(Mg,Fe)O$ and ringwoodite $(Mg,Fe)_2SiO_4$ (the minerals of ultrabasic rocks) to magnesiowustite $(Fe,Mg)O$ and stishovite SiO_2 (the minerals of basic rocks). Ca-perovskite $CaSiO_3$ and aluminous phases of Mg, Ca and Na are among the important inclusions in the lower mantle derived diamonds (Kaminsky 2017). Carbonate inclusions of $CaCO_3$ (calcite/aragonite), $CaMg(CO_3)_2$ (dolomite), $MgCO_3$ and $(Mg,Fe)CO_3$ (magnesite and Fe-magnesite), $Na_2Mg(CO_3)_2$ (eitelite), $(Na,K)_2Ca(CO_3)_2$ (nyerereite) and $NaHCO_3$ (nahcolite) are also reported. The interrelations between the transition zone and lower mantle diamond-producing melts and minerals may be presented in a form of the generalized composition diagrams, considering the effect of their ultrabasic-basic evolution (Litvin et al. 2016; Litvin 2017; Spivak and YuA 2019).

Meanwhile, the physicochemical mechanisms which control processes of the ultrabasic-basic evolution of the native mantle magmas and diamond-producing melts at the mantle depths are not evident from the analytical mineralogical data (Sobolev 1977; Dawson 1980; Logvinova et al. 2008; Klein-BenDavid et al. 2006; Zedgenizov et al. 2015) and the results of experimental investigation of the subsolidus transformations for the mantle minerals and their components (Akaogi 2007). It is evident, that the mechanisms can be revealed only in physicochemical experiments through studying the melting phase transformations of the mineral systems with the representative boundary compositions.

In the first place, this work is directed toward application of the experimental melting relations data for the upper mantle multicomponent system olivine–(clinopyroxene/omphacite)–corundum–coesite to substantiation of the mechanisms of physicochemical reactions of orthopyroxene and olivine to the process of ultrabasic-basic evolution of the native upper mantle magmas. This raises the necessity to considerate the experimental evidence for the upper magma evolution within the ultrabasic compositions (the simplexes A and B in Fig. 8.1). Another more important concern is in experimental study at 6 GPa of the olivine–diopside–jadeite–garnet system which melting relations have to be responsible for the ultrabasic-basic evolution of the upper mantle magmas (between the ultrabasic simplex A and basic simplexes C, D, E in Fig. 8.1).

Secondly, the challenge is to investigate the melting phase relations for the transition zone diamond-producing system $MgO–FeO–SiO_2–Na_2CO_3–CaCO_3–K_2CO_3$ in experiments at 20 GPa. The pressure provides the structural stability of ringwoodite $(Mg,Fe)_2SiO_4$ as the key rock-forming mineral of the deepest ultrabasic rocks of the transition zone.

Thirdly, of fundamental importance is the experimental estimation at 24 GPa of a physicochemical behavior of the chief lower mantle mineral bridgmanite $(Mg,Fe)SiO_3$ at the ultrabasic-basic evolution of the key system $MgO–FeO–SiO_2$. The system involves the key minerals of as the ultrabasic rocks (ferropericlase and bridgmanite) so basic ones (magnesiowustite and stishovite).

Hence much attention is given to the melting relations of the systems with olivine, ringwoodite and bridgmanite as the key minerals of ultrabasic rocks at the upper mantle, transition zone and lower mantle depths and estimation of their role in the ultrabasic-basic evolution of the Earth's mantle magmatic and diamond-producing materials.

8.2 Experimental, Theoretical and Analytical Methods

Melting relations of the upper mantle systems olivine–orthopyroxene–clinopyroxene–garnet at 4 GPa and olivine–diopside–jadeite–garnet at 6 GPa are studied in high pressure and temperature toroidal anvil-with-hole apparatus with isothermal cells in the Institute of experimental mineralogy (Litvin 1991; Litvin et al. 2008). The starting materials were gel mixtures of olivine, orthopyroxene, clinopyroxene and garnet (presented in Table 8.1), diopside $CaMgSi_2O_6$ and jadeite $NaAlSi_2O_6$. Melting relations were studied with the use of a graphite heater for generation of high temperature inside the strongly compressed cells from lithographic limestone. The starting materials have been introduced in Pt or $Pt_{60}Rh_{40}$ capsules being welded up. The capsules were protected from graphite heaters with MgO sleeves. Accuracies in pressure determinations were ±0.05 GPa from statistic evaluation. Uncertainties in the temperature measurements using the $Pt_{70}Rh_{30}$-$Pt_{94}Rh_{06}$ thermocouple were

Table 8.1 Compositions of boundary phases (in wt%) of ultrabasic system olivine (Ol)—orthopyroxene (Opx)—clinopyroxene (Cpx)—garnet (Grt) studied in experiments at 4 GPa

Oxides	Ol		Opx		Cpx		Grt	
	PM2258	EXPERIM.	PH1600	EXPERIM.	327	EXPERIM.	327	EXPERIM.
SiO_2	38.00	40.02	56.00	57.80	55.10	55.90	41.70	42.55
TiO_2	0.03		0.19		0.20		0.68	
Al_2O_3	0.05		0.92	0.49	3.88	4.28	21.10	24.07
Cr_2O_3	0.03		0.03		0.50		1.10	
FeO	13.60	14.35	9.62	7.26	4.13	4.68	8.84	8.48
MnO	0.13		0.19		0.12		0.32	
NiO	0.28							
MgO	47.60	45.63	33.60	33.34	17.10	16.88	21.00	20.93
CaO	0.07		0.86	0.81	15.10	15.66	4.68	3.97
Na_2O			0.23	0.30	2.77	2.60	0.11	
Total	99.70	100.00	101.6	100.00	98.90	100.00	99.40	100.00

The compositions of natural Ol (PM2258), Opx (PH1600), Cpx (327) and Grt (327) (Boyd and Danchin 1980) are used as prototypes

estimated as ±20 °C. Experimental samples were quenched with temperature change of about 300 °C/s between 1600 and 1000 °C in the sample zone.

Melting relations of the transition zone system $MgO–FeO–SiO_2–Na_2CO_3–CaCO_3–K_2CO_3$ at 20 GPa and lower mantle systems $MgO–FeO–CaO–SiO_2$ at 24 and 26 GPa were studied with the use of the multianvil apparatus in the Bavarian Institute of experimental geochemistry and geophysics of the University in Bayreuth, Germany (Frost et al. 2004). The powdered mixtures of oxides MgO, FeO, CaO, SiO_2, carbonates Na_2CO_3, $CaCO_3$, K_2CO_3 and ultrapure synthetic graphite were used as the starting materials. High temperature at the experimental sample was generated with the $LaCrO_3$ heater and measured with accuracy of ±50 °C. High pressure was maintained within the ±0.5–1.0 GPa limits of experimental error.

A fundamental investigation of the native mantle systems with complex multiphase compositions in physicochemical experiments would be highly profitable only in conjunction with theoretical methods of the physical chemistry of multicomponent systems (Rhines 1956; Palatnik and Landau 1964; Zakharov 1964). Among the methods is the concept of geometrical image of multicomponent system in a form of the complex tetrahedral diagram divided into the primitive tetrahedral simplexes ("simplex triangulation of diagrams"). In this work the method of two-dimensional polythermal sections of composition diagrams is used as most efficient in high-pressure study of melting relations of the multicomponent mantle systems. The method is appropriate for ternary phase diagrams and isothermal ternary sections of the quadruple tetrahedral diagrams. In this case the compositions for the melting diagrams of two-dimensional polythermal sections are determined from the boundary compositions of the multicomponent systems studied. The melting relations of the polythermal sections are capable to untangle the key physicochemical properties of the multicomponent mantle systems. The Rhines' phase rule (Rhines 1956) gives a rigorous control over the certainty in topological construction of the experimental diagrams of melting relation for polythermal sections of the mantle multicomponent systems. Physicochemical behaviour for the petrological systems at the regime of fractional crystallization have been considered repeatedly (Maaloe 1985; Litvin 1991). The crucial effect lies in the consecutive changes of the original general composition of the system owing to the fractional removal of crystallized solid phases from the residual melts at decreasing temperature. Under these circumstances, the original composition of the fractionated magma continuously changes and becomes coincident with the compositions of the residual melts (formed after permanent crystallization and fractional removal of the solid phases from the melts). The fractional magmatic evolution is effected principally over the univariant curves of the liquidus surface, and the invariant peritectic points could not interfere the process. Whereas the invariant eutectic points, being served as the thermal barriers, would terminate the evolution. It should be remembered that the original general compositions of the equilibrium systems are bound to be unchanged from start to finish of the melt crystallization.

The recovered experimental samples were examined using a CamScan M2300 (VEGA TS 5130MM) electron microscope. The compositions of the phases were analyzed with a Link INCA energy-dispersive microprobe. The SiO_2, albite, MgO, Al_2O_3, wollastonite and metallic Fe were used as standard materials. The analyzes were carried out at the Institute of Experimental Mineralogy.

8.3 Research on the Magmatic and Diamond-Producing Systems of the Earth's Mantle in Theory and High-Pressure Experiments

8.3.1 The Upper Mantle Orthopyroxene Peritectic Reaction at 4 GPa

Melting relations on the upper mantle ultrabasic multicomponent system olivine (Ol)–orthopyroxene (Opx)–clinopyroxene (Cpx)–garnet (Grt) have been initially studied at 4 GPa (Litvin 1991; Litvin et al. 2016). The general composition of the system was correlated to the natural one; the starting materials are given in the Table 8.1. The experimental melting relations for several polythermal sections, including the $Opx_{51}Ol_{09}Grt_{40}$–$Cpx_{51}Ol_{09}Grt_{40}$ one (Fig. 8.3), on the Ol–Opx–Cpx–Grt peridotite–pyroxenite system made it possible to construct the diagram of liquidus surface for the ultrabasic garnet-bearing substance of the upper mantle (Fig. 8.4). The ultrabasic system is characterized by invariant peritectic reaction of orthopyroxene and komatiitic melt at 1445–1465 °C, 4 GPa with formation of clinopyroxene. The composition of the primary invariant peritectic melt was found to be olivine-normative komatiitic (wt.%): SiO_2 48.7, Al_2O_3 7.0, MgO 24.0, FeO 11.7, CaO 5.8, Na_2O 2.8. Accordingly, the komatiitic melts have to be also generated at higher degrees of melting of

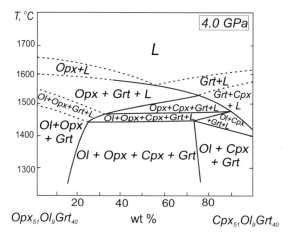

Fig. 8.3 Melting phase relations of the upper-mantle peridotite–pyroxenite system Ol–Opx–Cpx–Grt system for the polythermal section $Opx_{51}Ol_{09}Grt_{40}$–$Cpx_{51}Ol_{09}Grt_{40}$ at 4 GPa. Black points—for experimental compositions (Litvin 1991)

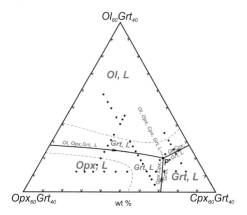

Fig. 8.4 Liquidus surface of peridotite–pyroxenite Ol–Opx–Grt–Cpx system in projection from Grt apex onto isothermal section $Ol_{60}Grt_{40}$–$Opx_{60}Grt_{40}$–$Cpx_{60}Grt_{40}$. Points for experimental compositions over polythermal sections studied (Litvin 1991)

the upper-mantle garnet lherzolite. As temperature decreases, the melt composition may follow over the univariant cotectic curve Ol + Opx + Grt + L to the invariant peritectic point Ol + Opx + Cpx + Grt + L where the orthopyroxene peritectic reaction opens the door to the univariant cotectic curve Ol + Cpx + Grt + L.

What's more it has became evident the equilibrium liquidus structure for the ultrabasic-basic multicomponent system olivine–(clinopyroxene/omphacite)–corundum–coesite (Fig. 8.5). Speculative analysis of the equilibrium liquidus diagram, which is defined as the complex of two ultrabasic (A and B) and three basic (C, D and E) simplexes, offers a clearer view of the existing physicochemical restrictions

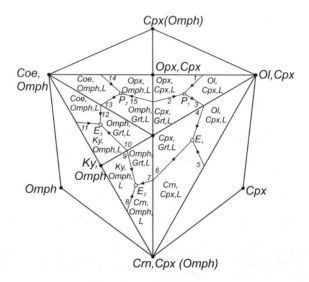

Fig. 8.5 Diagram-complex of ultrabasic-basic liquidus structure for upper mantle Ol–Opx–Crn–Coe system in projection onto isothermal section Ol,Cpx–Crn,Cpx(Omph)–Coe,Omph. Symbols: for ultrabasic compositions: P_1—peritectic point Ol + Opx + Grt + Cpx + L and E_1—eutectic point Ol + Crn + Grt + Cpx + L. Symbols for basic compositions: P_2—peritectic point Coe + Opx + Grt + Omph + L and eutectic points E_2—Crn + Ky + Grt + Omph + L and E_3—Ky + Coe + Grt + Omph + L. Numbers by univariant curves—ultrabasic: 1—Ol + Opx + Cpx + L, 2—Opx + Grt + Cpx + L, 3 and 4—Ol + Grt + Cpx + L, 5—Ol + Cpx + Crn + L, 6—Crn + Grt + Cpx + L; basic: 7—Crn + Grt + Omph + L, 8—Crn + Ky + Omph + L, 9 and 10—Ky + Grt + Omph + L, 11—Ky + Coe + Omph + L, 12 and 13—Coe + Grt + Omph + L, 14—Coe + Opx + Omph + L, 15—Opx + Grt + Omph + L. Arrows for directions of temperatures lowering along cotectic curves

for the upper mantle ultrabasic-basic magma evolution (Litvin 2017). Primarily it must be emphasized that the invariant peritectic point P_1 (where Opx is lost) and T-regressing univariant curve Ol + Cpx + Grt + L (denoted by 3) have capable of regulating the direction and limits for the magma evolution rigorously within the compositions of the ultrabasic peridotite–pyroxenite simplex A. Hence the univariant curve Ol + Cpx + Grt + L ($P_1 \rightarrow 3 \rightarrow 4 \rightarrow E_1$), being common for the simplexes A and B. Joins the peridotite–pyroxenite invariant peritectic point P_1 with the Ol–Crn–eclogitic eutectic point E_1. The possibility for equilibrium evolution of the ultrabasic komatiitic melt composition has a restriction exclusively within the simplex A because the compositions of the peridotite–pyroxenite system can not be changed outside their simplex. It should be realized that the fractional crystallization could in principle provide the evolution of the melt composition from the simplex A over the univariant curve Ol + Cpx + Grt + L into the simplex B with access of the eutectic point E_1. However this scenario does not hold under natural conditions as judged from the feasible absence of olivine eclogites among the upper-mantle ultrabasic xenoliths in kimberlites (MacGregor and Carter 1970; Sobolev 1977; Dawson 1980).

Of special significance is that the univariant curves Opx + (Cpx/Omph) + Grt + L (P_1 ← 2 ← max → 15 → P_2) between the ultrabasic A and basic E peritectic simplexes as well as (Cpx/Omph) + Crn + Grt + L (E_1 ← 6 ← max → 7 → E_2) between the ultrabasic B and basic C eutectic simplexes are characterized by the temperature maxima (max) at the piercing points between the adjacent simplexes (Fig. 8.5). The temperature maxima represent impassable thermal barriers for the ultrabasic-basic magmatic evolution of the upper mantle magmas under both the regimes of equilibrium and fractional crystallization. This leads to the conclusion that the peritectic reaction with loss of orthopyroxene may promote the evolution of the upper mantle primary magma exclusively within the ultrabasic peridotite–pyroxenite–(Ol–eclogite) compositions whereas the way to the basic eclogite compositions is overlapped by the thermal barriers in the piercing points.

8.3.2 The Upper Mantle Olivine Peritectic Reaction at 6 GPa

Encouraging perspective for a rigorous understanding the physicochemical mechanism of ultrabasic-basic evolution of the upper mantle magma has appeared with experimental discovery at higher 4.5 GPa pressures of the reaction of forsterite and jadeite components with the formation of pyropic garnet in the systems forsterite–jadeite, enstatite–nepheline and forsterite–diopside–jadeite (Gasparik and Litvin 1997; Litvin et al. 2000). This is of particular interest in the experimental study of melting relations on the ultrabasic-basic olivine–diopside–jadeite–garnet system which composition is in closer approximation to the native upper mantle matter (after the loss of orthopyroxene). Contents of Na_2O in clinopyroxenes of peridotites and pyroxenites averages between 1.5 and 4.5 wt.% and are noticeably growing in omphacites of eclogites (2.5–6.5 wt.%) and grospydites (4.5–9.9 wt.%), that indicates the commensurable increase of concentration of the jadeitic component in the ultrabasic-basic system (Dawson 1980). The contents of Na_2O in primary omphacitic inclusions in diamond are even higher (6.4–10.7 wt.%) (Sobolev 1977).

Experimental study at 6 GPa of melting relations on the multicomponent olivine–diopside–jadeite join was carried out using its polythermal sections Ol–Omph (where Ol = $Fo_{80}Fa_{20}$ and Omph = $Jd_{62}Di_{38}$) and $Ol_{90}Jd_{10}$–Cpx (where Cpx = $Di_{90}Jd_{10}$). Experimental results are presented in Table 8.2 and Figs. 8.6, 8.7 and 8.8. The quasi binary melting phase diagram of the polythermal section Ol–Omph in the Fig. 8.7 demonstrates that olivine, garnet and omphacite represent the liquidus phases. Therewith the garnet phase is a result of the reaction between olivine and jadeite components of the melts. The formation of the ultrabasic Cpx together with Grt in the vicinity to the solidus conditions is significant for the compositions enriched with Ol component. As the concentration of Jd component in the system increases, the reaction of Ol and Jd components has accomplished in the peritectic point P with complete loss of Ol and the transfer of the Di-Jd solid solutions compositions from the ultrabasic Cpx phases to the basic Omph phases.

Table 8.2 Experimental conditions at 6 GPa, chemical compositions of mineral phases after quenching and phase assemblies (at equilibrium approximation) used in construction of phase diagrams in Figs. 8.7 and 8.8

Run №.	Starting composition	T, °C	τ, min	Phase	Composition of experimental phase							Phase assembly
					SiO$_2$	Al$_2$O$_3$	FeO	MgO	CaO	Na$_2$O	Total	
Polythermal section Ol-Omph												
3168	Ol$_{65}$Di$_{13.3}$Jd$_{21.7}$	1625	10	Ol	41.39		10.24	47.54	0.20	0.07	99.44	Ol + Grt + L
				Grt	44.09	23.35	3.92	24.10	4.69	0.22	100.37	
				L	54.47	21.80	2.82	5.19	3.87	11.77	100.00	
3166	"	1440	10	Ol	42.36	0.22	5.32	52.78		0.01	100.69	Ol + Grt + L
				Grt	50.94	14.77	7.00	27.01	2.73	0.38	102.83	
				L	56.02	9.30	6.07	9.92	8.65	10.04	100.00	
3118 Figure 8.5a	Ol$_{56}$Di$_{26.4}$Jd$_{17.6}$	1390	10	Ol	40.84	0.04	14.2	45.99	0.05	0.08	101.21	Ol + Cpx + Grt + L
				Cpx	60.78	2.81	4.60	25.30	1.91	4.28	99.68	
				Grt	33.07	21.75	13.19	18.37	3.74	0.46	100.58	
				L	58.81	9.84	5.47	9.25	7.60	9.03	100.00	
3187	Ol$_{50}$Di$_{19}$Jd$_{31}$	1545	10	Ol	40.42		12.03	47.39	0.40		100.24	Ol + Grt + L
				Grt	47.16	21.87	1.37	28.17	2.00	0.03	100.60	
				L	52.81	17.64	3.16	11.61	5.38	9.40	100.00	
3165 Figure 8.5b	"	1450	10	Ol	41.78	0.22	6.97	51.44	0.13	0.06	100.60	Ol + Grt + L
				Grt	43.18	23.41	9.89	19.34	4.35	0.35	100.52	
				L	56.68	21.79	1.95	4.23	3.03	12.32	100.00	

(continued)

Table 8.2 (continued)

Run №.	Starting composition	T, °C	τ, min	Phase	Composition of experimental phase							Phase assembly
					SiO$_2$	Al$_2$O$_3$	FeO	MgO	CaO	Na$_2$O	Total	
3160	Ol$_{40}$Di$_{24}$Jd$_{36}$	1550	10	Grt	44.41	24.03	3.49	25.40	2.69	0.16	100.18	Grt + L
				L	Quenched carbonate and silicate dendritic microphases							
3180	”	1500	15	Ol	41.32		7.47	51.29			100.08	Ol + Grt + L
				Grt	43.02	23.98	7.57	21.67	4.67		100.07	
				L								
3119	”	1395	30	Ol	41.03	0.45	11.28	46.13	0.12	0.50	99.58	Ol + Cpx + L
				Cpx	54.17	10.10	2.90	20.19	8.40	3.24	99.00	
				Grt	41.67	21.17	9.82	19.44	4.03	0.69	96.82	
				L	55.05	14.18	1.99	16.66	5.21	6.92	100.0	L
3158	Ol$_{30}$Di$_{13.5}$Jd$_{38.5}$	1560	15	L	53.91	16.54	1.17	13.61	5.38	9.40	100.00	
3157	”	1500	10	Grt	44.65	23.82	3.76	34.72	3.15	0.36	100.44	Grt + L
				L	Quenched carbonate and silicate dendritic microphases							
3159	”	1420	20	Grt	43.26	22.44	9.42	21.17	3.11	0.45	99.85	Grt + L
				L	55.05	14.18	1.99	16'66	5.21	6.92	100.0	
3132 Figure 8.5c	”	1390	30	Omph	58.04	8.20	2.40	16.95	10.10	5.11	100.80	Omph + Grt + L
				Grt	43.33	23.32	2.54	24.74	3.60	0.24		
				L	Quenched carbonate and silicate dendritic microphases							

(continued)

Table 8.2 (continued)

Run №.	Starting composition	T, °C	τ, min	Phase	Composition of experimental phase							Phase assembly
					SiO$_2$	Al$_2$O$_3$	FeO	MgO	CaO	Na$_2$O	Total	
3183	Ol$_{20}$Di$_{30.4}$Jd$_{31.6}$	1470	10	L	Quenched carbonate and silicate dendritic microphases							L
3133 Figure 8.5d	"	1380	30	Omph	56.61	8.05	0.40	17.47	13.02	4.56	100.11	Omph + Grt + L
				Grt	44.34	24.03	0.91	25.88	4.56	0.20	99.92	
				L	53.39	15.30	0.41	12.98	6.55	8.88	100.00	
3184	Ol$_{20}$Di$_{30.4}$Jd$_{49.6}$	1540	10	L	53.87	13.46	1.98	15.76	5.26	9.67	100.00	L
3170	Ol$_{08}$Di$_{37.2}$Jd$_{54.8}$	1520	10	Omph	56.12	7.19	4.96	17.60	8.61	5.16	99.64	Omph + L
				L	55.49	15.55	0.66	12.48	7.05	8.77	100.00	
Polythermal section Ol-Cpx												
3179	Ol$_{43}$Di$_{27.4}$Jd$_{29.6}$	1520	10	Ol	39.09	0.19	11.54	44.71	0.25	0.09	95.87	Ol + L
				L	49.76	2.69	2.58	27.18	16.34	1.96	100.51	
3181	Ol$_{34}$Di$_{44.2}$Jd$_{21.8}$	1500	10	Cpx	57.38	8.82	3.15	16.38	10.13	4.31	100.17	Cpx + L
				L	Quenched carbonate and silicate dendritic microphases							

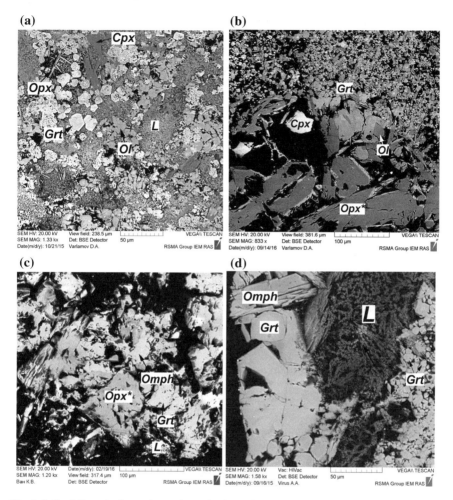

Fig. 8.6 Partially melted experimental samples being formed at 6 GPa after quenching and polishing. Univariant assemblies—ultrabasic Ol + Cpx + Grt + L: **a** sample 3118 (1400 °C), **b** sample 3164 (1400 °C) and basic Omph + Grt + L: **c** sample 3132 (1390 °C), **d** sample 3133 (1389 °C). Asterisk marks solid phases formed metastably due to quenching process

The polythermal sections Ol–Omph ($Jd_{62}Di_{38}$) and Ol–Cpx ($Di_{90}Jd_{10}$) are propagated, respectively, through the three Ol + L, Grt + L, (Cpx ↔ Omph) + L and the two Ol + l, (Cpx ↔ Omph) + L liquidus associations. This clearly demonstrates the principal peritectic liquidus structure of the ternary Ol–Di–Jd system (Fig. 8.8).

The constructed melting diagrams are evidenced that the invariant peritectic reaction of olivine and jadeitic components of melts with formation of garnet has proceeded. The ternary liquidus diagram (Fig. 8.8) is demonstrated that the figurative points of the ultrabasic compositions of the system can move over the newly formed regressive univariant curve Ol + Grt + L towards the quasi-invariant peritectic point

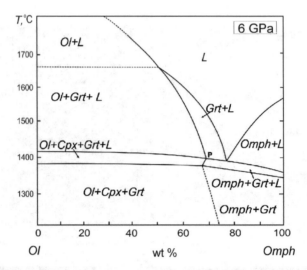

Fig. 8.7 Diagram of melting relations for polythermal section Ol–Omph of the ultrabasic-basic Ol–Di–Jd system. Black points for experimental compositions with sample numbers. Analytical data are in the Table 8.2

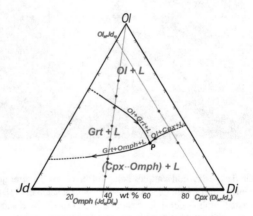

Fig. 8.8 Liquidus surface of ultrabasic-basic system Ol–Di–Jd. Positions of the polythermal sections Ol–Jd$_{62}$Di$_{38}$ and Ol$_{90}$Jd$_{10}$–Di$_{90}$Jd$_{10}$ are marked by dotted lines. Positions of the experimental points of the polythermal sections are presented by their projections onto the liquidus surface. Analytical data are in the Table 8.2

P at 1380–1420 °C, 6 GPa (to which the other univariant curve Ol + Cpx + L is also attained). Olivine components are completely reacted with the Jd-component of melt in the peritectic assemblage Ol + (Cpx/Omph) + Grt + L that is followed with garnet and omphacite assembly formation, and then the system composition becomes basic and evolves along the univariant curve Grt + Omph + L. This has the effect of creating the mechanism of physicochemical reaction capable to provide the ultrabasic-basic evolution of the upper mantle magmas (Litvin et al. 2017).

8.3.3 The Transition Zone Ringwoodite Peritectic Reaction at 20 GPa

Experimental study of the subsolidus transformations in the system Mg_2SiO_4–Fe_2SiO_4 disclosed that the ringwoodite solid solution $(Mg_2SiO_4 \cdot Fe_2SiO_4)_{ss}$ is limited one due to its ferrous component Fe_2SiO_4 is unstable at pressure above 17 GPa and disproportionates into wustite FeO and stishovite SiO_2 (Ito and Takahashi 1989). Therewith, the Fe–Mg–solid solutions, depleted in Mg_2SiO_4 component, have endured a similar decomposition to oxides. It results in formation of two stishovite-bearing subsolidus assemblages: (1) ringwoodite $(Mg,Fe)_2SiO_4$ + ferropericlase-magnesiowustite solid solutions $[(Mg,Fe)O \leftrightarrow (Fe,Mg)O]_{ss}$ + stishovite SiO_2 and (2) magnesiowustite $(Fe,Mg)O$ + stishovite SiO_2. The fact that the component Fe_2SiO_4 is unstable at the depths of the transition zone is appropriate to take properly into account and present the boundary compositions of the diamond-producing system under study as Mg_2SiO_4–$(2FeO + SiO_2)$–Na_2CO_3–$CaCO_3$–K_2CO_3–C. There is a good probability that experimental melting relations of this diamond-producing system are capable of revealing the peritectic reaction of ringwoodite and Fe-component richer melts (the chance of such reaction is high in the transition zone carbonate-free material).

The schematic subsolidus diagram of the system MgO–FeO–SiO_2–Carb* (the symbol Carb* indicates the composition of the carbonate constituent Na_2CO_3–$CaCO_3$–K_2CO_3) includes the phase assemblages which are stable at the experimental conditions at 20 GPa and 1000–1700 °C (Fig. 8.9). In a triangulation subdivision of the system, it is taken into consideration the unlimited periclase-wustite solid solutions $(MgOFeO)_{ss}$ (FPer/MWus) as well as limited for ferroringwoodite $(Mg,Fe)_2SiO_4$ (FRwd) up to 55–65 mol% (Ito and Takahashi 1989; Matsuzaka et al. 2000) and ferroakimotoite $(Mg,Fe)SiO_3$ (FAki) up to 20–40 mol% (Tomioka et al. 2002).

The main objective is to determine the physicochemical behavior of ringwoodite $(Mg,Fe)_2SiO_4$ by melting relations of the diamond-producing Mg_2SiO_4–$(2FeO \cdot SiO_2)$–Carb*–C system at its polythermal section $[(Mg,Fe)_2SiO_4$ + Carb*]–$[(2FeO \cdot SiO_2)$ + Carb*)] in experiments at 20 GPa. It is anticipated a determination of the physicochemical mechanisms capable of providing an ultrabasic-basic evolution of the diamond-producing melts and formation of the corresponding ringwoodite- and stishovite-bearing subsolidus mineral assemblages. Experimental results are presented in Table 8.3 and Fig. 8.10. The quasi binary melting phase diagram of the polythermal section $[(Mg,Fe)_2SiO_4$ + Carb*]–$[(2FeO \cdot SiO_2)$ + Carb*)] in Fig. 8.11 demonstrates that ferroringwoodite FRwd and magnesiowustite MWus are the

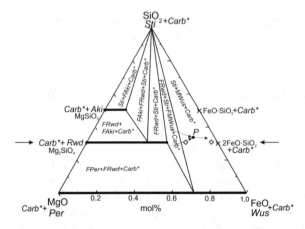

Fig. 8.9 Preliminary subsolidus associations of the MgO–FeO–SiO$_2$–Carb* system at 20 GPa and 1100 °C (based on available experimental data and results of this study). Carb*—simplified notation for the carbonate CaCO$_3$—Na$_2$CO$_3$—K$_2$CO$_3$ system; the notation being common for all phase assemblages of the system is disposed beyond the diagram; rhombuses—for the starting compositions of the series of experiments at 20 Gpa and 1000–1700 °C; P—peritectic point; dotted lines with arrows—direction of temperature lowering for cotectic curves, crosses—oxide compositions FeO·SiO$_2$ and 2FeO·SiO$_2$ for the unstable components «FeSiO$_3$» and «Fe$_2$SiO$_4$», correspondingly. Position of the polythermal section (Mg$_2$SiO$_4$)$_{70}$Carb*$_{30}$–(2FeO·SiO$_2$)$_{70}$Carb*$_{30}$ is shown with arrows

liquidus phases. Under the solidus conditions, the ferroringwoodite in the invariant peritectic reaction with Fe-rich melt decomposes into magnesiowustite MWus and stishovite Sti. With the participation of the Ca-carbonate component, the attendant reaction of formation of the Ca-perovskite CaPrv proceeds concurrently. Thus, the complete loss of ferroringwoodite takes place in the peritectic point P (its position is shown at the Fig. 8.11) with subsequent transfer from the ultrabasic ferroringwoodite-bearing assemblages to the basic ones with magnesiowustite and stishovite (Table 8.3).

8.3.4 The Lower Mantle Bridgmanite Peritectic Reaction at 26 GPa

Experimental data at pressures higher 24 GPa of the subsolidus phase relationes in the lower-mantle system MgSiO$_3$–FeSiO$_3$ are indicative of the limited bridgmanite solid solution (MgSiO$_3$·FeSiO$_3$)$_{ss}$ and decomposition of the boundary ferrous component FeSiO$_3$ into the oxides wustite FeO and stishovite SiO$_2$. In similar way, the Fe–Mg–solid solution phases depleted in MgSiO$_3$ component can also decompose into oxides (FeO·MgO)$_{ss}$ and SiO$_2$ (Irifune and Tsuchiya 2007). The oxides are involved into formation of two stishovite-bearing subsolidus associations: (1) bridgmanite (Mg,Fe)SiO$_3$ + ferropericlase-magnesiowustite solid solutions

Table 8.3 Conditions and results of experiments on study of phase relations for the system MgO–FeO–SiO$_2$–(Na,Ca,K)CO$_3$ at 20 GPa

Sample#	T, °C	T, min	Experimental results								
			Phase	SiO$_2$	FeO	MgO	CaO	Na$_2$O	K$_2$O	CO$_2$*	Sum
				wt%							
I. (Fe$_2$SiO$_4$)$_{52.5}$(Mg$_2$SiO$_4$)$_{17.5}$(Carb*)$_{30}$											
S6873-1	1700	30	L	10.47	9.79	8.70	18.57	18.69	12.47	21.32	100.00
			Rwd**	34.42	34.73	27.02	1.55	0.95	1.32	–	99.99
S6880-1	1500	20	L	7.26	10.63	11.64	11.97	6.16	8.80	43.54	100.00
			Rwd	33.86	35.02	25.87	1.90	1.50	1.82	–	99.98
			MWus	0.49	87.47	5.50	2.25	2.55	1.73	–	99.98
			Sti	98.15	1.27	n.d.	0.14	n.d.	n.d.	–	99.56
S6877-1	1200	40	L	12.64	5.40	7.79	18.33	9.32	12.72	33.52	100.00
			Rwd	33.78	30.02	32.40	1.63	1.11	1.06	–	99.99
			MWus	0.55	92.68	5.33	0.49	0.23	0.32	–	99.84
			Sti	95.93	1.82	0.29	0.84	0.63	0.47	–	99.99
			CaPrv	48.57	1.64	0.55	46.32	0.92	1.79	–	99.97
			Mag	0.43	2.40	40.27	0.35	0.16	0.22	56.00	100.00
			K-Arg	0.73	1.43	6.72	12.27	5.31	17.82	55.69	100.00
S6886-1	1000	60	MWus	2.93	87.49	2.79	1.36	2.08	3.35	–	99.99
			Sti	97.12	1.20	0.16	0.14	0.61	0.77	–	99.98
			CaPrv	48.70	1.27	0.24	46.26	0.99	2.13	–	99.59
			Mag	1.81	3.27	42.25	0.90	1.18	1.69	48.91	100.00
			Na,K-Arg	1.44	1.68	1.03	20.21	9.59	8.57	57.48	100.00

(continued)

Table 8.3 (continued)

Sample#	T, °C	t, min	Experimental results								
			Phase	SiO$_2$	FeO	MgO	CaO	Na$_2$O	K$_2$O	CO$_2$*	Sum
				wt%							
II. (Fe$_2$SiO$_4$)$_{66.5}$(Mg$_2$SiO$_4$)$_{3.5}$(Carb*)$_{30}$											
S6873-2	1700	30	L	11.66	18.62	1.94	10.26	6.66	8.89	41.85	100.00
			MWus	0.82	97.43	1.13	0.22	0.34	n.d.	–	99.95
			Sti	98.43	1.39	n.d.	n.d.	n.d.	n.d.	–	99.82
S6875-2	1400	10	L	3.19	5.74	2.89	19.62	9.58	11.72	47.26	100.00
			MWus	0.44	98.15	0.98	0.15	0.26	n.d.	–	99.98
			Sti	96.92	2.28	n.d.	0.36	0.15	0.21	–	99.92
S6877-2	1200	40	L	11.28	4.39	3.44	17.12	6.00	10.69	47.08	100.00
			MWus	0.50	97.26	1.46	0.39	0.16	0.17	–	99.98
			Sti	95.27	1.29	0.32	1.39	0.85	0.87	–	99.99
			K-Arg	0.74	3.45	3.01	21.20	9.49	13.53	48.59	100.00
S6886-2	1000	60	MWus	0.62	92.06	0.79	0.20	1.05	0.41	–	99.99
			Sti	97.88	1.06	0.33	0.18	n.d.	0.53	–	99.99
			Mag	n.d.	4.83	39.94	1.33	1.10	3.60	49.21	100.00
			K-Cal	0.58	0.89	0.20	38.67	3.04	14.62	42.01	100.00
			Na, K-Arg	0.95	1.13	0.64	19.51	15.72	12.18	49.87	100.00

Calculated data are marked as **, the quenched phase—as *

(a)   (b)

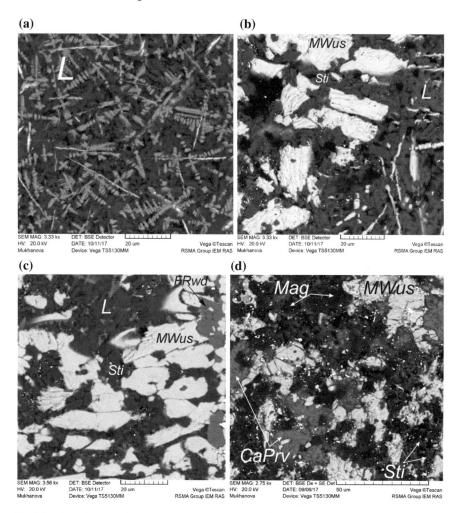

Fig. 8.10 SEM images of experimental samples of the polythermal section $(Mg_2SiO_4)_{70}Carb*_{30}$—$(2FeO \cdot SiO_2)_{70}Carb*_{30}$ at 20 GPa after quenching: **a** sample S6873-1 at 1700 °C; **b** sample S6875-2 at 1400 °C; **c** sample S6880-1 at 1500 °C; **d** sample S6886-1 at 1000 °C

$[(Mg,Fe)O \leftrightarrow (Fe,Mg)O]_{ss}$ + stishovite SiO_2 and (2) magnesiowustite $(Fe,Mg)O$ + stishovite SiO_2.

A triangulation subdivision of the subsolidus diagram for the lower mantle system periclase Per–stishovite Sti–wustite Wus–Ca-perovskite CaPrv is given at Fig. 8.12. The boundary Per–Sti–Wus join including the solid solutions as limited ferrobridgmanite $(Mg,Fe)SiO_3$ (FBrd) as unlimited periclase-wustite $(MgO \cdot FeO)ss$ (FPer/MWus) has the controlling significance.

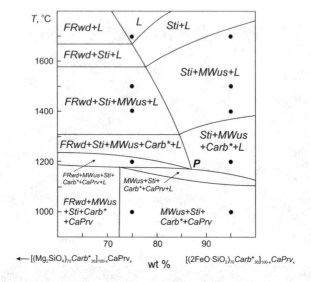

Fig. 8.11 Preliminary melting phase diagram for the polythermal section $[(Mg_2SiO_4)_{70}Carb^*_{30}]_{100-x}CaPrv_x–[(2FeO·SiO_2)_{70}Carb^*_{30}]_{100-x}CaPrv_x$ at 20 GPa. Experimental compositions are marked with black points. P—peritectic point L + Rwd + MWus + Sti + CaPrv + Carb* with the key reaction L + Rwd = Sti + MWus

Physicochemical behaviour of ferrobridgmanite $(Mg,Fe)SiO_3$ is bound to be find by melting relations of the native mantle and diamond-producing $MgSiO_3$–$(FeO·SiO_3)$–$(\pm Carb^*)$ systems in experiments at 26 GPa (Table 8.4). It is anticipated a determination of the physicochemical mechanisms capable of providing the ultrabasic-basic evolution of the diamond-producing melts and formation of the corresponding bridgmanite- and stishovite-bearing subsolidus mineral assemblages (Fig. 8.13).

The main objective is to study in experiments at 26 GPa the physicochemical behavior of ferrobridgmanite $(Mg,Fe)SiO_3$ by evaluation of melting relations of the representative native lower mantle system periclase Per–stishovite Sti–wustite Wus–Ca-perovskite CaPrv and the diamond-producing system Per–Sti–Wus–CaPrv–Carb*–C at their polythermal sections. It would be expected the establishment of the physicochemical mechanisms capable of providing the ultrabasic-basic evolution of the lower mantle magmatism and diamond-producing melts with formation of the subsolidus (ferropericlase + bridgmanite)- and (magnesiowustite + stishovite)-bearing mineral assemblages (Litvin and Spivak 2019). Experimental results for the melting relations at the polythermal section $(MgO)_{49}(FeO)_{21}CaPrv_{30}$–$(SiO_2)_{49}(FeO)_{21}CaPrv_{30}$ of the native lower mantle system are presented in the Fig. 8.14. A quenched material of the complete melt is demonstrated by Fig. 8.15. Ferropericlase FPer and stishovite Sti (Fig. 8.15c) are identified as the liquidus phases. With decreasing temperature, the liquidus phases are joined by FBrd within the field FPer + FBrd + L (Fig. 8.15a) and then the near-solidus field

Table 8.4 Experimental results, conditions, experimental phases compositions, and estimations of equilibrium mineral associations at polythermal section $(MgO)_{49}(FeO)_{21}(CaSiO_3)_{30}$–$(SiO_2)_{49}(FeO)_{21}(CaSiO_3)_{30}$ of root system of th lower mantle MgO–FeO–SiO_2–$CaSiO_3$ at 26 GPa

# Sampl.	T, °C	T, min	Experimental results							
			Phase association	Phase	MgO	FeO	SiO₂	CaO	Sum	
					wt%					
$(MgO)_{39.2}(SiO_2)_{9.8}(FeO)_{21}(CaSiO_3)_{30}$										
H4074-1a	2500	5	L	L	32.31	27.88	28.53	10.62	99.34	
H4060a	2200	10	L + (Per/Wus)ss	L	20.79	27.16	32.78	19.13	99.87	
				FPer	64.38	34.51	0.49	0.48	99.80	
H4065a	2100	10	L + (Per/Wus)ss + FBrd	L	22.24	21.47	21.00	35.22	99.92	
				MWus	43.06	55.61	0.28	0.34	99.29	
				FBrd	33.19	11.86	54.38	0.54	99.97	
H4070-1c	2000	20	L + (Per/Wus)ss + FBrd +CaPrv	L	32.50	34.66	9.26	23.26	99.67	
				MWus	42.89	56.22	0.16	0.59	99.86	
				FBrd	35.79	6.28	57.37	0.42	99.83	
				CaPrv	0.67	0.93	50.50	47.84	99.95	
H4072-1a	1800	30	(Per/Wus)ss+ FBrd + CaPrv	MWus	43.55	55.04	0.49	0.31	99.39	
				FBrd	34.86	7.70	56.83	0.38	99.76	
				MWus	9.39	88.51	1.04	0.51	99.52	
				CaPrv	0.49	0.66	49.37	48.80	99.32	
H4073-1c	1650	30	(Per/Wus)ss+ FBrd + CaPrv	FBrd	31.70	10.08	55.26	2.84	99.88	
				MWus	11.70	86.73	0.81	0.69	99.93	
				CaPrv	0.52	0.63	55.52	47.96	99.62	

(continued)

Table 8.4 (continued)

# Sampl.	T, °C	T, min	Experimental results Phase association	Phase	MgO wt%	FeO	SiO$_2$	CaO	Sum
(MgO)$_{24.5}$(SiO$_2$)$_{24.5}$(FeO)$_{21}$(CaSiO$_3$)$_{30}$									
H4073-1b	2200	5	L	L	19.86	25.88	40.85	13.09	99.68
H4065b	2100	10	L + (Wus/Per)$_{ss}$ + Sti	L	29.66	13.64	27.62	29.01	99.94
				MWus	19.83	79.47	0.48	0.05	99.83
				Sti	0.29	0.72	98.91	–	99.92
H4070-1c	2000	20	L + (Wus/Per)$_{ss}$ + FBrd + Sti	L	28.40	7.22	6.71	57.51	99.83
				MWus	11.55	86.63	1.30	0.45	99.93
				FBrd	32.88	11.23	55.47	0.32	99.90
				Sti	0.17	0.18	99.96	0.12	100.43
H4073-1b	1650	30	(Wus/Per)$_{ss}$ + FBrd + Sti + CaPrv	MWus	8.57	88.03	2.11	0.91	99.61
				FBrd	33.29	8.05	58.04	1.09	100.47
				Sti	0.01	0.48	98.94	0.18	99.61
				CaPrv	1.53	0.91	49.11	48.25	99.80
(MgO)$_{9.8}$(SiO$_2$)$_{39.2}$(FeO)$_{21}$(CaSiO$_3$)$_{30}$									
H4074-1c	2500	5	L + Sti	L	9.91	30.51	41.26	17.99	99.67
				Sti	0.05	1.89	97.60	0.52	100.06
H4060c	2200	10	L + Sti + (Wus/Per)$_{ss}$	L	10.25	4.97	56.12	28.45	99.78
				Sti	–	2.13	97.14	0.72	99.99
				MWus	11.29	86.97	0.91	0.49	99.65

(continued)

Table 8.4 (continued)

# Sampl.	T, °C	τ, min	Experimental results							
			Phase association	Phase	MgO	FeO	SiO$_2$	CaO	Sum	
					wt%					
H4070-1c	2000	20	$L + Sti + (Wus/Per)_{ss}$	L	14.33	8.68	42.33	34.50	99.83	
				Sti	–	0.48	98.92	0.08	99.48	
				$MWus$	7.86	83.89	0.98	0.39	93.62	
H4073-1a	1650	30	$(Wus/Per)_{ss} + FBrd + Sti + CaPrv$	$FBrd$	31.81	12.33	55.46	0.18	99.78	
				$MWus$	8.58	88.35	2.06	0.66	99.66	
				Sti	0.06	0.81	98.93	0.69	100.49	
				$CaPrv$	0.21	1.91	49.61	47.77	99.50	

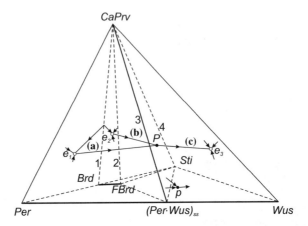

Fig. 8.12 Subsolidus and liquidus structures of the original lower-mantle system Per–Wus–Sti–CaPrv at 26 GPa. The subsolidus volume (Per–Sti–Wus–CaPrv) is divided by triangulation procedure (dotted lines) onto the simplex phase volumes and indicated as **1**: Per–(Brd + FBrd)–(Per/Wus)ss–CaPrv; **2**: (Brd + FBrd)–Sti–CaPrv; **3**: FBrd–Sti–(Per/Wus)ss–CaPrv and **4**: (Per/Wus)ss–Sti + Wus–CaPrv. The Brd–FBrd solid solutions are shown with solid thick line. The general liquidus topology of the system (solid thin lines) is determined with the univariant cotectic curves (**a**) L + (Per/Wus)ss + FBrd + CaPrv (at ultrabasic simplex compositions) and (**b**) L + Sti + FBrd + CaPrv (at basic simplex compositions) which are running through quasi-invariant peritectic point **P** to the univariant curve (**c**) L + Sti + (Wus/Per)ss + CaPrv (at basic simplex compositions). The peritectic point P: L + (Wus/Per)ss + FBrd + Sti + CaPrv is founded on the reaction L + FBrd = Sti + (Wus/Per)ss known as the effect "stishovite paradox" (Litvin 2014). Also shown are the invariant points of the boundary ternary systems as eutectic—**e1** L + Per + Brd + CaPrv, **e2** L + Brd + Sti + CaPrv, **e3** L + Sti + Wus + CaPrv and peritectic **p**: L + Brd = Sti + MWus. Arrows—for directions of temperature lowering at univariant curves

FPer + FBrd + CaPrv + L (Fig. 8.15c) is formed. The Sti-bearing assemblies are presented by the fields MWus + Sti + L (Fig. 8.15d) and the near-solidus field FPer + FBrd + Sti + L. At the solidus conditions, the quasi-invariant assemblage FBrd + MWus + Sti + CaPrv + L (Fig. 8.15d) with peritectic reaction FBrd + L → FPer/MWus + Sti is formed. After a complete reactional loss of FBrd in the peritectic point P, the solidus univariant curve MWus + Sti + CaPrv + L came into being. The subsolidus phase field FPer + FBrd + CaPrv is composed of the most representative rock-forming minerals for the primary lower mantle ultrabasic material. With increasing SiO_2 content in the system, the basic assemblages FBrd + FPer/MWus + Sti + CaPrv and MWus + Sti + CaPrv are formed.

The diamond-producing system periclase Per–stishovite Sti–wustite Wus–Ca-perovskite CaPrv–carbonate Carb*–C carbon (where the symbol Carb* is for the Mg–Fe–Ca–Na–carbonate constituent) (Fig. 8.15) has a similar topology with the lower mantle system periclase Per–stishovite Sti–wustite Wus–Ca-perovskite CaPrv (Fig. 8.12). The diagram for the melting relations at the polythermal section $(MgO)_{20}(FeO)_{15}CaPrv_{15}Carb*_{50}$–$(SiO_2)_{20}(FeO)_{15}CaPrv_{15}Carb*_{50}$ of the diamond-producing system is (Fig. 8.16) is based on experimental data (Litvin and Spivak

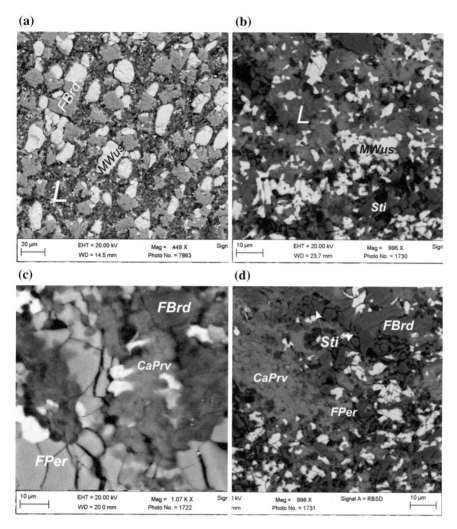

Fig. 8.13 SEM images of experimental samples of the polythermal section $(MgO)_{49}(FeO)_{21}(CaSiO_3)_{30}$–$(SiO_2)_{49}(FeO)_{21}(CaSiO_3)_{30}$ at 26 GPa after quenching. Starting compositions: **a, c** $(MgO)_{39.2}(FeO)_{21}(SiO_2)_{9.8}(CaSiO_3)_{30}$; **b** $(MgO)_{24.5}(FeO)_{21}(SiO_2)_{24.5}(CaSiO_3)_{30}$; **d** $(SiO_2)_{39.2}(FeO)_{21}(MgO)_{9.8}(CaSiO_3)_{30}$. FPer = (Per·Wus)ss]; MWus = (Wus·Per)ss

2019). Crystallization of diamonds and genetically associated minerals under conditions of the ultrabasic-basic evolution of the lower mantle diamond-parental silicate-carbonate melts is attended with the bridgmanite-and-melt peritectic reaction resulted in formation of the magnesiowustite-and-stishovite assemblage (Fig. 8.17) that is the effect of "stishovite paradox" (Litvin 2014).

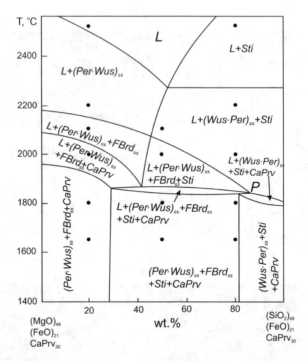

Fig. 8.14 Melting phase diagram for the polythermal section $(MgO)_{49}(FeO)_{21}(CaSiO_3)_{30}$–$(SiO_2)_{49}(FeO)_{21}(CaSiO_3)_{30}$ at 26 GPa. Experimental compositions are marked with black points. P—quasi invariant peritectic point L + FBrd + [(Per/Wus)ss ↔ (Wus/Per)ss] + Sti + CaPrv with the key reaction L + FBrd = Sti + (Wus/Per)ss characterizes the effect of "stishovite paradox" (Litvin 2014)

8.4 Discussion and Concluding Remarks

The genetic ratios between the native ultrabasic and basic rocks have been consistently of chief interest for the upper mantle petrology and geochemistry (Yoder and Tilley 1962; O'Hara and Yoder1967; O'Hara 1968; Ringwood 1975; Yoder 1976). The ratios have eluded intelligibility in the absence of a knowledge of the peritectic reactions of orthopyroxene and olivine with their complete eliminations in the native upper mantle multicomponent ultrabasic-basic systems. It has been possible to reveal the mechanisms for the reactions exclusively in physicochemical experiments under high pressures and temperatures. The multysimplex composition of the native multicomponent systems calls for the regime of fractional magma crystallization in order that the both peritectic mechanisms for orthopyroxene and olivine can be realized. The fractional crystallization is extremely significant by virtue of the fact that contents of the relatively fusible components of Fe and Na increase appreciably in the upper mantle eclogitic clinopyroxenes, garnets and their melts in relation to the peridotitic ones (Dawson 1980; Sobolev 1977). The contents of jadeitic component $NaAlSi_2O_6$ in clinopyroxenes of garnet lherzolites and websterites are variable

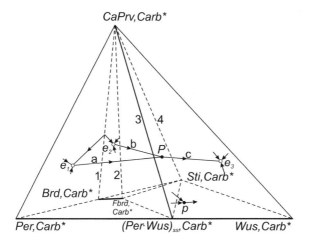

Fig. 8.15 Subsolidus and liquidus structures of the lower-mantle diamond-producing system Per–Wus–Sti–CaPrv–Carb* at 26 GPa. The subsolidus volume (Per–Sti–Wus–CaPrv–Carb*) is divided by triangulation procedure (dotted lines) onto the simplex phase volumes indicated as **1**: Per–(Brd + FBrd)–(Per/Wus)ss–CaPrv–Carb*; **2**: (Brd + FBrd)–Sti–CaPrv–Carb*; **3**: FBrd–Sti–(Per/Wus)ss–CaPrv–Carb* and **4**: (Per/Wus)ss–Sti + Wus–CaPr–Carb*. The Brd–FBrd solid solutions are shown with solid thick line. The general liquidus topology of the system (solid thin lines) is determined with the univariant cotectic curves **a** L + (Per/Wus)ss + FBrd + CaPrv + Carb* (at ultrabasic simplex compositions) and **b** L + Sti + FBrd + CaPrv + Carb* (at basic simplex compositions) which are running through quasi-invariant peritectic point P to the univariant curve **c** L + Sti + (Wus/Per)ss + CaPrv + Carb* (at basic simplex compositions). The peritectic point P: L + (Wus/Per)ss + FBrd + Sti + CaPrv + Carb* is founded on the reaction L + FBrd + Carb* = Sti + (Wus/Per)ss + Carb*. Also shown are the invariant points of the boundary ternary systems as eutectic—**e1**: L + Per + Brd + CaPrv + Carb*, **e2**: L + Brd + Sti + CaPrv, **e₃** L + Sti + Wus + CaPrv + Carb* and peritectic **p**: L + Brd = Sti + MWus + Carb*. Arrows—for directions of temperature lowering at univariant curves

within 10–35 wt% whereas in eclogitic omphacites may be as much as 15–70 wt%. Some overlaps of the data are attributable to the difficulties in distinguishes between pyroxenites and olivine-eclogites (Dawson 1980). All this is compatible with the scenario of fractional evolution of the ultrabasic-basic magma when jadeitic component being distributed between clinopyroxene/omphacite minerals and melts. At the same time the fractional removal together with clinopyroxene of the Na-free phases of olivine and garnet from the ultrabasic melts and then the garnet together with omphacite from the basic melts makes the residual melts richer in jadeitic component. In consequence of this, the jadeitic component concentration can increase as in the residual melts so in the fractionating clinopyroxene and omphacite phases.

As a consequence of the joint action of olivine-jadeite peritectic reaction and fractional crystallization of the residual melts, the system olivine–diopside–jadeite–garnet with high content of the boundary jageitic component is newly formed. It would be expected that its liquidus structure has assumed a decisive importance in the ultrabasic-basic evolution of the upper mantle magmatism. It is valid to say that

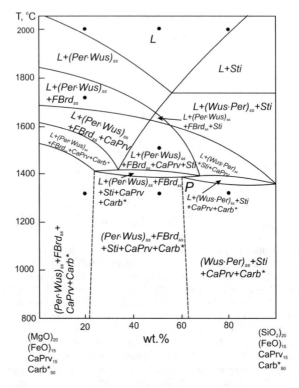

Fig. 8.16 Melting phase relations for the polythermal section $(MgO)_{20}(FeO)_{15}CaPrv_{15}Carb*_{50}$—$(SiO_2)_{20}(FeO)_{15}CaPrv_{15}Carb*_{50}$ at 26 GPa (Litvin and Spivak 2019)

the regime of fractional magma evolution has culminated in the rise of the olivine–diopside–jadeite–garnet system as the new ultrabasic-basic simplex which is not be taken into account in the equilibrium complex of ultrabasic and basic simplexes (Figs. 8.1 and 8.5). Melting relations at the boundary ternary joins for the olivine-diopside–jadeite–garnet system based on the available experimental data (Davies 1964; Davies and Shairer 1965; Gasparik and Litvin 1997) are schematically displayed by the development of its tetrahedral simplex in Fig. 8.8. The liquidus structures of the boundary joins at the development make possible the constructing of the generalized liquidus structure for the diagram of the utrabasic-basic tetrahedral simplex olivine–diopside–jadeite–garnet (Fig. 8.7). It is able to verify that the univariant curve Ol + Cpx + Grt + L being passed from the boundary eutectic point E (Ol + Cpx + Grt + L) of the boundary ultrabasic Ol–Di–Grt join through the simplex volume has connected with the peritectic point P (Ol + Cpx/Omph + Grt + L) of the ultrabasic-basic Ol–Di–Jd join. The reaction of "olivine garnetization" is developing with growing the jadeitic component concentration at the fractionary-motivated varying melt compositions at the Ol + Cpx + Grt + L curve and brought

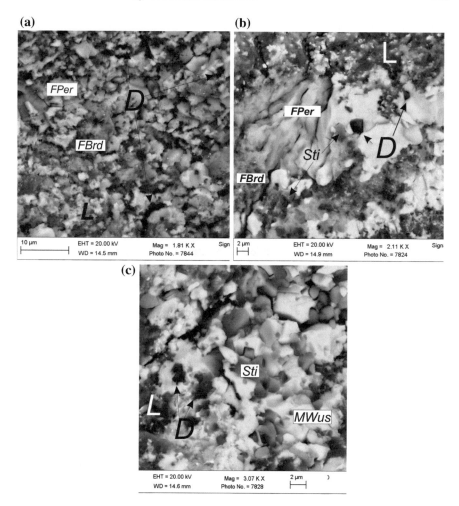

Fig. 8.17 SEM images of experimental samples demonstrative for crystallization of diamonds and associated phases in the melts-solutions oversaturated with dissolved carbon of graphite source in the polythermal section $(MgO)_{30}(FeO)_{20}Carb^*_{50}–(SiO_2)_{30}(FeO)_{20}Carb^*_{50}$ at 26 GPa after quenching. Starting compositions: **a** $(MgO)_{12.6}(FeO)_{12}(SiO_2)_{5.4}Carb^*_{30}G_{40}$; **b** $(MgO)_{09}(FeO)_{12}(SiO_2)_9Carb^*_{30}G_{40}$; **c** $(MgO)_{5.4}(FeO)_{12}(SiO2)_{12.6}Carb^*_{30}G_{40}$

to completion in the peritectic point P. As a result of the reaction, the olivine components die out, and ultrabasic compositions of the clinopyroxenes transfer into basic omphacitic.

Hence the qualitative changes being able to producing new physicochemical potentialities for the compositional evolution of the upper mantle geochemical and petrological systems have contributed with the jadeitic component. The liquidus structure of the system olivine–diopside–jadeite–garnet, being came into existence due to fractional crystallization of the ultrabasic magmas, could play a crucial role as a physico-chemical "bridge" between the ultrabasic peridotite–pyroxenitic and basic eclogitic compositions of the material of the upper mantle garnet-peridotitic facies. All this governs the ultrabasic-basic magmatism evolution and petrogenesis of the combined series of the olivine-normative peridotite–pyroxenite and silica-normative eclogite-grospydite rocks of the Earth's upper mantle.

The physicochemical mechanisms, which are produced at the peritectic points for orthopyroxene and olivine, make it possible to construct the syngenesis phase diagrams for diamonds and genetically associated mineral phases of ultrabasic and basic parageneses under conditions of fractional crystallization of the mantle diamond-producing systems (Litvin 2017). The syngenesis diagram for the upper mantle peridotite Prd–carbonate Carb–diamond D (Fig. 8.18) includes in its boundary composition the starting ultrabasic material (Prd + Carb), which melts experience a fractional crystallization. The resulted basic subsolidus assemblage (Ky + Coes-Ecl + Carb) is placed within square brackets with arrow as the symbol of fractional crystallization. The sequence of the phase fields reflects the action of the physicochemical mechanisms which are responsible for the continuous ultrabasic-basic transfer of diamond producing silicate-carbonate growth melts-solutions with a consecutive formation of minerals of the peridotitic and eclogitic parageneses. The syngenesis phase diagram for diamond and minerals of ultrabasic and basic parageneses under conditions of fractional crystallization of the ultrabasic lower-mantle diamond-producing system FPer–FBrd–CaPrv–Carb–D, resulting in formation of the MWus + Sti + CaPrv + Carb + D subsolidus assemblage, is presented in the Fig. 8.19.

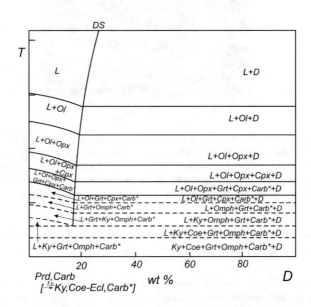

Fig. 8.18 Syngenesis phase diagram for diamond and minerals of ultrabasic and basic parageneses under conditions of fractional crystallization of the upper-mantle diamond-producing system peridotite Prd–Carb–D

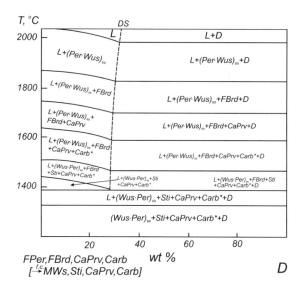

Fig. 8.19 Syngenesis phase diagram for diamond and minerals of ultrabasic and basic parageneses under conditions of fractional crystallization of the lower-mantle diamond-producing system Per–FBrd–aPrv–Carb–D

It can be seen that the possibility of the ultrabasic-basic magmatic evolution occurring at the deeper Earth's horizons—as transition zone as lower mantle—can be solved in the context of relatively simpler physicochemical scenarios. Under the transition zone conditions, the peritectic reaction of ringwoodite and Fe-richer magma with formation of the magnesiowustite and stishovite assemblage became a moving force of the ultrabasic-basic magma evolution. In the case of the lower mantle, the peritectic reaction of bridgmanite and Fe-richer magma with formation of the magnesiowustite and stishovite phases has to approach the problem. Nonetheless, the ultrabasic-basic evolution in both of the cases have been only made effective with the regime of fractional crystallization of the deep mantle magmas.

It is symptomatic that the peritectic reactions of olivine, orthopyroxene, ringwoodite and bridgmanite with coexisting melts at the corresponding upper mantle, transition zone and lower mantle depths reveal themselves in the diamond-and-inclusions-producing reactions generated by the multicomponent silicate–(±oxide)–carbonate–carbon systems. These peritectic reactions are responsible for the ultrabasic-basic evolution of the diamond-parental melts and genesis of diamonds and mineral phases associated with diamonds in primary inclusions and diamond-bearing peridotite, pyroxenite and eclogite rocks in accordance with the mantle-carbonatite theory (Litvin 2017; Spivak and Litvin 2019).

Acknowledgements This work is supported by the Program of the Russian Academy of Sciences № I.08.P "Physics of condensed substances and materials of new generation", and IEM RAS projects AAAA-A18-118020590140-7 and AAAA-A18-118021990093-9.

References

Akaogi M (2007) Phase transformations of minerals in the transition zone and upper part of the lower mantle. In: Ohtani E (ed) Advances in high-pressure mineralogy. Geological Society of America Special Paper, vol 421, pp 1–13

Bobrov AV, Litvin YuA (2011) Mineral equilibria of diamond-forming carbonate-silicate systems. Geochem Int 49(13):1267–1363

Bobrov AV, Dymshits AM, Litvin YuA (2009) Condition of magmatic crystallization of Na-bearing majoritic garnets in the Earth mantle: evidence from experimental and natural data. Geochem Int 47(10):951–965

Boyd FR, Danchin RV (1980) Lherzolites, eclogites, and megacrysts from some kimberlites of Angola. Am J Sci A280(2):528–548

Davies BTC (1964) The system diopside-forsterite-pyrope at 40 kbar. Carnegie Inst Wash Yb 63:165–171

Davies BTC, Shairer JF (1965) Melting relations in the join Di-Fo-Py at 40 kbar and 1 atm. Carnegie Inst Wash Yb 64:183–187

Dawson JB (1968) Recent researches on kimberlite and diamond geology. Econ Geol 63:504–511

Dawson JB (1980) Kimberlites and their xenoliths. Springer, Berlin, p 252

Dymshits AM, Bobrov AV, Litvin YuA (2015) Phase relations in the system $(Mg,Ca)_3Al_2Si_3O_{12}$–$Na_2MgSi_5O_{12}$ at 7.0 and 8.5 GPa and 1400–1900 °C. Geochem Int 53(1):9–18

Frost DJ, Poe BT, Tronnes RG, Libske C, Duba F, Rubie DC (2004) A new large-volume multianvil system. Phys Earth Planet Inter 143:507–514

Gasparik T, Litvin YuA (1997) Stability of $Na_2Mg_2Si_2O_7$ and melting relations on the forsterite-jadeite join at pressures up to 22 GPa. Eur J Mineral 9:311–326

Harte B (2010) Diamond formation in the deep mantle: the record of mineral inclusions and their distribution in relation to mantle dehydration zone. Mineral Mag 74(2):189–215

Irifune T, Tsuchiya T (2007) Mineralogy of the Earth–phase transitions and mineralogy of the lower mantle. In: Treatise on Geophysics. Elsevier, pp 33–62

Ito E, Takahashi E (1989) Postspinel transformations in the system Mg_2SiO_4–Fe_2SiO_4 and some geophysical implications. J Geophys Res 94:10637–10646

Kaminsky FV (2012) Mineralogy of the lower mantle: a review of "super-deep" mineral inclusions in diamond. Earth-Sci Rev 110(1–4):127–147

Kaminsky FV (2017) The Earth's lower mantle. Composition and structure. Springer Nature, 331 p

Klein-BenDavid O, Wirth R, Navon O (2006) TEM imaging and analysis of microinclusions in diamonds: a close look at diamond-growing fluids. Am Miner 91:353–365

Litvin YA (1991) Physicochemical study of melting relations of the deep-seated Earth's substance. Nauka, Moscow, 312 p (in Russian)

Litvin YA (2014) The stishovite paradox in genesis of ultradeep diamonds. Dokl Earth Sci 455(1):274–278

Litvin YA (2017) Genesis of diamonds and associated phases. Springer, Berlin, 137 p

Litvin YuA, Spivak AV (2019) Genesis of diamonds and paragenetic inclusions under lower mantle conditions: the liquidus structure of the parental system at 26 GPa. Geoch Int 57(2):134–150

Litvin VY, Gasparik T, Litvin YA (2000) The enstatite-nepheline system in experiments at 6.5–13.5 GPa: an importance of $Na_2Mg_2Si_2O_7$ for melting of the nepheline-normative mantle. Geochem Int 38(1):100–107

Litvin YA, Litvin VY, Kadik AA (2008) Experimental characterization of diamond crystallization in melts of mantle silicate-carbonate-carbon system at 7.0–8.5 GPa. Geochem Int 46:531–553

Litvin YuA, Spivak AV, Kuzyura AV (2016) Fundamentals of the mantle carbonatite concept of diamond genesis. Geochem Int 54:839–857

Litvin YA, Kuzyura AV, Limanov EV (2017) Olivine garnetization—the key to ultrabasic—basic evolution of upper-mantle magmatism: evidence from melting relations of olivine–clinopyroxene–jadeite system at 6 GPa. 9th High Pressure Mineral Physics Seminar (HPMPS-9), 24–28 Sept 2017, Saint Malo, France

Logvinova AM, Wirth R, Fedorova EN, Sobolev NV (2008) Nanometre-sized mineral and fluid inclusions in cloudy Siberian diamonds: new insights on diamond formation. Eur J Mineral 20:1223–1233

Maaloe S (1985) Principles of igneous petrology. Springer, 374 p

MacGregor ID, Carter JL (1970) The chemistry of clinopyroxenes and garnets of eclogite and peridotite xenoliths from Roberts Victormine, South Africa. Phys Earth Planet Inter 3:391–397

Marakushev AA (1984) Nodules of peridotites in kimberlites as indicators of the deep-seated structure of the lithosphere. In: Lectures of Soviet geologists at the XXYII session of the International Geological Congress. Petrology. Nauka, Moscow, pp 153–160 (in Russian)

Meyer HOA, Boyd FR (1972) Composition and origin of crystalline inclusions in natural diamonds. Geochim Cosmochim Acta 36:1255–1274

O'Hara MJ (1968) The bearing of phase equilibria studies in synthetic and natural systems on the origin and evolution of basic and ultrabasic rocks. Earth-Sci Rev 4:69–133

O'Hara MJ, Yoder HS (1967) Formation and fractionation of basic magmas at high pressures. Scott J Geol 3:67–117

Palatnik LS, Landau AI (1964) Phase equilibria in multicomponent systems. Holt, Rinehart and Winston Inc., New York, p 454

Rhines FN (1956) Phase diagrams in metallurgy: their development and application. McGraw-Hill, New York, p 348

Ringwood AE (1975) Composition and petrology of the Earth's mantle. McGraw-Hill, New York, p 618

Ringwood AE, Major A (1971) Synthesis of majorite and other high pressure garnets and perovskites. Earth Planet Sci Lett 12:411–418

Sobolev NV (1977) Deap-seated inclusions in kimberlites and the problem of the composition of the upper mantle. American Geophysics Union, Washington, p 279

Spivak AV, Litvin YA (2019) Evolution of magmatic and diamond-forming systems of the Earth's lower mantle. Springer, Berlin, 95 p

Tomioka N, Fujino K, Ito E, Katsura T, Sharp T, Kato T (2002) Microstructures and structural phase transition in (Mg, Fe)SiO$_3$ majorite. Eur J Mineral 14:7–14

Wang W (1998) Formation of diamonds with minerals inclusions of "mixed" eclogite and peridotite parageneseis. Earth Planet Sci Lett 160:831–843

Yoder HS (1976) Generation of basaltic magma. National Academy of Sciences, Washington

Yoder HS, Tilley CE (1962) Origin of basalt magmas: an experimental study of natural and synthetic rock systems. J Petrol 3:342–353

Zakharov AM (1964) Phase state diagrams of quadruple systems. Metallurgia, Moscow, p 240

Zedgenizov DA, Shatsky VS, Panin AV, Evtushenko OV, Ragozin AL, Kagi H (2015) Evidence for phase transitions in mineral inclusions in superdeep diamonds of the Sao Luiz deposit (Brazil). Rus Geol Geophys 56(1–2):384–396

Chapter 9
Formation of K–Cr Titanates from Reactions of Chromite and Ilmenite/Rutile with Potassic Aqueous-Carbonic Fluid: Experiment at 5 GPa and Applications to the Mantle Metasomatism

V. G. Butvina, S. S. Vorobey, O. G. Safonov, and G. V. Bondarenko

Abstract Magnetoplumbite (yimengite-hawthorneite, HAWYIM), crichtonite (lindsleyite-mathiasite, LIMA) and hollandite (priderite) minerals are exotic titanate phases, which formed during metasomatism at the conditions of high alkali activity, especially K, in the fluids in the upper mantle peridotites. The paper presents data on experiments on formation of K-end-members priderite, yimengite and mathiasite, as the result of the interaction of chromite, chromite + rutile and chromite + ilmenite assemblages in the presence of a small amount of silicate material with H_2O–CO_2–K_2CO_3 fluids at 5 GPa and 1200 °C. Cr-bearing Ba-free priderite, characteristic for metasomatized Cr-rich harzburgites, was firstly synthesized. The experiments demonstrated the principal possibility of the formation of the titanates in the reactions of chromite with alkaline aqueous-carbonic fluids and melts. However, the formation of these phases does not proceed directly on chromite, but requires additional titanium source. The relationship between titanates is found to be a function of the activity of the potassium component in the fluid/melt. Priderite is an indicator of the highest potassium activity in the mineral-forming medium. Titanates in the run products are constantly associated with phlogopite. Experiments prove that the formation of titanates manifests the most advanced or repeated stages of metasomatism in mantle peridotites. Association of titanates with phlogopite characterizes a higher activity of the potassium component in the fluid/melt than the formation of phlogopite alone. The examples from natural associations, reviewed in the paper, well illustrate these conclusions.

V. G. Butvina (✉) · O. G. Safonov · G. V. Bondarenko
D.S. Korzhinskii Institute of Experimental Mineralogy, Russian Academy Sciences, Academician Ossipian str. 4, Chernogolovka, Moscow Region, Russia 142432
e-mail: butvina@iem.ac.ru

S. S. Vorobey · O. G. Safonov
Geological Department, Lomonosov Moscow State University, Vorob'evy Gory, Moscow, Russia 119899

201

Y. Litvin and O. Safonov (eds.), *Advances in Experimental and Genetic Mineralogy*, Springer Mineralogy, https://doi.org/10.1007/978-3-030-42859-4_9

Keywords K-Cr titanates · Yimengite · Priderite · Mathiasite · Potassic fluid · Peridotite · Mantle metasomatism · High pressure experiment

9.1 Introduction

The term "mantle metasomatism" (Lloyd and Bailey 1975; Harte and Gurney 1975; Bailey 1982, 1987; Menzies and Hawkesworth 1987) combines the processes of transformation of mantle rocks under the influence of external fluids and/or melts, regardless of their origin and composition. The metasomatic processes in the upper mantle are responsible not only for great variability of parageneses in the mantle rocks, but also for formation of magmas specific in composition, such as kimber-lites, carbonatites, lamproites, kamafugites and others (e.g., Haggerty 1987). The type of mantle metasomatism, expressed in the formation of new phases, which are non-characteristic of the peridotites and eclogites, Harte (1983) referred to as the "modal mantle metasomatism". This process is commonly expressed in the forma-tion of amphiboles, phlogopite, apatite, carbonates, sulfides, titanite, ilmenite, rutile. A number of unique mineral phases form as the products of metasomatism in the mantle. Among them are minerals of magnetoplumbite (yimengite-hawthorneite, HAWYIM), crichtonite (lindsleyite-mathiasite, LIMA) and hollandite (priderite) groups (Haggerty 1991), i.e. rare K–Ba titanates, enriched in large-ion lithophile (LILE), high field strength (HFSE), light rare earth (LREE) elements, U and Th.

Yimengite K $(Cr, Ti, Mg, Fe, Al)_{12}O_{19}$ is the end-member of the magnetoplumbite group with general formula $AM_{12}O_{19}$ (Haggerty 1991). The 12-coordinated site A in the structure of this mineral is located in perovskite-like layers (AMO_3) and con-tains large cations (K, Ba and LILE), while small cations M (Ti, Cr, Fe, Mg, Zr, Nb, V, Zn) are located in 4 to- 6-coordinated polyhedra in spinel-like layers (Grey et al. 1987). Yimengite forms a limited solid solution with hawthorneite Ba (Cr, Ti, Mg, Fe, Al)$_{12}$O$_{19}$ (Haggerty et al. 1989). Natural minerals of yimengite-hawthorneite series (HAWYIM) usually do not correspond to an ideal formula, being characterized by significant variations component populations in both sites and vacancies in the struc-ture. Yimengite was first described in kimberlite dykes of the Shandong province, China (Dong et al. 1983) in association with olivine, chromian pyrope, chromite, phlogopite, ilmenite, chromian diopside, apatite, zircon, moissanite. As a product of modifications of chromite xenocrysts, yimengite was identified in the heavy con-centrates of the kimberlites of the Guaniamo area, Venezuela (Nixon and Condliffe 1989), and Float-Ouellet, Australia (Kiviets et al. 1998). In all cases, yimengite contains BaO (up to 3.4 wt% in yimengite from Venezuela), indicating a solid solu-tion with hawthorneite (Grey et al. 1987; Haggerty et al. 1989; Peng and Lu 1985). The compositional characteristics of chromite, after which yimengite forms, indicate that they belong to the assemblage of diamondiferous garnet harzburgites (Nixon and Condliffe 1989). Yimengite was discovered as the inclusions in diamonds (Sobolev et al. 1988, 1998; Bulanova et al. 2004), where it is also associated with the typical minerals of harzburgite paragenesis, i.e. chromite, chromian subcalcic garnet and enstatite. Inclusions of yimegite described by Bulanova et al. (2004) contain elevated Rb, Cs, Sr concentrations. According to some authors, yimengite is a product of reactions of diamond-bearing harzburgites at the base of the lithospheric continental mantle at depths of about 150 km with fluids enriched in K, HFSE, LREE. This is confirmed by the discovery of yimegite and hawthorneite in metasomatic veins, which cross harzburgite xenolith from a kimberlite pipe Bultfontein (South Africa) along with phlogopite, K-richterite, the LIMA phases, armalcolite, rutile, ilmenite (Haggerty et al. 1986).

Mathiasite is titanate of the crichtonite group $AM_{21}O_{38}$, where the position of A is characterized by isomorphism Ba $\leftrightarrow$ K (lindsleyite-mathiasite, LIMA). Firstly, lindsleyite and mathiasite were identified by Haggerty (1975) in the De Beers kimberlite pipe, South Africa. Subsequently, the minerals were described in the metasomatized peridotite xenoliths from other kimberlite pipes of South Africa (Erlank and Rickard 1977; Smyth et al. 1978; Jones et al. 1982; Konzett et al. 2000, 2013). In the metasomatized peridotites, it occurs both as end-members of the LIMA series and as intermediate members of the solid solution. Characteristic minerals associated with the LIMA are phlogopite, diopside, K-richterite, Nb–Cr-rutile, Mg–Cr–Nb-ilmenite and Mg–Cr spinel (Haggerty et al. 1983). The LIMA minerals are described as inclusions in xenocrysts of chromian pyrope (Rezvukhin et al. 2018) and in diamonds (Sobolev and Yefimova 2000) from kimberlites.

Priderite is titanate of hollandite group, solid solution in the system of $A^{2+}B^{2+}Ti_7O_{16}$–$A^+_2B^{2+}Ti_7O_{16}$–$A^{2+}B^{3+}_2Ti_6O_{16}$–$A^+_2B^{3+}_2Ti_6O_{16}$. The site A is occupied by Ba and K, as well as Na, Pb, Sr, Ca and REE; the site B holds Mg, Fe^{2+}, Fe^{3+}, Al, Cr, substituting Ti. Priderite was firstly described by Prider (1939) and Norrish (1951) in the Kimberly lamproites, Western Australia, and was subsequently found as a typomorphic mineral of leucite lamproites (Jaques et al. 1989). It is known in metasomatized peridotite xenoliths from kimberlites (Konzett et al. 2013; Giuliani et al. 2012; Naemura et al. 2015; Haggerty 1987), as well as diamond inclusions (Jaques et al. 1989). Chrome-dominant variety of priderite is found in the metasomatized peridotites only (Haggerty 1987, 1991; Konzett et al. 2013; Giuliani et al. 2012; Naemura et al. 2015).

Experimental data on the stability of K–Ba-titanates are limited to several studies. Podpora and Lindsley (1984) synthesized lindsleyite and mathiasite with compositions given by Haggerty et al. (1983) at 2 GPa/1300 °C and 2.2 GPa/900 °C. Foley et al. (1994) studied stability of priderite, its Fe^{3+} and Fe^{2+}-bearing varieties, as well as HAWYIM and LIMA phases in experiments on synthesis from oxide and simple titanate mixtures at 3.5, 4.3 and 5.0 GPa. At these pressures, priderite is stable up to temperatures of 1500 °C, HAWYIM—to 1150–1300 °C and LIMA—to 1200–1350 °C (Foley et al. 1994). There is no data on the synthesis of Cr-rich priderite, which is characteristic for metasomatized peridotites. Synthetic hawthorneite and yimengite in the system TiO_2–ZrO_2–Cr_2O_3–Fe_2O_3–MgO–BaO–K_2O are stable up to 15 GPa and 1400–1500 °C, and synthetic LIMA solid solutions are stable up to 11 GPa and 1400–1600 °C (Konzett et al. 2005). The experimental data indicate a very wide PT-range of K–Ba-titanates stability confirming the possibility of coexistence of these phases with diamond in the subcontinent upper mantle in the regions of generation of kimberlites and lamproites.

However, it is evident that not only temperature and pressure, but also specific chemical conditions are responsible for the stability of K–Ba-titanates. These minerals are formed when the ability to concentrate K and LILE in phlogopite and potassium richterite is exhausted. Formation of minerals of the HAWYIM, LIMA groups and priderite characterizes the highest degree of metasomatic changes in the conditions of high activity (concentration) of alkaline components, especially potassium in the fluids, noticeably larger than is necessary for the formation of phlogopite and potassium richterite (Safonov and Butvina 2016). The formation of these minerals is usually associated with peridotite reactions with alkali-rich fluids (melts) with low SiO_2 activity (Konzett et al. 2013; Giuliani et al 2012). The inclusions of such fluids are found in the MARID assemblages (Konzett et al. 2014), which represent the closest to those assemblages, in which K–Ba-titanates were identified. Thus, these minerals can be considered as indicators of the activity of high-alkaline aqueous-carbonic fluids or carbonate-salt melts in the upper mantle. However, there are no experimental or calculated data that would answer the question of how high the concentrations of alkali-salt components in fluids should be for the appearance of these minerals.

The paper reports results of experimental study of reactions of chromite, chromite + rutile and chromite + ilmenite forming priderite, yimengite and mathiasite with the participation of potassic aqueous-carbonic fluid at the upper-mantle P-T conditions.

9.2 Starting Materials, Experimental and Analytical Procedures

9.2.1 Starting Materials

As starting materials for experiments, mixtures (1:1 or 2:1 wt. ratios) of natural chromite with ilmenite or synthetic TiO_2 powder were used. The chromite with composition $(Mg_{0.49-0.54}Fe_{0.50-0.54}Mn_{0.01-0.02}Zn_{0.01-0.02})(Al_{0.17-0.20}Cr_{1.55-1.61}Fe_{0.10-0.22}Ti_{0.03-0.07})O_4$ (Table 1) was picked from a garnet lherzolite xenolith from the Pionerskaya kimberlite pipe, Arkhangelsk region. Ilmenite with composition $Fe_{0.98}Mg_{0.01}Mn_{0.06}Ti_{0.93}Al_{0.01}Nb_{0.01}O_3$ is a xenocrystal from the Udachnaya kimberlite pipe, Yakutia. Mixture of K_2CO_3 with oxalic acid (9:1; 7:3; 5:5; 3:7 and 1:9 ratios by weight) were used as a starting fluid component. A mixture chromite + TiO_2 was mixed with the fluid mixture as 4:1 and 9:1 by weight, whereas chromite + ilmenite as 9:1 by weight (Table 2). The starting mixtures were contained in platinum lense-like capsules with 0.2 mm wall thickness. The capsules were welded using the pulse Ar welding stage PUK-04, that allowed avoiding a loss of volatiles from the capsules.

Table 1 Chemical compositions (wt. %) of the initial minerals used in experiments on the synthesis of K-Cr titanates

Oxides	Chromite			Ilmenite	
TiO_2	1,96	1,23	2,8	46,81	45,10
Cr_2O_3	57,2	58,7	56,6	-	-
FeO*	25,7	26,4	22,9	44,25	43,39
Al_2O_3	4,10	4,24	4,80	0,78	0,25
MnO	0.75	0,18	0,58	2,99	2.65
MgO	10.10	9,49	10,50	0,58	0,49
Nb_2O_5	-	-	-	1,00	1,66
Sum	99,72	100,16	98,19	93,54	96,41
	per 4 O			**per 3 O**	
Ti	0,05	0,03	0,07	0,93	0,93
Cr	1,57	1,61	1,55	-	-
Fe^{3+}	0,15	0,15	0,10	0,09	0,07
Fe^{2+}	0,52	0,61	0,56	0,92	0,91
Al	0,18	0,17	0,20	0,03	0,05
Mn	0,02	0,01	0,02	0,06	0,06
Mg	0,53	0,49	0,54	0,03	0,02
Nb^{5+}	-	-	-	0,01	0,01

*$FeO=FeO+Fe_2O_3$

9.2.2 Experimental Procedure

The high-pressure high-temperature experiments were performed at 5 GPa and 1200 °C at the Korzhinskii Institute of Experimental Mineralogy using a toroidal anvil-with-hole apparatus of uni-axial compression by a 500-ton hydraulic press (Litvin 1991). High-pressure cells were manufactured from limestone and arranged into the space between the upper and lower anvils. Tubular heaters from graphite (length 8 mm, diameter 7.5 mm, wall thickness 0.75 mm) were disposed in the central spaces of the cells. The welded Pt capsules (diameter 4.0 mm, height 2.5 mm) with starting materials are placed at the heater centers between the electro-insulating holders being made from pressed mixtures of MgO and BN in the ratio 3:1. The temperature versus current power was calibrated using a $Pt_{70}Rh_{30}/Pt_{94}Rh_{06}$ thermocouple. The run temperatures were regulated using a MINITERM-300.31 controller with accuracy of ±0.5 °C. The quenching rate is close to 300 °C/s for the starting temperatures within 1600–1000 °C. The pressure versus press force calibration was performed against standard phase transitions in Bi (at 2.7 and 7.7 GPa) and Ba (5.5 GPa) with accuracy of 0.01–0.05 GPa (Hall 1971; Litvin et al. 1983). The accuracies of the HP-HT-experiments are estimated as ±0.1 GPa and ±20 °C, when the

P and T gradients within the cells and experimental samples are taken into account. Duration of experiments is varied (Table 2).

Oxygen fugacity was not specially controlled in the experiments being suggested that it was controlled by starting mineral phases, first of all, by chromite. The chromite composition allows estimation of $\log f_{O2}$ on the basis of the Fe^{3+}/Fe^{2+} ratio in this mineral. We used the dependence of the deviation of the oxygen fugacity of the system from the QFM buffer ($\Delta \log f_{O2}$) on the ratio Fe^{3+}/Fe^{2+} in chromite given by Nikitina et al. (2010). The Fe^{3+}/Fe^{2+} ratios in chromite were estimated from the crystal chemical formulas. The obtained values $\Delta \log f_{O2}$ (Table 3) indicate the oxygen fugacity in the experiments at 1.1–1.6 logarithmic unit below the QFM buffer.

Table 2 Run conditions and products of experiments

Run #	Starting mixture, (wt. ratio)	Fluid, (wt. ratios)	Fluid content in the system, %	Time, h.	Synthesis of priderite, yimengite, mathiasite, phlogopite
Sp1	Chromite	K_2CO_3	30	21	-,-,-, phl
Sp2	Chromite: rutile (1:1)	K_2CO_3: oxalic acid (9:1)	20	23	+,-,-, phl
A1	Chromite: rutile (1:1)	K_2CO_3: oxalic acid (9:1)	10	20	+,-,-, phl
A2	Chromite: rutile (2:1)	K_2CO_3: oxalic acid (9:1)	10	24	+,-,-, phl
B1	Chromite: ilmenite (1:1)	K_2CO_3: oxalic acid (9:1)	10	22	+,+,-, phl
B1-1	Chromite: ilmenite (1:1)	K_2CO_3: oxalic acid (7:3)	10	22	-,+,+, phl
B1-2	Chromite: ilmenite (1:1)	K_2CO_3: oxalic acid (5:5)	10	22	-,+,-, phl
B1-3	Chromite: ilmenite (1:1)	K_2CO_3: oxalic acid (3:7)	10	22	-,-,-, phl
B1-4	Chromite: ilmenite (1:1)	K_2CO_3: oxalic acid (1:9)	10	22	-,-,-, phl
B2	Chromite: ilmenite (2:1)	K_2CO_3: oxalic acid (9:1)	10	20	+,+,-, phl

Table 3 Representative analyses of chromite in the run products

№	Sp2-1	A1-1	A2-6	B1-11	B2-4
Ox/Mineral	Chr	Chr	Chr	Chr	Chr
TiO_2	2,67	3,64	3,61	4,62	4,16
Cr_2O_3	58,12	57,76	56,01	53,43	58,25
FeO	18,60	17,99	20,23	24,67	21,22
Al_2O_3	5,14	4,55	5,73	5,10	5,61
K_2O	0,24	0,20	0,10	0,38	0,20
MnO	0,60	0,29	0,81	0,39	0,61
MgO	11,51	12,42	11,92	9,61	11,48
ZnO	0,15	0,12	0,10	0,12	0,10
Sum	97,03	96,97	98,51	98,30	101,63
	Formula quantities per 3 cations				
Ti	0,07	0,09	0,09	0,12	0,10
Cr	1,57	1,56	1,49	1,45	1,51
Fe^{3+}	0,09	0,08	0,11	0,13	0,08
Fe^{2+}	0,44	0,43	0,46	0,58	0,51
Al	0,21	0,18	0,23	0,21	0,22
K	0,01	0,01	0,00	0,02	0,01
Mn	0,02	0,01	0,02	0,01	0,02
Mg	0,59	0,63	0,60	0,49	0,56
Zn	0,00	0,00	0,00	0,00	0,00
φ_{Sp}*	0,170	0,157	0,193	0,183	0,136
$\Delta logf_{O2}$**	-1,3	-1,4	-1,1	-1,2	-1,6

*$Fe^{3+}/(Fe^{3+}+Fe^{2+})$ ratio in chromite
** $\Delta logf_{O2} = (logf_{O2} - logf_{O2}^{QFM})$ (Nikitina et al., 2010)

9.3   Analytical Methods

9.3.1   Microprobe Analyses

Each run sample was embedded in epoxy and polished. After preliminary examination in reflected light, the microscopic features of run products Microprobe analyses of minerals were performed using CamScan MV2300 (VEGA TS 5130MM) electron microscope equipped with EDS INCA Energy 350 and Tescan VEGA-II XMU microscope equipped with EDS INCA Energy 450 and WDS Oxford INCA Wave 700 at the Korzhinskii Institute of Experimental Mineralogy. Analyses were performed at 20 kV accelerating voltage with a beam current up to 400 pA, spot size 115–150 nm and a zone of "excitation" of 3–4 μm diameter. Counting times was 100 s for all elements. The ZAF matrix correction was applied.

9.3.2 Raman Spectroscopy

The Raman spectra were obtained using the Renishaw RM1000 micro-scope/spectrometer equipped with the diode-pumped modular laser 532 nm. The typical parameters of measurements are following: laser output power 22 mW, slit 50 mm, collection time 100 s. The alignment of the spectrometer was checked before run by taking spectra of high-purity monocrystalline Si.

9.4 Experimental Results

9.4.1 Phase Relations

Natural Cr-rich K–Ba-titanates are closely related to chromite. Therefore, in order to find out a possibility for formation of any of these phases in direct interaction of chromite with the potassic fluid, an experiment with a mixture of chromite with K_2CO_3 (run Sp1; Table 2) has been performed. However, no any potassium-bearing phases were identified in the products of this experiment. The principal conclusion from this experiment is necessity of additional Ti-bearing phases which associated with chromite for the formation of titanates.

In fact, interaction of chromite with the K_2CO_3–H_2O–CO_2 fluid in presence of rutile (runs Sp2, A1, A2; Table 2) resulted in formation of priderite. It forms individual anhedral and subhedral grains, as well as tetragonal prisms and di-tetragonal di-pyramids up to 40 μm in size (Figs. 1 and 2a). Priderite also forms inclusions in rutile (Fig. 2b). Formation of priderite results in changes in the chromite composition, which becomes poorer in FeO + Fe_2O_3 and Cr_2O_3, but richer in TiO_2 compared to the starting chromite (Table 1). Using composition of phases in the run Sp2 products

Fig. 1 Tetragonal crystals of priderite (run A, Table 2) in the system chromite-rutile-K_2CO_3–H_2O–CO_2 at 5 GPa and 1200 °C

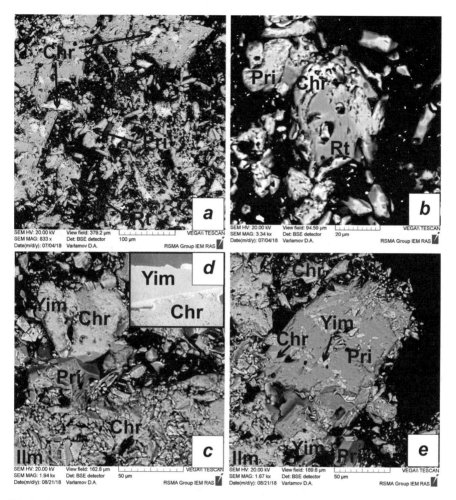

Fig. 2 Run products of the experiment B1 (Table 2) in the system chromite-rutile/ilmenite-K_2CO_3–H_2O–CO_2 at 5 GPa and 1200 °C. **a, b** Experiments A1 and A2 (Table 2): priderite, chromite, rutile; **c, e** experiment B1 (Tables 2 and 5): chromite, ilmenite, priderite, yimengite; **d** intergrowth of yimengite and chromite from the kimberlite sill Prospect 039, Guaniamo province, Venezuela (Nixon, Condliffe, 1989)

(Table 5), the following equilibrium represents the formation of priderite:

$$24.50Chr_2 + 5.79TiO_2 + K_2CO_3 + 0.46O_2 + 1.10Cr_2O_3$$
$$= 1.17Pri + 9.45Chr + 14.57Chr_1 + CO_2 \tag{1}$$

The equilibrium demonstrates formation of priderite with variations of the chromite composition. Slight oxidation is needed to accommodate Fe^{3+} in the forming priderite. Some excess of Cr_2O_3 demonstrates either non-stoicheometry of priderite or formation of additional oxides in the run products. In fact, some grains of unidentified Cr and Ti-bearing oxides were detected in the run products.

Other titanates are absent in the runs with rutile at any chromite/rutile ratios. The possible reason is an insufficient amount of Fe, first of all, Fe^{3+}. This suggestion is supported by the experiments in the chromite-ilmenite-K_2CO_3-H_2O-CO_2 system (runs B1, B1-1, B1-2, B1-3, B1-4, B2; Table 2). Products of these experiments contain yimengite along with priderite. The titanates are associated with chromite, ilmenite, as well as with newly formed rutile (compositions see in Table 4). Anhedral or subhedral grains of priderite of 10–100 µm in size locally contain inclusions of chromite and ilmenite. Yimengite is also found as inclusions in priderite (Fig. 2e). This phase forms intergrowths with chromite, which closely resemble natural examples (Nixon and Condliffe 1989) (Fig. 2c, d). The chromite/ilmenite ratio does not influence on the titanate crystallization.

However, their formation is affected by the $K_2CO_3/(H_2O + CO_2)$ ratio in the starting fluid mixture at the constant fluid content in the system. Experiments of the series B1, B1-1, B1-2, B1-3, B1-4 and B2 (Table 2) were performed using starting mixtures with variable wt. $K_2CO_3/(H_2O + CO_2)$ ratios and allowed evaluation of a sequence of the formation of various K–Cr-titanates and their assemblages on the fluid composition (Table 2). The assemblage priderite + yimengite form at the ratio $K_2CO_3/(H_2O + CO_2) = 9/1$ (Table 2). Using composition of phases in the run B1 products (Table 2), a number of equilibria could be written to illustrate the formation of this titanate assemblage, for example:

$$Rt + 0.11Chr + 0.12K_2CO_3 + 0.21Ilm_1 = 0.11Pri + 0.05Yim$$
$$+ 0.03Cr_2O_3 + 0.13Ilm_2 + CO_2 \qquad (2)$$

The equilibrium demonstrates formation of priderite and yimengite with variations of the ilmenite composition. Again, slight excess of Cr_2O_3 demonstrates either non-stoicheometry of titanates or formation of additional oxides in the run products. In fact, some grains of Cr-bearing (0.2 a.p.f.u.) rutile were detected in the run products.

Decrease of the ratio $K_2CO_3/(H_2O + CO_2)$ to 7/3 leads to disappearance of priderite, but yimengite actively forms as subhedral hexagonal crystals of 10–100 µm in size (Fig. 3). K-poor titanate, mathiasite, appears along with yimengite. Yimengite only forms at $K_2CO_3/(H_2O + CO_2) = 5/5$. Using representative composition of phases in the run B1-1 products (Table 2), a number of equilibria could be written to illustrate the formation of this titanate assemblage, for example:

$$5.29Chr + 9.87Ilm + 1.08K_2CO_3 + 7.61TiO_2$$
$$= Yim_1 + 1.14Yim_2 + 0.68Cr_2O_3 + 8.09Ilm_3 \qquad (3)$$

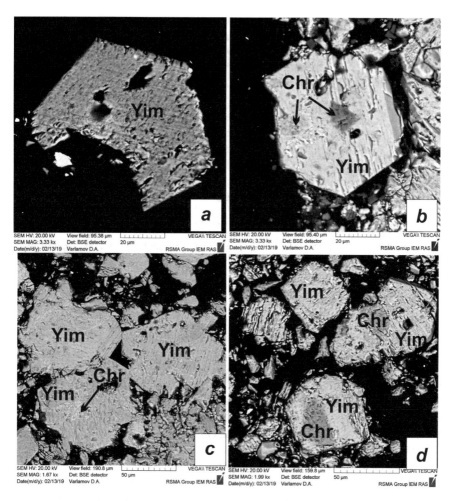

Fig. 3 Products of the experiment B1-1 and B1-2 (Table 2) in the system chromite-ilmenite-K_2CO_3–H_2O–CO_2 at 5 GPa and 1200 °C: hexagonal crystals of yimengite

The equilibrium demonstrates formation of yimengite with variations of the ilmenite composition (it becomes more magnesian). The excess of Cr_2O_3 and TiO_2 demonstrates possible non-stoicheometry of titanates or formation of additional non-detectible oxides in the run products.

No titanates are observed at lower $K_2CO_3/(H_2O + CO_2)$ ratios (Table 2).

The products of all experiments contain some phlogopite, which was formed, probably, due to a presence of various silicate phases – inclusions or intergrowths in the starting ilmenite and chromite. Usually, it forms aggregates between grains of starting and newly formed phases being closely associated with K-bearing titanates (Fig. 4). However, phlogopite is present even in run products, which do not contain titanates (experiments Sp1, B1-3 and B1-4; Table 2).

Fig. 4 Products of experiment B2 (Table 2) in the system chromite-ilmenite-K_2CO_3–H_2O–CO_2 at 5 GPa and 1200 °C containing phlogopite

9.4.2 Phase Compositions

Representative microprobe analyses of synthesized titanates are given in Tables 4 and 5. The ratios of Fe ($Fe^{2+} + Fe^{3+}$), Ti and Cr in formulas of priderite synthesized in systems both with rutile and ilmenite differ insignificantly (Fig. 5). Compositions of yimengite produced in the reactions of chromite and ilmenite with the fluid form the trend, which reflect the isomorphism ($Fe^{2+} + Fe^{3+}$) + Ti $\leftrightarrow$ Cr at the nearly constant ($Fe^{2+} + Fe^{3+}$)/Ti ratio (Fig. 5). Yimengite is characterized by a relatively low content of Al_2O_3 and MgO. It contains up to 3.5 wt% Nb_2O_5, despite the fact that coexisting priderite does not contain this component (Tables 4 and 5).

Phlogopite in the run products contains 0.6–1.9 wt% TiO_2 and 1.8–2.7 wt% Cr_2O_3 reflecting its formation in the reactions involving both chromite and ilmenite or rutile. In the series of experiments with variable K_2CO_3/($H_2O + CO_2$) ratio (series B in Table 2), the highest concentrations of these components are detected in phlogopites forming in the reactions with fluids with high K_2CO_3/($H_2O + CO_2$) ratio (9:1 and 7:3) suggesting more active involvement of chromite and ilmenite in the phlogopite-forming reactions.

9.4.3 Raman Spectra of Titanates

The presence of priderite, yimengite and mathiasite was confirmed using Raman spectroscopy. Raman spectra of Cr-bearing priderite within the range of 150–1200 cm^{-1} (Fig. 6) are characterized by three intense bands at 159, 359 and 692 cm^{-1} (Fig. 6 **I**), close to those for the spectra of natural K–Cr-priderite (Konzett et al. 2013; Naemura et al. 2015). However, in contrast to the spectra of natural K–Cr-priderite,

Table 4 Representative analyses of priderite, yimengite and mathiasite

№№	Sp-4	Sp-5	A1-5	A2-7	B1-12	B1-3	B1-1-2	B1-1-12
Минерал	Pri	Pri	Pri	Pri	Yim	Yim	Yim	Ma
TiO_2	69,45	69,03	69,83	69,21	34,29	38,34	36,11	51,16
Cr_2O_3	15,62	16,45	19,00	17,03	25,31	21,9	23,14	2,29
FeO*	0,83	0,5	0,52	0,42	23,09	26,3	24,4	32,42
Al_2O_3	0,91	0,89	0,79	0,70	1,97	1,68	1,56	0,39
K_2O	11,26	11,39	10,95	11,26	4,78	4,88	5,15	1,89
MnO	0,37	0,03	0,15	0,06	1,79	1,80	1,68	2,82
MgO	0,68	0,51	0,30	0,40	2,20	1,82	2,27	5,36
Nb_2O_5	0,00	0,00	0,00	0,12	2,94	3,44	2,35	2,00
Сумма	99,12	98,80	101,54	99,20	96,37	100,16	96,66	98,33
	Per 16 O				**Per 19 O**			**Per 38 O**
Ti	6,23	6,22	6,13	6,22	4,04	4,34	4,23	12,05
Cr	1,47	1,56	1,75	1,61	3,14	2,60	2,85	0,57
Fe^{3+}	0,07	0,05	0,05	0,04	2,24	2,41	2,47	8,48
Fe^{2+}	-	-	-	-	0,48	0,57	0,39	-
Al	0,13	0,13	0,11	0,10	0,36	0,30	0,29	0,14
K	1,71	1,74	1,63	1,72	0,96	0,94	1,02	0,75
Mn	0,04	0,00	0,01	0,01	0,24	0,23	0,22	0,75
Mg	0,12	0,09	0,05	0,07	0,51	0,41	0,53	2,50
Nb^{5+}	0,00	0,00	0,00	0,01	0,24	0,27	0,19	0,28

*$FeO=FeO+Fe_2O_3$

the main bands in the spectra of the synthetic priderite are noticeably shifted to higher wave numbers. This is probably due to the lack of Ba in the synthetic priderite.

Raman spectra of yimengite (Fig. 6 **II**) show intense bands at 378, 508, 682 and 757 cm^{-1}, consistent with the spectra of the synthetic HAWYIM solid solution (Konzett et al. 2005). In comparison to the HAWYIM solid solution (Konzett et al. 2005), the bands in the spectra of the synthesized yimengite are shifted to higher wave numbers. As in the case of priderite, the shift of the bands, apparently, is associated with the absence of Ba in the composition of yimengite.

Raman spectra of mathiasite (Fig. 6 **III**) are characterized by intense peaks at 258, 332, 456, 587, 682 cm^{-1}, consistent with the spectra of the synthetic LIMA solid solution (bands at 243, 327, 433, 560 and 661–702 cm^{-1}) (Konzett et al. 2005). The shift by 5–20 cm^{-1} of all bands can be explained by different contents of the end-members of lindsleyite, loveringite (Ca-containing end-member) and mathiasite in the crichtonite phase.

Table 5 Representative compositions (a.p.f.u.) of coexisting phases in some run products

Run #	B1				
a.p.f.u./mineral	Yim	Pri	Ilm_1	Ilm_2	Rt
Ti	4,76	6,15	0,93	0,98	0,85
Cr	2,87	1,56	-	0,01	0,20
Fe^{3+}	0,25	0,15	0,08	0,03	-
Fe^{2+}	3,30	0,10	0,95	0,70	-
Al	0,35	0,06	0,02	-	-
K	1,01	1,77	-	-	-
Mg	0,47	0,06	0,04	0,28	-

Run #	Sp2			B1-1		
a.p.f.u./mineral	Chr_1	Chr_2	Pri	Yim_1	Yim_2	Ilm_3
Ti	0,10	0,14	6,23	4,34	4,25	0,97
Cr	1,47	1,46	1,47	2,99	3,31	0,02
Fe^{3+}	0,02	-	0,11	0,68	0,42	0,04
Fe^{2+}	0,31	0,46	-	2,61	2,74	0,74
Al	0,31	0,26	0,13	0,69	0,66	-
K	-	-	1,71	1,00	1,01	-
Mg	0,79	0,68	0,12	0,67	0,67	0,23

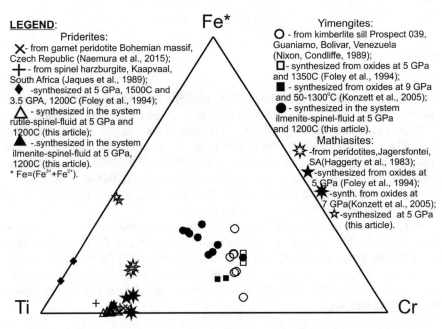

LEGEND:

Priderites:
✗- from garnet peridotite Bohemian massif, Czech Republic (Naemura et al., 2015);
╋- from spinel harzburgite, Kaapvaal, South Africa (Jaques et al., 1989);
◆ -synthesized at 5 GPa, 1500C and 3.5 GPa, 1200C (Foley et al., 1994);
△ - synthesized in the system rutile-spinel-fluid at 5 GPa and 1200C (this article);
▲ -.synthesized in the system ilmenite-spinel-fluid at 5 GPa, 1200C (this article).
* $Fe=(Fe^{3+}+Fe^{2+})$.

Yimengites:
○ - from kimberlite sill Prospect 039, Guaniamo, Bolivar, Venezuela (Nixon, Condliffe, 1989);
□- synthesized from oxides at 5 GPa and 1350C (Foley et al., 1994);
■ - synthesized from oxides at 9 GPa and 50-1300°C (Konzett et al., 2005);
● - synthesized in the system ilmenite-spinel-fluid at 5 GPa and 1200C (this article).

Mathiasites:
✶ -from peridotites,Jagersfontei, SA(Haggerty et al., 1983);
★-synthesized from oxides at 5 GPa (Foley et al., 1994);
✱-synth. from oxides at 7 GPa(Konzett et al., 2005);
☆-synthesized at 5 GPa (this article).

Fig. 5 Diagram of Ti – Fe – Cr illustrating the variation of the composition of synthesized yimengite, mathiasite and priderite in comparison with the compositions of natural minerals

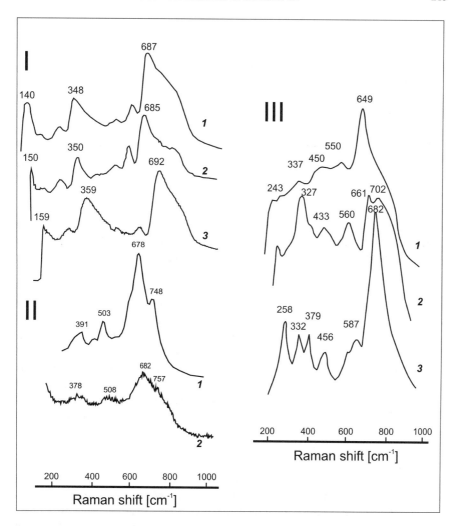

Fig. 6 Raman spectra of the synthesized phases: (I-1) K-Cr priderite from metasomatized peridotites from kimberlites of South Africa (Konzett et al. 2013); (I-2) K-Ba-Cr priderite from inclusions in chromites of Bohemian Massif garnet peridotites (Naemura et al. 2015); (I-3) synthesized K-Cr priderite (analysis Sp-4, Table 4); (II-1) solid solution yimengite-hawthorneite synthesized from the oxides at 12 GPa and 1400 °C (ex. JKW88, see Konzett et al. 2005); (II-2) the synthesized yimengite (analysis B1-12, Table 4); (III-1) mathiasite synthesized from the oxides at 7 GPa and 1300 °C (ex. JKW43, see Konzett et al. 2005); (III-2) lindsleyite synthesized from the oxides at 7 GPa and 1300 °C (ex. JKW43, see Konzett et al. 2005); (III-3) synthesized mathiasite (analysis B1-1-12, Table 4)

9.5 Discussion

Experimental data on synthesis of K–Ba-titanates demonstrate that there are virtually no restrictions on their stability at the upper-mantle and, possibly, transition zone PT-conditions (Podpora and Lindsley 1984; Foley et al. 1994; Konzett et al. 2005). The main restriction is imposed by the compositional characteristics of the mantle protolith and fluids, whose reactions produce these minerals.

Petrological and mineralogical studies of mantle xenoliths show that the predominant type of mantle protolith for the formation of Cr-bearing K–Ba-titanates are peridotites enriched in Cr_2O_3, but rather poor in Al_2O_3. These are subcratonic depleted harzburgites. However, K–Ba-titanates are known in lherzolite assemblages, as well (e.g. Rezvukhin et al. 2018). The formation of Cr-rich K–Ba-titanates is usually associated with reactions of chromite in peridotites with fluids or melts enriched in potassium and incompatible elements (e.g. Haggerty et al. 1983; Haggerty 1983; Nixon and Condliffe 1989; Sobolev et al. 1998; Bulanova et al. 2004; Rezvukhin et al. 2018). The phase relations in the run products clearly showed this process (Figs. 2a–e and 3b, d). However, experiments also showed that these phases did not form after chromite directly, but demanded an additional source of titanium. This source cannot be depleted peridotites usually containing less than 0.1 wt% TiO_2 (e.g. Rudnick et al. 1998). Many authors pay attention to elevated TiO_2 content (up to 7 wt%) of chromite associated with K–Ba titanates (Zhou 1986; Nixon and Condliffe 1989; Bulanova et al. 2004), which is atypical for chromite in the subcratonic peridotites. Of course, this compositional feature is related to the preliminary interaction of chromite with metasomatizing fluids/melts, which introduce Ti and incompatible elements. Such interaction results in not only the enrichment of chromite in titanium, but also in the formation of ilmenite and rutile, which subsequently serve as reagents for the formation of titanates (Konzett et al. 2000; Almeida et al. 2014). For example, lindsleyite-mathiasite reaction on Cr-bearing ilmenite, which is itself a product of the metasomatism, has been described in xenoliths from South African kimberlites (Konzett et al. 2000). It cannot be excluded that the suppliers of titanium for the K–Ba-titanate-forming reactions may be garnet and clinopyroxene, which are capable to contain high concentrations of both titanium and chromium, especially at high pressures (e.g. Zhang et al. 2003). However, this conclusion requires experimental confirmation, since there are no experiments on crystallization of K–Ba-titanates in the presence of chromium and titanium-bearing silicates. Natural assemblages also do not provide evidence for the formation of these phases after garnet and pyroxenes, although the inclusions of the crichtonite group minerals in chromium pyropes are known (Wang et al. 1999; Rezvukhin et al. 2018). It is possible that the absence of such evidence is explicable by the specificity of mantle metasomatism. During the processes of modal mantle metasomatism of peridotites with the participation of alkaline fluids, garnet, as a rule, disappears quickly during the phlogopite and/or potassium richterite-forming reactions (Safonov and Butvina 2016). In addition to limited substitution into hydrous aluminosilicates (phlogopite and amphibole), Cr from garnet forms a new chromium-bearing spinel. This spinel can also serve a reagent for the formation of Cr-bearing K–Ba-titanates in the subsequent stages of metasomatic transformations.

Phlogopite and potassium richterite are typical products of modal mantle metasomatism, during which K–Ba-titanates form. Their formation is determined not only by the potassium content in the system, but also by the ratio of its content to the activity of water in fluids. Hydrous aluminosilicates are not only active containers of potassium and other LILE, but also able to accommodate Cr and Ti. Thus, phlogopite and potassium richterite can act as contenders for K–Ba-titanates. Cr and Ti-rich phlogopite was identified in the products of all experiments (Table 2 and Fig. 4). It appears even in experiments where K-titanates are not detected (experiments B1-3 and B1-4). These experiments are characterized by the lowest $K_2CO_3/(H_2O + CO_2)$ ratios in the starting fluid mixture (3:7 and 1:9, respectively). This result is consistent with experimental data on the interaction of peridotites with K_2CO_3 and KCl-bearing fluids (Edgar and Arima 1984; Thibault and Edgar 1990; Safonov and Butvina 2013; Sokol et al. 2015), which demonstrate that phlogopite is formed in a wide range of H_2O/salt ratios, from undersaturated water-salt fluids to hydrous salt melts. Thermodynamic calculations show that phlogopite in peridotite assemblages is formed at water activities down to values of 0.1, and potassium activity is the leading factor determining the formation of this mineral (Safonov and Butvina 2016). In experiments with higher $K_2CO_3/(H_2O + CO_2)$ ratios in starting fluids, phlogopite coexists with Cr-bearing K-titanates. This means that the crystallization of titanates together with phlogopite is determined not so much by the activity of water as by the activity of the potassium component of the fluid. The close natural association of titanates with phlogopite and/or K-amphibole indicates that the formation of HAWYIM and LIMA minerals and priderite is possible at excess of potassium and other LILE, when the mineral capacity of the system relative to these components exceeds the possibility for mica and amphibole formation (Konzett et al. 2005; Safonov and Butvina 2016). Reactions of phlogopite with relic Cr-rich spinel can form K-bearing titanates at the advanced stage of metasomatic processes (e.g. Almeida et al. 2014).

High potassium activity in the mineral-forming medium, necessary for the formation of Cr-bearing K–Ba-titanates, corresponds to the highest degrees of metasomatism. Such conditions can be created either with a continuous and intensive interaction of ultrapotassic fluids/melts with rocks, or during a multi-stage process with an increasing effect. For example, two-stage metasomatic process is described by Konzett et al. (2013) in the spinel harzburgites from kimberlites of South Africa. The first stage of metasomatic transformation results in the formation of phlogopite, K-amphibole, titaniferous phase (rutile and srilankite) and crichtonite phase (mathiasite). The subsequent stage of metasomatism is manifested by the decomposition of these minerals to form new generations of amphibole, phlogopite, clinopyroxene, olivine, various rare Ti and Zr-bearing minerals. In this association, Cr-rich priderite replaces mathiasite. The authors believe that the second stage of metasomatism was due to the interaction of rocks with ultra-alkaline fluids/melts with low silica activity (Konzett et al. 2013), which is typical for alkaline carbonatite melts. The relics of such melts have been repeatedly described as polymineral inclusions in kimberlite minerals, including diamonds, minerals of peridotite xenoliths in kimberlites, as well as in

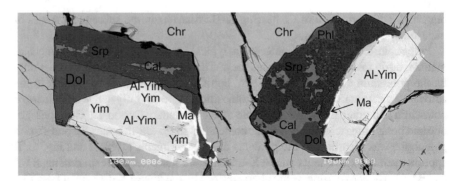

Fig. 7 Polyphase inclusions in chromite: yimengite, Al-rich yimengite, mathiasite, dolomite, calcite, serpentine and phlogopite. BSE image

minerals of alpine-type orogenic peridotites. Associations of these inclusions consist of mixtures of carbonate and silicate minerals with phosphate, sulfide, chloride, sulfate and other phases reflecting the complex composition of the trapped carbonatite melts. Among phases composing the polymineral inclusions, K–Ba titanates were described. Priderite has been discovered in carbonate-silicate inclusions in chromite of garnet peridotite in the Bohemian massif (Naemura et al. 2015) and in ilmenites of the Bultfontein kimberlite pipe, South Africa (Giuliani et al. 2012). The crichtonite group minerals were found in carbonate-bearing polyphase inclusions in chromium pyropes from the International kimberlite pipe, Yakutia (Rezvukhin et al. 2018). Figure 7 shows large polyphase inclusions in chromite from garnet lherzolite xenolith from the Obnazhennaya kimberlite pipe, Yakutia. They contain yimengite. Subhedral outlines of the yimengite crystals manifest a free growth of mineral within the inclusions. Zoning in these crystals (Fig. 7) is resulted from the alternation of zones with different Al/Cr ratios. Yimengite is associated with dolomite, phlogopite, serpentine and calcite (probably, products of a later reaction inside the inclusions on cooling $6Dol + 4H_2O + 4SiO_2 = 2Srp + 6Cal + 6CO_2$). The presence of phlogopite and yimengite indicates a high primary content of potassium in the inclusions. It is evident that the inclusions are relics of solidified alkaline hydrous carbonate-silicate melts, the interaction of which with the host chromite led to the formation of titanates.

The experiments confirmed the possibility of joint formation of different titanates as a result of the interaction of chromite and ilmenite with potassic aqueous carbonate fluid/melt with different $K_2CO_3/(H_2O + CO_2)$ ratios. Associations containing two different titanates are known in metasomatized xenoliths. In these cases, reaction relationships between titanates are usually observed. Replacement of yimengite by priderite were described in xenoliths from Chinese kimberlites (Zhou 1986). In the above-mentioned xenoliths of spinel harzburgites from South African kimberlites (Konzett et al. 2013), priderite was produced during the repeated stage of metasomatism and replaced the LIMA phases. The authors conclude that the second stage of metasomatism was due to the interaction of rocks with fluids characterized by higher potassium activity (Konzett et al. 2013). Almeida et al. (2014) concluded that priderite-bearing xenoliths record an action of more potassic metasomatizing melts than mathiasite-bearing xenoliths. These observations and conclusions are in good agreement with the observed sequence of appearance of various titanates depending on the $K_2CO_3/(H_2O + CO_2)$ ratio. Priderite appears at a higher ratio, which corresponds to higher potassium activity. In the above-mentioned polyphase inclusions in chromite from garnet lherzolite xenolith, crystals of yimengite are replaced by thin bay-shaped edges of mathiasite (Fig. 7). Following to the experiments, the association yimengite + mathiasite appears during the reaction of chromite and ilmenite with the fluid with higher $K_2CO_3/(H_2O + CO_2)$ ratio than yimengite alone. With application to the polyphase inclusions, the replacement of yimengite by mathiasite can be interpreted as the result of an accumulation of potassium component of the fluid/melt by crystallization within the inclusion and interaction with the chromite host.

9.6 Conclusions

Experiments on the interaction of chromite + rutile and chromite + ilmenite associations in the presence of a small amount of silicate material with $H_2O–CO_2–K_2CO_3$ fluids at 5 GPa revealed the following features of crystallization of titanate minerals and allowed interpretation of their associations in metasomatized mantle peridotites.

(1) The principal possibility of the formation of minerals of crichtonite and magnetoplumbite groups and priderite in the reactions of chromite with alkaline aqueous-carbonic fluids and melts is confirmed. Such substances are considered as main agents of potassium metasomatism, leading to the formation of titanates in the upper mantle (Konzett et al. 2013; Rezvukhin et al. 2018).

(2) The formation of these phases does not proceed directly on chromite (e.g. Haggerty et al. 1983; Haggerty 1983; Nixon and Condliffe 1989), and requires additional titanium source. They are rutile and ilmenite, which are themselves usually are products of modal metasomatism of peridotites. This experimental fact demonstrates that the formation of titanates marks probably the most advanced or repeated stages of metasomatism in mantle peridotites.

(3) This is also proved by the relationships of titanates with phlogopite. Association of titanates with phlogopite is characterized by a higher activity of the potassium component in the fluid/melt than the formation of phlogopite alone. Such conditions can again be created at the most advanced or repeated stages of mantle metasomatism.

(4) The relationship between titanates is also a function of the activity of the potassium component in the fluid/melt. Priderite is an indicator of the highest potassium activity in the mineral-forming medium. The above examples from natural associations (Zhou 1986; Konzett et al. 2013; Almeida et al. 2014) well illustrate this conclusion.

Acknowledgements The work was financially supported by the governmental projects AAAAA18-118020590148-3 and AAAA-A18-118020590140-7 of the Korzhinskii Institute of Experimental Mineralogy.

References

Almeida V, Janasi V, Svisero D, Nannini F (2014) Mathiasite-loveringite and priderite in mantle xenoliths from the Alto Paranaíba Igneous Province, Brazil: genesis and constraints on mantle metasomatism. Open Geosci 6(4):614–632

Bailey DK (1982) Mantle metasomatism—Continued chemical change within the Earth. Nature 296:525–580

Bailey DK (1987) Mantle metasomatism–perspective and prospect. In: Fitton JG, Upton BGJ (eds) Alkaline igneous rocks, vol 30. Geological Society Special Publication, pp 1–13

Bulanova GP, Muchemwa E, Pearson DG, Griffin BJ, Kelley SP, Klemme S, Smith CB (2004) Syngenetic inclusions of yimengite in diamond from Sese kimberlite (Zimbabwe)—evidence for metasomatic conditions of growth. Lithos 77(1–4):181–192

Dong Z, Zhou J, Lu Q, Peng Z (1983) Yimengite, K(Cr, Ti, Fe, Mg)$_{12}$O$_{19}$, a new mineral from China. Kexue Tongbao Bull Sci 15:932–936 (in Chinese)

Edgar AD, Arima M (1984) Experimental studies on K-metasomatism of a model pyrolite mantle and their bearing on the genesis of uitrapotassic magmas. In: Proceedings of 27th International Geological Congress Petroleum (Igneous and metamorphic rocks), vol 9, pp 509–541

Erlank AJ, Rickard RS (1977) Potassic richterite bearing peridotites from kimberlite and the evidence they provide for upper mantle metasomatism. (Abstract) Second International Kimberlite Conference, Santa Fe, New Mexico

Foley S, Hofer H, Brey G (1994) High-pressure synthesis of priderite and members of lindsleyite-mathiasite and hawthorneite-yimengite series. Contrib Mineral Petrol 117:164–174

Giuliani A, Kamenetsky VS, Phillips D, Kendrick MA, Wyatt BA, Goemann K (2012) Nature of alkali-carbonate fluids in the sub-continental lithospheric mantle. Geology 40(11):967–970

Grey IE, Madsen IC, Haggerty SE (1987) Structure of a new upper-mantle magnetoplumbite-type phase, Ba(Ti$_3$Cr$_4$Fe$_4$Mg)O$_{19}$. Am Mineral 72:633–636

Haggerty SE (1975) The chemistry and genesis of opaque minerals in kimberlites. Phys Chem Earth 9:295–307

Haggerty SE (1983) The mineral chemistry of new titanates from the Jagersfontein kimberlite, South Africa: implications for metasomatism in the upper mantle. Geochim Cosmochim Acta 47(11):1833–1854

Haggerty SE (1987) Metasomatic mineral titanates in upper mantle xenoliths. In: Nixon PH (ed) mantle xenoliths. Wiley, Chichester, pp 90–671

Haggerty SE (1991) Oxide mineralogy of the upper mantle. In: Lindsley DH (eds) Oxide minerals: Petrologic and magnetic significance. Rev Mineral 25: 355–416

Haggerty SE, Smyth JR, Erlank AJ, Rickard RS, Danchin RV (1983) Lindsleyite (Ba) and mathiasite (K): two new chromium-titanates in the crichtonite series from the upper mantle. Am Mineral 68:494–505

Haggerty SE, Erlank AJ, Grey IE (1986) Metasomatic mineral titanate complexing in the upper mantle. Nature 319(6056):761–763

Haggerty SE, Grey IE, Madsen IC, Criddle AJ, Stanley CJ, Erlank AJ (1989) Hawthorneite, Ba[Ti$_3$Cr$_4$Fe$_4$Mg]O$_{19}$: a new metasomatic magnetoblumbite-type mineral from the upper mantle. Am Mineral 74:668–675

Hall HT (1971) Fixed points near room temperature: accurate characterization of the high pressure environment. National Bureau of Standards. US. Spec. Publ. №. 326, Washington, DC, p 313

Harte B (1983) Mantle peridotites and processes—The kimberlite sample. In: Hawkesworth CJ, Norry MJ (eds) Continental basalts and mantle xenoliths. Shiva, Cheshire, pp 46–91

Harte B, Gurney JJ (1975) Ore mineral and phlogopite mineralization within ultramafic nodules from the Matsoku kimberlite pipe, Lesotho, vol 74. Carnegie Inst Yearbook, Washington, pp 528–536

Jaques AL, Hall AE, Sheraton JW, Smith CB, Sun SS, Drew RM, Foudoulis C, Ellingsen K (1989) Composition of crystalline inclusions and C-isotopic composition of Argyle and Ellendale diamonds. In: Jaques AL, F'erguson J, Green DH (eds) Kimberlites and related rocks 2: their crust/mantle setting, diamonds, and diamond exploration. Blackwells, Melbourne, pp 966–989

Jones AP, Smith JV, Dawson JB (1982) Mantle metasomatism in 14 veined peridotites from Bultfontein Mine, South Africa. J Geol 435–453

Kiviets GB, Phillips D, Shee SR, Vercoe SC, Barton ES, Smith CB, Fourie LF (1998) 40Ar/39Ar dating of yimengite from Turkey Well kimberlite, Australia: the oldest and the rarest. In: 7th International Kimberlite Conference, pp 432–434

Konzett J, Armstrong RA, Günther D (2000) Modal metasomatism in the Kaapvaal craton lithosphere: Constraints on timing and genesis from U-Pb zircon dating of metasomatized peridotites and MARID-type xenoliths. Contrib Mineral Petrol 139(6):704–719

Konzett J, Yang H, Frost DJ (2005) Phase relations and stability of magnetoplumbite- and crichtoniteseries phases under upper-mantle P-T conditions: an experimental study to 15 GPa with implications for LILE metasomatism in the lithospheric mantle. J Petrol 46(4):749–781

Konzett J, Wirth R, Hauzenberger C, Whitehouse M (2013) Two episodes of fluid migration in the Kaapvaal Craton lithospheric mantle associated with Cretaceous kimberlite activity: evidence from a harzburgite containing a unique assemblage of metasomatic zirconium-phases. Lithos 182:165–184

Konzett J, Krenn K, Rubatto D, Hauzenberger C, Stalder R (2014) The formation of saline mantle fluids by open-system crystallization of hydrous silicate–rich vein assemblages—evidence from fluid inclusions and their host phases in MARID xenoliths from the central Kaapvaal Craton, South Africa. Geochim Cosmochim Acta 147:1–25

Litvin YA (1991) Physico-chemical studies of the melting of the earth's deep matter. Science, Moscow, 312 p (in Russian)

Litvin YA, Livshits LD, Karasev VV, Chudinovskikh LT (1983) On reliability of experiments and P-T measurements at studies of physico-chemical equilibria in the solid-phase apparatuses. Phys Techn High Press 14:50–56

Lloyd FE, Bailey DK (1975) Light element metasomatism of the continental mantle: the evidence and the consequences. Phys Chem Earth 9:389–416

Menzies MA, Hawkesworth CJ (1987) Mantle metasomatism. Academic Press, London

Naemura K, Shimizu I, Svojtka M, Hirajima T (2015) Accessory priderite and burbankite in multiphase solid inclusions in the orogenic garnet peridotite from the Bohemian Massif, Czech Republic. Mineral Petrol 110:20–28

Nikitina LP, Goncharov AG, Saltykova AK, Babushkina MS (2010) The redox state of the continental lithospheric mantle of the Baikal-Mongolia region. Geochem Int 48(1):15–40

Nixon PH, Condliffe E (1989) Yimengite of K-Ti metasomatic origin in kimberlitic rocks from Venezuela. Min Mag 53:305–309

Norrish K (1951) Priderite, a new mineral from the leucite lamproites of the West Kimberley area, Western Australia. Min Mag 73:1007–1024

Peng Z, Lu Q (1985) The crystal structure of yimengite. Sci Sinica (Ser B) 28:882–887

Podpora C, Lindsley DH (1984) Lindsleyite and mathiasite: synthesis of chromium-titanates in the crichtonite (A1M21O38) series. EOS Trans Am Geophys Union 65:293

Prider RT (1939) Some minerals from the leucite-rich rocks of the west Kimberley area, Western Australia. Min Mag 25:373–387

Rezvukhin DI, Malkovets VG, Sharygin IS, Tretiakova IG, Griffin WL, O'Reilly SY (2018) Inclusions of crichtonite-group minerals in Cr-pyropes from the Internatsionalnaya kimberlite pipe,

Siberian Craton: crystal chemistry, parageneses and relationships to mantle metasomatism. Lithos 308:181–195

Rudnick RL, McDonough WF, O'Connell RJ (1998) Thermal structure, thickness and composition of continental lithosphere. Chem Geol 145:399–415

Safonov OG, Butvina VG (2013) Interaction of model peridotite with H_2O-KCl fluid: experiment at 1.9 GPa and its implications for upper mantle metasomatism. Petrology 21(6):599–615

Safonov OG, Butvina VG (2016) Indicator reactions of K and Na activities in the upper mantle: natural mineral assemblages, experimental data, and thermodynamic modeling. Geochemistry 54(10):858–872

Smyth JR, Erlank AJ, Rickard RS (1978) A new Ba–Sr–Cr–Fe titanate mineral from a kimberlite nodule, vol 59. EOS American Geophysical Union, 394 p

Sobolev NV, Yefimova ES (2000) Composition and petrogenesis of Ti-oxides associated with diamonds. Int Geol Rev 42(8):758–767

Sobolev NV, Yefimova ES, Kaminsky FV, Lavrentiev YG, Usova LV (1988) Titanate of complex composition and phlogopite in the diamond stability field. In: Sobolev NV (ed) Composition and processes of deep seated zones of continental lithosphere. Nauka, Novosibirsk, pp 185–186

Sobolev NV, Yefimova ES, Channer DM DeR, Anderson PFN, Barron KM (1998) Unusual upper mantle beneath Guaniamo, Guyana Shield, Venezuela: evidence from diamond inclusions. Geol 26:971–974

Sokol AG, Kruk AN, Chebotarev DA, Pal'yanov YN, Sobolev NV (2015) Conditions of phlogopite formation upon interaction of carbonate melts with peridotite of the subcratonic lithosphere. Doklady Earth Sci 462(2):638–642

Thibault Y, Edgar AD (1990) Patent mantle-metasomatism: inferences based on experimental studies. Proc Indian Acad Sci-Earth Planet Sci. 99:21–37

Wang L, Essene EJ, Zhang Y (1999) Mineral inclusions in pyrope crystals from Garnet Ridge, Arizona, USA: implications for processes in the upper mantle. Contrib Mineral Pet 135:164–178

Zhang RY, Zhai SM, Fei YW, Liou JG (2003) Titanium solubility in coexisting garnet and clinopyroxene at very high pressure: the significance of exsolved rutile in garnet. Earth Planet Sci Lett 216:591–601

Zhou J (1986) LIL-bearing Ti-Cr-Fe oxides in Chinese kimberlites. In 4th International Kimberlite conference, pp 100–102

Chapter 10
How Biopolymers Control the Kinetics of Calcite Precipitation from Aqueous Solutions

L. Z. Lakshtanov, O. N. Karaseva, D. V. Okhrimenko, and S. L. S. Stipp

Abstract This work is a continuation and extension of our studies of the kinetics of recrystallization and precipitation of calcite in the presence of biopolymers, as well as precipitation of calcite on the surface of biogenic calcite (chalk). Our results show that the dependence of calcite precipitation rate on supersaturation does not obey the parabolic law that is characteristic for calcite growth. Instead, the dependence was exponential, which indicates surface nucleation-mediated growth. Rate of calcite precipitation on chalk surfaces at the lowest supersaturation studied in this work is 3 orders of magnitude lower than on pure calcite surfaces. The presence of non-polar organic compounds, such as biomarkers, decreases surface energy but does not change the mechanism of calcite precipitation. On the contrary, the presence of adsorbed polysaccharides and polyamino acids on the surface determines the mechanism of calcite growth and precipitation. This is confirmed by the results of measuring ζ-potential as well as the induction period for homogeneous nucleation during CO_2 degassing from solutions saturated at $pCO_2 = 1$ atm with respect to calcite or various chalks.

Keywords Calcite · Chalk · Recrystallization · Polysaccharides · Polyamino acids · Surface nucleation · Inhibition

10.1 Introduction

The kinetics of calcite growth and nucleation is of fundamental importance for understanding the processes that control many natural and anthropogenic geochemical

L. Z. Lakshtanov (✉) · O. N. Karaseva
Institute of Experimental Mineralogy RAS, Chernogolovka, Russia 142432
e-mail: leonidlak@gmail.com

L. Z. Lakshtanov · D. V. Okhrimenko · S. L. S. Stipp
Department of Chemistry, Nano-Science Center, University of Copenhagen, Universitetsparken 5, 2100 Copenhagen, Denmark

Y. Litvin and O. Safonov (eds.), *Advances in Experimental and Genetic Mineralogy*, Springer Mineralogy, https://doi.org/10.1007/978-3-030-42859-4_10

systems. It is well known that calcite has a great impact on the evolution of such processes as the diagenesis of marine sediments (Bathurst 1975), deterioration of agricultural soils (Buol et al. 1973), the effectiveness of water treatment methods (Reddy 1978).

Inhibitors of calcite precipitation are of great importance for geochemistry and biomineralization. The inhibitory effect could explain the extremely slow rate of recrystallization of some types of chalk, which can consist of more than 90% calcite. Persistence of microfossils suggests that something inherent in chalk dramatically reduces the rate of recrystallization (Marsh 2003). In particular, there are evidences that exogenous biopolymers, especially polysaccharides and polyamino acids, significantly inhibit the precipitation of calcium carbonate (Borman et al. 1982; Sand et al. 2014; Belova et al. 2012).

It is widely believed that growth inhibition is caused by adsorption of organic compounds at active growth centers on mineral surfaces. Adsorption leads to a reduction in the effective surface free energy and therefore reduces the energy barrier for nucleation. Adsorbed molecules can make it difficult for crystalline lattice ions to attach during growth and to act as obstacles to the advancement of growth steps (called step-pinning) (Cabrera and Vermilyea 1958).

Adsorbed biopolymers can also change the mechanism of calcite crystallization, inducing surface nucleation-mediated growth (Dove and Hochella 1993; Karaseva et al. 2018). "When polymer is introduced to the system during the growth, it inhibits step movement by pinning at a number of kink sites along the step length. A calcite surface, wetted in the presence of an organic polymer, activates growth controlled by two-dimensional nucleation" (Karaseva et al. 2018).

Fossil polysaccharides extracted from chalk are still active. They have been shown to have a strong influence on the morphology of calcite crystals, their dissolution and growth (Borman et al. 1982; Sand et al. 2014). "It is reasonable to assume that the polysaccharides, inherent on the surface of chalk particles, are responsible for extremely low recrystallization rates" (Lakshtanov et al. 2015).

A major unresolved problem in many industries, including desalination and water treatment (Lattemann and Höpner 2008), power generation (Zhao and Liu 2004), oil and gas extraction (Crabtree et al. 1999; Wang et al. 2005), is the formation of scale, which results in significant operational losses as a result of heat flow disruption, reduction of the effective crosssection in pipes, decreased pressure and shortened equipment life.

This work is a continuation and extension of efforts to understand the interaction of organic compounds with calcite and to discover the reasons for the extremely low rate of chalk recrystallization. We studied calcite precipitation in the presence of the biopolymers, alginate and polyaspartate, as well as calcite precipitation on natural chalk samples, using the constant composition method. We examined the relationship between the precipitation rate and the supersaturation of the solution to establish the active mechanism of calcite precipitation on the surface of calcite particles with adsorbed biopolymers, as well as on chalk particles. The data obtained were correlated with the surface concentration of C–O bonds, as well as with the measured values of ζ-potential. We used solvent extraction to remove some non-polar organic compounds and then determined the rate of calcite precipitation and ζ-potential for samples after hydrocarbon removal.

Surface free energy is a key parameter that controls nucleation and growth, and in the case of three-phase contact (calcite–biopolymer–aqueous solution), the effective free surface energy is a function of several parameters. To estimate that, we determined the free energy of the interface calcite–biopolymer–solution by the vapor adsorption method and compared with that obtained from the calcite precipitation kinetics. In addition, using optical spectroscopy (UV–vis) and potentiometry, we measured the induction period of homogeneous nucleation of calcite during CO_2 degassing from solutions saturated with respect to calcite and various chalk samples at $pCO_2 = 1$ atm, and correlated the data with the surface concentration of C–O bonds, as well as with the measured values of ζ-potential of the same samples.

10.2 Materials

Calcite powder (Sigma) was used for all experiments. The powder was recrystallized by several treatments with deionized water saturated with CO_2 (Stipp and Hochella 1991). Calcite crystal size was ~10 μm and surface area, 0.2 m^2/g. All solutions were prepared with Milli-Q deionized water (specific resistance 18.2 MOhm). In experiments with calcite precipitation we used freshly prepared solutions of 0.1 M $CaCl_2$, 0.1 M Na_2CO_3 and 0.1 M $NaHCO_3$ with concentrations of Ca^{2+}, CO_3^{2-} and HCO_3^- measured by atomic absorption spectroscopy (PerkinElmer AAS Analyst 800) and titration (Metrohm 809 Titrano). Initial solutions of biopolymers were prepared with alginic acid sodium salt (Alg, 0.5 g/l) or poly(α, β)-DL-aspartic acid sodium salt (pAsp, 68 mg/l).

Calcite samples with adsorbed polymers were prepared for precipitation experiments and measurements of vapor adsorption. Calcite powder, weighing 2 g was placed in 50 ml of calcite saturated solution at pH 8.5 and a determined amount of biopolymer solution. After 2 or 24 h, the suspension was filtered out, washed with a saturated calcite solution and freeze dried.

Most of the chalk samples were of Maastrichtian age and were provided by Maersk Oil and Gas A/S as fragments of core samples drilled from water-saturated (2-3, 2-4 and 7-1) and gas-saturated (10-4) zones in the North Sea Basin. Geological evidence suggests that none of the samples have ever been in contact with oil. Some chalk samples were obtained from the Aalborg quarry (Aalborg, Denmark). The chalk was ground in an agate mortar and sieved to retain the <60 μm fraction. This was again ground and used in precipitation experiments. The chalk particles were 0.5–5 μm diameter. Using X-ray photoelectron spectroscopy, we obtained the surface concentration of carbon compounds from ratios of peak intensities (Table 10.1).

Samples 10-4 and 7-1 contained a higher proportion of C–C or C–H bonds and a lower proportion of C–O bonds, which indicates a higher concentration of non-polar organic compounds on the surface of these chalk samples.

Table 10.1 Atomic percent of carbon, derived from the XPS peak intensity ratios from the chalk samples

Chalk	Zone	CO_3	C–O	CC or CH	O=C–O	C–O/CO_3	CC or CH/CO_3
2-3	WS	57.34	27.53	15.13	5	0.48	0.26
2-4	WS	47.60	36.29	16.10	5.1	0.76	0.34
Aalborg	Quarry	65.00	13.70	17.70	3.7	0.21	0.27
10-4	GS	28.60	12.90	52.60	5.9	0.45	1.84
7-1	WS	55.90	14.20	23.80	6.1	0.25	0.42

The results of fitting the C1s peak with four organic carbon contributions: adventitious carbon and hydrocarbons (C–C and C–H), alcohol groups (C–O(H)), carboxylic groups (O=C–O) and carbonate (CO_3) (Belova et al. 2012; Skovbjerg et al. 1984; Okhrimenko et al. 2014)

To investigate the effect of non-polar organic compounds on the rate of calcite precipitation, we separated Sample 10-4 into two parts. One of the parts was investigated after the removal of non-polar organic compounds. To remove the organic material, we placed 3 g into a Soxhlet extractor with 100 ml of a boiling mixture of dichloromethane (CH_2Cl_2, 93 vol.%; Lab-Scan, purity > 99.8%) and methanol (CH_3OH, 7 vol.%; Sigma Aldrich, purity > 99.9%). After 12 h, the chalk was collected and dried. In the text, this sample is called "10-4 treated". It was shown earlier (Okhrimenko et al. 2014) that extracts from similar chalk samples contain high concentrations of oil biomarkers, including steranes, diasteranes, triaromatic steroids and trypentacyclichopanes but the concentration of components in crude oil (polycyclic aromatic hydrocarbons) is significantly lower. The extraction of these compounds decreased the peak intensity of C–C/C–H XPS C1s in the high resolution spectra (Okhrimenko et al. 2014). The peak corresponding to the C–O bond remained unchanged, which indicates that the polysaccharides were not removed from the chalk surface during the extraction. They are tightly bound to the chalk particles.

10.3 Calcite Precipitation Experiments

To study the kinetics of calcite precipitation, we used the constant composition method described earlier (Manoli and Dalas 2002; Lakshtanov et al. 2011, 2015; Montanari et al. 2016; Karaseva et al. 2018). The experiments were carried out in a thermostated cell at 25 °C. The working solutions contained the chosen amounts of $CaCl_2$ and $NaHCO_3$ to achieve the required supersaturation with respect to calcite. The total volume of the solution for each experiment was 60 ml. Ionic strength was maintained with 0.1 M NaCl solution. Solution pH was brought to 8.5 by adding 0.1 M HCl or NaOH solutions.

For initiation of calcite precipitation, the calcite or chalk powder (0.05–0.15 g) was introduced into the vessel. When calcite started to precipitate, the pH of the solution decreased, which was compensated by addition of $CaCl_2$ and $Na_2CO_3/NaHCO_3$ solutions using a peristaltic pump (Ismatech). Calcite precipitation rate was determined from the amount of added titrant solutions. To check the constancy of the working solution composition, we took samples from time to time and analyzed the Ca^{2+} concentration using atomic absorption spectroscopy. Supersaturation during the experiment was constant within $\pm 5\%$.

Supersaturation, S, is defined as

$$S = \left(\frac{IAP}{K_{sp}}\right)^{1/2} = \left(\frac{a_{Ca^{2+}} a_{CO_3^{2-}}}{K_{sp}}\right)^{1/2}, \tag{10.1}$$

where IAP represents the ion activity product, a_{ion}, the activity of each ion, and K_{sp} represents the solubility product of calcite. We used the PHREEQC speciation program (Parkhurst 1995) with the PHREEQC database, to calculate the supersaturation with respect to calcite.

The precipitation rate, R, was determined from

$$R = \frac{[Ca]_{titr}}{mA}\frac{dV}{dt} = \frac{[Ca]_{titr}}{mA}R', \tag{10.2}$$

where R' represents the rate of solution addition, $[Ca]_{titr}$, the concentration of Ca in the titrant solution, m, the initial mass of the calcite seed, A, the specific surface area and V, the volume of added titrant.

10.4 Induction Period

The induction period was determined by two independent methods: by monitoring the pH drop (Metrohm microelectrode) at a rate of more than 0.1 pH in 10 log (t) minutes and by measuring the change in turbidity of the solution using a spectrophotometer (AvaLight-DH-S-BAL) in a $10 \times 10 \times 40$ mm cuvette with a magnetic stirrer (stirring speed 300 rpm).

The method for determining surface energy by measuring vapor adsorption, as well as the method for determining ζ-potential from streaming potential measurements, are described in detail in our previous studies (Karaseva et al. 2018; Lakshtanov et al. 2015).

10.5 Results and Discussion

10.5.1 Calcite Precipitation Rate in the Presence of Biopolymers

The driving force for calcite precipitation (crystal growth) is the difference in the Gibbs free energy, between initial (supersaturated) and final (equilibrium) indices:

$$\Delta G = -kT\upsilon \ln S, \tag{10.3}$$

where k is Boltzmann's constant, υ represents the number of ions in the dissolved substance formula, T, temperature and S, supersaturation.

In our experiments, we did not observe any evidence of homogeneous nucleation so precipitation occurred only on the seed crystals.

We define the precipitation rate as:

$$R = Kf(S), \tag{10.4}$$

where K represents the apparent growth rate constant and $f(S)$, a function of super-saturation that depends on the growth mechanism (Nielsen and Toft 1984; Verdoes et al. 1992; Van der Leeden et al. 1993):

$$f(S) = (S - 1) \tag{10.5}$$

for linear growth,

$$f(S) = (S - 1)^2 \tag{10.6}$$

for spiral growth and

$$f(S) = (S - 1)^{2/3} S^{1/3} \exp(-B_{2D}/3\ln S) \tag{10.7}$$

for two-dimensional, nucleation-mediated growth.

B_{2D}, the free energy barrier to create a new step, is:

$$B_{2D} = \frac{\beta \kappa^2 a}{2k^2 T^2}, \tag{10.8}$$

where β represents the shape factor, κ, the effective edge free energy and a, the molecular area.

In most experiments, four supersaturation values were used: 2.40 ± 0.20, 3.02 ± 0.20, 3.87 ± 0.15 and 4.50 ± 0.15, corresponding to a concentration of Ca (total Ca equal to total carbonate content) in a working solution of 2.0, 2.7, 3.3 and 4.0 mM.

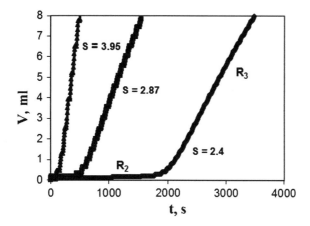

Fig. 10.1 Typical plot from a constant composition experiment with Alg (alginate), showing the trend for the added volume of titrants (i.e. precipitation rate is represented by the slope)

The reproducibility of the measured precipitation rates, which was an average of the three measurements, was ±5%.

When calcite precipitates on calcite particles in systems with an adsorbed biopolymer, it can be expected that the formation of growth steps and their advancement is inhibited by the adsorbed polymer molecules and growth begins with the formation of surface nuclei (Dove and Hochella 1993). Within a certain period of time, t, after the beginning of the experiment, the initial growth rate, R_2, remains very low and increases only slightly with time (Fig. 10.1).

After the initial period of slow growth, the duration of which depends on supersaturation, the rate of precipitation, R_3, increases to about the initial value of R_0 for a pure system without biopolymers. This behavior was the same in all experiments. Depending on supersaturation and polymer concentration, the inhibition period ranged from 60 to 1000 s for the Alg experiments and from 30 to 2300 s for the experiments with pAsp (Karaseva et al. 2018).

To analyze the experimental data, we plot the dependence of ln (R/f(S)) on 1/ln S, which, according to Eqs. (10.5) and (10.7), should give a straight line for the exponential growth, typical when calcite precipitates in the presence of an adsorbed polymer. Figure 10.2 shows that the experimental data for calcite precipitation rate, plotted as ln (R/f(S))(f(S)) from Eq. (10.7) depending on 1/ln S, is successfully approximated by straight lines for experiments with both Alg and pAsp.

"Obtained for the exponential rate law, these straight lines indicate the change of rate determining mechanism for calcite precipitation in the presence of polymers from spiral growth to 2D nucleation mediated growth" (Karaseva et al. 2018).

Images obtained using atomic force microscopy (AFM, Fig. 10.3a) and scanning electron microscopy (SEM, Fig. 10.3b), show the formation of two-dimensional surface nuclei at Stage R_2, which leads to calcite growth at Stage R_3, observed with SEM (Fig. 10.4).

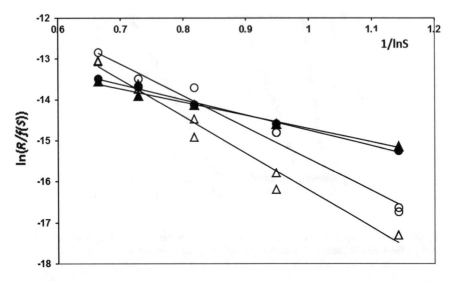

Fig. 10.2 $\ln(R/f(S))$ (from Eq. 10.7) plotted as a function of $1/\ln S$. Calcite seed with pAsp adsorbed from solutions of 0.32 mg L^{-1} (open triangles) and 32 mg L^{-1} (filled triangles); calcite seed with Alg adsorbed from 2 mg L^{-1} solution (open circles) and 20 mg L^{-1} (filled circles)

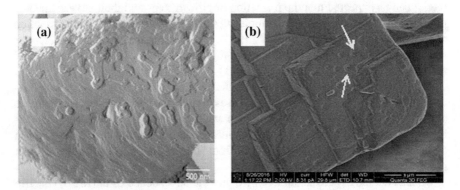

Fig. 10.3 **a** AFM amplitude image and **b** SEM image of a calcite sample grown at $S = 2.4$ in the presence of 0.32 mg L^{-1} pAsp, taken at the beginning of precipitation experiment (Stage R_2)

The slope of the lines in Fig. 10.2 is $-B_{2D}/3$, from which, using Eq. (10.8), we can estimate the free edge energy, κ, and the free interface energy, γ, from

$$\gamma = \kappa \bar{V}^{-1/3}, \tag{10.9}$$

where $\bar{V}$ represents molecular volume.

It is also possible to estimate the number of molecules of CaCO$_3$, n_{2D}, that participate in the elementary step of surface nucleus formation (Nielsen 1964; Verdoes et al. 1992):

(a)   **(b)**

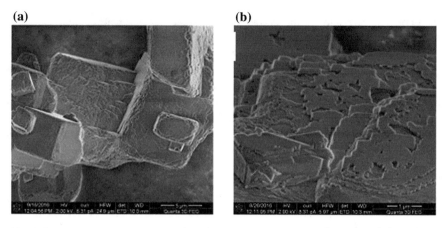

Fig. 10.4 Scanning electron microscopy images of calcite samples grown at $S = 2.4$ (Stage R_3) in the presence of **a** 2 mg L^{-1} Alg and **b** 0.32 mg L^{-1} pAsp

Table 10.2 Results from the regression analysis of the function ln $(R/f(S))$ versus 1/ln S and the related parameters

Polymer	C (10^{-3} g L^{-1})	Slope	B_{2D}	γ (mJ m^{-2})	N_{2D} ($S = 4.5$–2.4)
Alg	2	-7.7 ± 0.4	23.1	101	10–30
	20	-3.7 ± 0.2	11.1	71	5–14
pAsp	0.32	-7.7 ± 0.3	23.1	108	10–30
	32	-3.2 ± 0.2	9.6	65	4–13

$$n_{2D} = B_{2D}/\ln^2 S. \tag{10.10}$$

The results of the regression analysis for the dependencies presented in Fig. 10.2 and the corresponding calculated parameters are shown in Table 10.2.

The critical radius, r_{2D}, for 2D nucleation is half the size of a nucleus for homogeneous nucleation (Cubillas and Anderson 2010):

$$r_{2D} = \gamma \bar{V}/kT \ \ln S, \tag{10.11}$$

Thus, r_{2D} can be estimated to be 1.0–1.7 nm at low and 0.6–1.0 nm at high polymer concentration for S in the range from 4.5 to 2.4.

Using n_{2D} (Table 10.2), we can also calculate r_{2D}:

$$r_{2D} = \left(\bar{V}^{2/3} n_{2D}/\pi\right)^{1/2}, \tag{10.12}$$

which is in the range of 0.7–1.4 nm for low and 0.4–0.9 nm for high polymer concentrations for the same S, so the size of the critical nucleus obtained from the calculated n_{2D} values is slightly lower than the predicted value. One possible explanation might be that the combination "active surface center - adsorbed atom" (e.g. carboxyl group—Ca^{2+}) can be considered stable and, therefore, the active center itself can be a part of a critical nucleus (Milchev and Stoyanov 1976; Torrent et al. 1993). This is not taken into account in Eq. (10.12). "Obviously, this effect is expected to be more pronounced at high polymer concentration, where there is a smaller difference between the theoretical values for r_{2D} and the results obtained from precipitation kinetic experiments. We are aware that the results obtained here are of more or less formal physical sense but we provide them for comparison with similar studies (Verdoes et al. 1992; Gomez-Morales et al. 1996)" (Karaseva et al. 2018).

As can be seen from Table 10.2, all parameters are very similar for both polymers. The free interface energy, γ, for both polymers is comparable and lies between the theoretically estimated values for calcite (120 mJ/m^2) and vaterite (90 mJ/m^2) (Nielsen and Söhnel 1971; Söhnel and Mullin 1978). The number of CaCO$_3$ units in a two-dimensional nucleus, n_{2D}, from this work is very close to the result obtained by Verdoes et al. (1992) for the precipitation of CaCO$_3$ in the presence of polyacrylic acid and polymaleic and polystyrene sulfonic acid copolymer. Thus, we can conclude that our analysis of calcite precipitation in the presence of biopolymers, considering two-dimensional nucleation-mediated growth, provides realistic physical parameters (Karaseva et al. 2018).

Theoretically, the expected value of γ for calcite is 120 mJ/m^2 (Söhnel and Mullin 1978). For heterogeneous nucleation of calcite on the surface of calcite with an adsorbed polymer (with surface coverage, θ), the effective free interface energy, γ^{eff}, would consist of two parts. The first one is the free energy of the calcite–water interface (γ_{cw}, for surface areas that do not contain an adsorbed polymer). When calcite precipitates on an adsorbed polymer, three interface boundaries must be considered: calcite–water (γ_{cw}), adsorbed polymer–water (γ_{pw}), and calcite–adsorbed polymer (γ_{cp}) (Karaseva et al. 2018; Lakshtanov et al. 2015). Thus, the effective free interface energy, γ^{eff}, can be written as:

$$\gamma^{eff} = \gamma_{cw}(1 - \theta) + \gamma'\theta, \tag{10.13}$$

where

$$\gamma' = \gamma_{cw} + h\gamma_{cp} - h\gamma_{pw}, \tag{10.14}$$

And h represents the nucleus shape factor (1/2 for a hemisphere) (Giuffre et al. 2013) or with the Langmuir equations,

$$\theta = \frac{K_{ads}C}{1 + K_{ads}C}, \tag{10.15}$$

we can rewrite Eq. (10.13):

$$\gamma^{eff} = \gamma_{cw}\left(\frac{1}{1 + K_{ads}C}\right) + \gamma'\left(\frac{K_{ads}C}{1 + K_{ads}C}\right), \tag{10.16}$$

From Eq. (10.16), it can be seen that the effective free interface energy, γ^{eff}, varies from γ_{cw} for the calcite–water interface at a low polymer concentration (at low surface coverage), to the energy for the three-phase interface at high polymer concentration. This free energy of the three-phase interface is a combination of three γ values for simple two-phase interfaces (Eq. 10.14) (Karaseva et al. 2018).

Table 10.2 shows that the effective free interface energy, γ^{eff}, decreases with increasing concentration of polymer or with surface coverage, θ. Thus, the free energy of the three-phase interface γ' is lower than that of the calcite–water interface, γ_{cw}, which in turn means that $\gamma_{\text{cp}} < \gamma_{\text{pw}}$. In other words, to minimize the change in free energy, the polymer prefers to remove its interface with water and instead create a new polymer–calcite nucleus interface. Both polymers adsorbed on the calcite surface decrease the effective interface energy, γ^{eff}, reducing the energy barrier for nucleation of calcite (Karaseva et al. 2018). This is in agreement with the Gibbs adsorption equation, which shows that the adsorption of dissolved matter leads to decrease of surface free energy:

$$\Gamma = -\frac{1}{RT}\left(\frac{d\gamma}{dC}\right),\qquad(10.17)$$

where Γ represents the adsorption density and C, the concentration of the dissolved substance.

Alg and pAsp are strongly adsorbed on calcite, probably with rather difficult desorption (Manoli and Dalas 2002; Lakshtanov et al. 2011). Thus, we can expect that γ_{cp} is quite small and the difference in the energy of interaction between the polymer and the crystal, γ_{cp}, for different carboxylated polymers is insignificant (Giuffre et al. 2013). Thus, the energy barrier is mainly controlled by the hydrophilicity of the polymer through γ_{pw}. From Eqs. (10.8), (10.13) and (10.14), we see that the higher value of γ_{pw} contributes to the reduction of the energy barrier for nucleation. Then the activation barrier can become very low or disappear completely and calcite can form a continuous layer on the polymer (Lakshtanov et al. 2015).

10.6 Calcite Precipitation on Chalk Samples

We were interested in the initial period (0.5–3 min depending on the supersaturation) of precipitation, when calcite was precipitated on the initial surface of the chalk particles. Subsequently, calcite already grew on freshly precipitated calcite.

Figure 10.5 depicts calcite precipitation rates on various chalk samples. For comparison, the calcite precipitation rate on calcite (Sigma) seeds (green triangles, Lakshtanov et al. 2011; Karaseva et al. 2018) is also shown.

When calcite is precipitated on pure calcite, the dependence of the precipitation rate on relative supersaturation (Fig. 10.5) is satisfactorily described by the equation with the apparent reaction order (line slope) n = 1.8, which practically corresponds to the parabolic dependence of the precipitation rate on supersaturation. This means

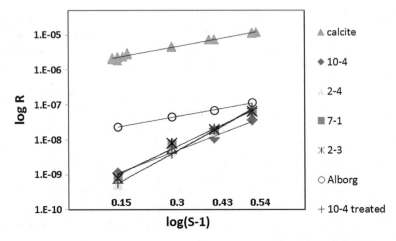

Fig. 10.5 Rate of calcite precipitation on the various chalk samples as a function of relative super-saturation. The data for pure calcite are from previous work, Lakshtanov et al. (2011), Karaseva et al. (2018) for comparison. The lines were calculated by linear regression of the data

that the rate-determining stage is surface diffusion (Burton et al. 1951). Figure 10.5 shows that, at low supersaturation, the rate of calcite precipitation on particles of all chalk samples, with the exception of Aalborg chalk, is 3 orders of magnitude lower than on the surface of pure calcite. For chalk Aalborg, which has been exposed to groundwater for 10,000 years, the precipitation rate is lower by 2 orders of magnitude. In addition, for all chalk samples (except for Aalborg chalk), the apparent reaction order n ranged from 3.7 to 5.4, which is much higher than 2. This indicates that calcite precipitation on the surface of chalk particles is caused (at least mainly) by growth due to surface nucleation (Eqs. 10.4 and 10.7). Calcite precipitation on the Aalborg chalk particles is described by almost parabolic dependence (n = 1.7), although it is much slower than on pure calcite. As a result of extraction of non-polar organic material from chalk 10-4, order of reaction n increased from 3.7 to 5.4, which indicates growth as a result of surface nucleation (Lakshtanov et al. 2015).

To analyse data on the rate of calcite precipitation on North Sea chalk samples with n > 2, we plot ln (R/f(S)) as a function of 1/ln S. If calcite growth on chalk particles obeys an exponential dependence, then according to Eqs. (10.4) and (10.7), these dependences should be described by straight lines that is clearly seen in Fig. 10.6 for all chalk samples.

These straight lines indicate a change in the mechanism of calcite precipitation on chalk surfaces from parabolic growth on pure calcite to growth due to surface nucleation (Lakshtanov et al. 2015). We observed a similar situation in the experiments described above with biopolymers adsorbed on calcite (Karaseva et al. 2018).

Using Eqs. (10.8) and (10.9), from the slope $B_{2D}/3$ of the lines, we can calculate the free interface energy γ. The calculation results for chalk samples before and after solvent extraction are shown in Table 10.3. For comparison, the free interface energy γ^v obtained by measuring the vapour adsorption is also indicated. To analyse

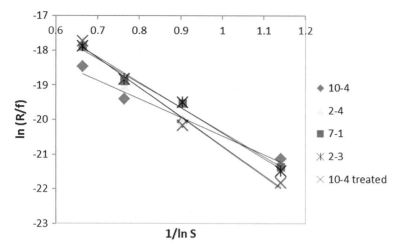

Fig. 10.6 ln ($R/f(S)$) (Eq. 10.7) plotted versus 1/ln S for calcite precipitation on the various chalk samples

Table 10.3 Interfacial free energy chalk-water (γ) obtained from the regression analysis of the function ln ($R/f(S)$) as a function of 1/ln S; γ^v obtained using the vapour adsorption method (Okhrimenko et al. 2014)

Chalk	γ (mJ m^{-2})		γ^v (mJ m^{-2})	
	Original	Treated	Original	Treated
2-3	99	–	–	–
2-4	106	–	–	–
7-1	96	–	153	176
10-4	84	106	55	80
Aalborg	–	–	291	263

changes in surface energy as a result of treatment, non-polar organic compounds were extracted from chalk samples in a Soxlet extractor using CH_2Cl_2/CH_3OH.

The values of the free interface energy obtained by the vapour adsorption method are the lowest for chalk with the highest content of non-polar organic matter (10-4) and the highest for chalk Aalborg, which, according to XPS, is characterized by the lowest content of organic matter. Since for the Aalborg chalk, the slope of the corresponding line in Fig. 10.5 is very close to 2, it can be stated that the precipitation of calcite on the surface of the Aalborg chalk particles is not the result of growth mediated by surface nucleation. Thus, unlike other chalk samples, free surface energy cannot be determined using Eqs. (10.7) and (10.8).

According to the results of measurements of vapour adsorption, after partial removal of non-polar organic matter from the chalk samples, free surface energy increases for all chalk samples of the North Sea, but decreases for Aalborg chalk.

This occurs as a result of the removal of organic matter, which makes the chalk surface less hydrophobic (Lakshtanov et al. 2015). The behaviour of Aalborg chalk is probably due to a significant difference in surface composition and low reproducibility of measurements of vapour adsorption (Okhrimenko et al. 2014). According to XPS data (Okhrimenko et al. 2014), Aalborg chalk contains twice as much silicon as amorphous silica and much less surface organic carbon (compounds with C–C and C–H bonds) than in the North Sea samples. The impact of the biosphere, organic acids from overlying soils, organic compounds from organisms living on the surface of the soil, and the presence of bacteria in groundwater could have a significant impact on the initial surface composition of the Aalborg chalk (Lakshtanov et al. 2015). The work of wetting remained virtually unchanged after the extraction of organic matter. Bearing in mind the low reproducibility of the results for Aalborg chalk, differences in surface energy before and after extraction can be considered insignificant within 5%. Therefore, we concluded that Aalborg chalk is insensitive to solvent extraction (Lakshtanov et al. 2015).

From the data in Figs. 10.5 and 10.6, it can be seen that, despite the very different origin of the chalk samples from completely different geographical areas and depths, from the water and gas zones, the behavior of all chalk samples of the North Sea was almost identical. The values of free surface energy γ obtained from the analysis of the kinetics of precipitation are comparable for all chalk samples and are in the range between the theoretically calculated values for calcite ($120 \, mJ/m^2$) and laterite ($90 \, mJ/m^2$) (Nielsen and Söhnel 1971; Söhnel and Mullin 1978). Thus, the analysis of the kinetics of calcite precipitation on the surface of chalk particles in accordance with the growth mechanism due to surface nucleation gives realistic physical quantities.

The slope of the straight line in Fig. 10.4 for chalk 10-4, the sample with the highest content of organic matter (Table 10.1), increased after extraction of non-polar organic matter, approaching the slope of the lines for other chalk samples. Accordingly, the thermodynamic barrier of nucleation and the free surface energy of chalk 10-4 also increased after extraction. Thus, the extraction of non-polar organic matter changed the surface energy, but did not affect the calcite precipitation mechanism, which is still determined by the growth due to surface nucleation (Lakshtanov et al. 2015). Polysaccharides, which are also present on the surface of the chalk, interact with calcite much more strongly than hydrocarbons, and cannot be extracted by our method. However, they can be oxidized to some extent with H_2O_2. Non-polar organic compounds adsorbed on the surface of calcite reduce the effective free energy of the interface, reducing the energy barrier to the nucleation of calcite (Lakshtanov et al. 2015).

10.7   Electrokinetics

10.7.1   Calcite in the Presence of Biopolymers

For a deeper understanding of the interaction of biopolymers with calcite surface, we have estimated ζ-potential on calcite samples in the presence of biopolymers. The measurements were carried out in a solution saturated with calcite with 0.001 M NaCl in equilibrium with atmospheric CO_2. Figure 10.7 presents the ζ-potential determined from measurements of the streaming potential on the samples studied. ζ-potential for pure calcite at pH < 8.5 is slightly positive and becomes negative at a higher pH. This is consistent with the model of calcite–water interface (Van Cappellen et al. 1993; Pokrovsky et al. 2000), which predicts the point of zero charge at pH = 8.2.

Calcite samples in the presence of biopolymers have a much negative ζ-potential than for pure calcite that decreases with increasing pH. There is much evidence in the literature that ζ-potential of calcite, including natural samples (e.g. chalk), as well as synthetic calcite, reflects the adsorption effect of organic substances, reducing the positive surface charge of calcite and making it negative (Cicerone et al. 1992; Vdović and Bišćan 1998; Vdović 2001).

The behavior of ζ-potential for calcite with adsorbed biopolymers compared to pure calcite indicates an insignificant contribution of electrostatics to the interaction energy. Since at pH > 8.5 the surface of calcite and the deprotonated carboxyl groups of the polymers are negatively charged and polymers are adsorbed regardless of their charge, electrokinetic measurements prove the presence of a specific coordination interaction between the carboxyl groups of biopolymers and the surface of calcite (Karaseva et al. 2018). Specific adsorption is the main reason for the strong inhibition

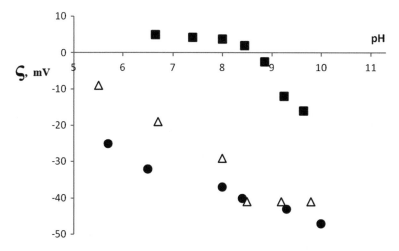

Fig. 10.7 ζ-potential of pure calcite seed overgrown in a pure system without polymers (squares), in the presence of 2 mg L^{-1} Alg (circles) and 0.32 mg L^{-1} pAsp (triangles)

of calcite precipitation by these organic compounds. Moreover, at pH > 8, when the carboxyl groups of the polymers are completely dissociated, ζ-potential of calcite with adsorbed Alg or pAsp becomes the same (Fig. 10.7). This may indicate a close adsorption density for both polymers, as well as a very similar amount of surface ionogenic groups. From this, we can conclude that the electrokinetic measurements are consistent with the results of inhibition of calcite precipitation by biopolymers described above (Karaseva et al. 2018).

10.7.2 Chalk Samples

Streaming potential measurements on chalk samples were carried out in solutions saturated with respect to the corresponding chalk with the addition of 0.001 M NaCl. The streaming potential was also measured on chalk samples 2-3, which were treated with H_2O_2 (2-3t) (Belova et al. 2012) to remove polysaccharides.

The surface of the chalk samples is characterized by a much more negative ζ-potential than pure calcite (Fig. 10.8). A similar behavior shows the ζ-potential of calcite with adsorbed biopolymers (Fig. 10.7). This is seen in Fig. 10.9, where the ζ-potential for calcite and chalk samples is plotted as a function of the relative surface concentration of C–O bonds. The presence of C–O bonds on the surfaces of chalk particles is due to the presence of polysaccharides that are produced by algae in order to control the morphology of their coccoliths (Belova et al. 2012). Thus, ζ-potential is a measure of the concentration of polysaccharides on the surface of chalk particles. The behavior of chalk 2-3 is indicative, when ζ-potential becomes less negative after treatment with H_2O_2, approaching that for pure calcite (Fig. 10.9).

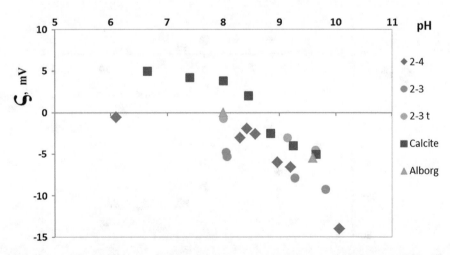

Fig. 10.8 pH dependence of ζ-potential for calcite and the chalk samples. Chalk 2-3t had been treated with H_2O_2 in an attempt to remove the polysaccharides (Belova et al. 2012)

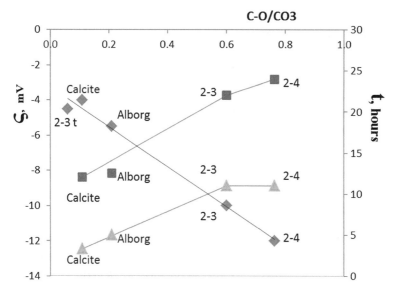

Fig. 10.9 ζ-potential (rhombs) of calcite and chalk samples at pH 9.6 as a function of the relative surface concentration of C–O bonds. Chalk 2-3t had been treated with H_2O_2 (Belova et al. 2012) and induction time for homogeneous nucleation during CO_2 degassing from solutions saturated with respect to calcite as well various chalks, monitored by pH drop (squares) and by turbidity (triangles)

ζ-potential of Aalborg chalk is also close to ζ-potential of calcite, which is in agreement with the low content of organic matter on the surface (Table 10.1). As for calcite with adsorbed biopolymers, the behavior of ζ-potential for chalk samples indicates an insignificant contribution of electrostatics to the interaction energy and thus, a specific coordinative interaction between the functional groups of polysaccharides and the calcite surface (Lakshtanov et al. 2015).

10.8 Induction Period of Homogeneous Nucleation

Figure 10.9 also shows the values of the induction period for homogeneous nucleation as a result of CO_2 degassing from solutions saturated with respect to calcite, as well as for different chalk samples at 100% partial pressure, i.e. $CO_2 = 1$ atm, then placed in atmospheric conditions ($pCO_2 = 10^{-3.5}$). This so-called fast controlled precipitation (FCP) method is often used to assess the effectiveness of scale inhibitors (Gauthier et al. 2012; Hamdi et al. 2016).

During CO_2 degassing, the supersaturation of the solution with respect to calcite, S, is not constant, but increases (for calcite, $S = 1$–5.9 at $t = 0$ to t_{ind}, for chalk 2-3, $S = 1$–8.1 at $t = 0$ to t_{ind}), so the measured values of the induction period cannot be used in the equations of nucleation theory to calculate, for example, the free surface energy of the resulting phase. However, these induction period values can be used to compare between different samples of chalk and calcite. The induction period identified by the pH jump is significantly higher than that measured from turbidity, which is explained by much slower degassing. There is almost 2 times less contact area of the solution with the air because of the pH-microelectrode introduced into cuvette. However, the curves for the induction period data (Fig. 10.9), obtained by different methods, have the same shape.

Figure 10.9 shows that the induction period for homogeneous nucleation for all samples, as well as the values of ζ-potential, correlate well with the relative surface concentration of C–O bonds, i.e. with the concentration of polysaccharides on the surface of the samples. For chalk samples from the North Sea Basin (Samples 2-3 and 2-4), the induction period is much longer than for pure calcite or for Aalborg chalk, which is explained by a significantly higher surface concentration of organic matter, in particular, polysaccharides, on the surface of these samples.

As shown above, adsorbed biopolymers reduce the effective free surface energy, γ^{eff}, and consequently, for high supersaturation, decrease the nucleation rate, i.e. increase the induction period, as we have observed in the experiments.

10.9 Free Surface Energy Determined by Measurements of Vapor Adsorption

The obtained values of the total surface energy, γ_s, and the free energy of calcite–water interface, γ_{sw}, are given in Table 10.4. The relative error in determining the amount of adsorbed vapor as well as the work of wetting is 2%. The total surface energy γ_s determined for a sample of pure calcite is in good agreement with our early data (Lakshtanov et al. 2015), but higher than that by Janczuk and Bialopiotrowicz (1986) for a similar system, 170 mJ/m^2. The free interface energy, γ_{sw}, was calculated from the equation:

$$\gamma^{SW} = \gamma^S + \gamma^L - W_A, \tag{10.18}$$

where W_A is work of adhesion.

We estimated the free surface energy of calcite–water to be 129.3 mJ/m^2. This value is close to 120 mJ/m^2 (Söhnel and Mullin 1982). It should be noticed that the choice of the reference vapor affects the results for the total surface energy and the distribution of its polar and dispersion components. During polymer adsorption on calcite, the total surface energy decreases, the distribution of the dispersion and polar components changes and the polar component decreases while the dispersion component grows (Karaseva et al. 2018). At a higher concentration of biopolymer and with an increase of adsorption period, this effect was more evident. Decrease of the free interface energy is shown in Fig. 10.10.

Table 10.4 Surface energies for calcite and calcite with adsorbed Alg and pAsp

Sample	Conc. (g/L)	Time of adsorption days	Surface energy γ^s (mJ/m^2)			Interface energy γ^{sw} (mJ/m^2)
			Dispersive	Polar	Total	
Pure calcite	–	–	2.9	328.3	331.2	129.3
Calcite + Alg	0.013	2	81.5	211.9	293.4	74.0
	0.1	2	12.3	331.6	343.9	123.9
	0.17	2	76.9	204.3	281.3	68.0
	0.013	24	53.0	261.8	314.8	88.5
	0.1	24	91.0	199.2	290.2	72.3
	0.1	24	90.0	246.3	336.3	96.3
	0.17	24	39.3	262.4	301.8	84.6
	0.17	24	18.3	327.1	345.4	119.9
Calcite + pAsp	0.0043	2	33.0	260.3	293.3	82.0
	0.0087	2	49.7	176.7	226.4	43.5
	0.025	2	36.0	218.3	254.2	60.1
	0.0435	2	31.1	236.0	267.1	68.4
	0.0435	2	36.8	285.8	322.7	97.3
	0.0043	24	18.9	234.8	253.7	67.1
	0.0087	24	25.6	159.9	185.6	30.5
	0.025	24	56.0	164.5	220.4	40.2
	0.0435	24	93.9	102.4	196.3	34.1
	0.87	24	1.8	947.5	949.4	569.8

We observed that the effect is more significant for pAsp than for Alg, as well as at higher adsorption time. For example, the free energy of the solid–water interface decreases from 129.3 mJ/m^2 for pure calcite to 30.5 mJ/m^2 at pAsp concentration of 8.7 mg/L and an adsorption time of 24 h. However, with an increase at pAsp concentration to 870 mg/L, the surface energy is 569.8 mJ/m^2, which indicates a significant adsorption of water. Probably at high pAsp concentrations, biopolymer completely covers calcite surface and as a result, the calcite–water interface disappears. In this case, the interaction of water molecules occurs not with the surface of calcite, but mainly with pAsp, which is able to retain water and form hydrogels (Meng et al. 2015); high water absorption leads to incorrect surface energy values.

The effective free energy of calcite–water interface obtained from vapor adsorption measurements differs from that obtained from the kinetics of calcite precipitation. As noted above, the effective free interface energy γ^{eff} for calcite nucleation on polymer consists of two parts. The free energy of the calcite–water interface, γ_{cw}, for surfaces areas not containing adsorbed polymers is the first part, and the second part is the free energy of the three-phase interface, γ' (calcite–polymer–water). In the method of the vapor adsorption, γ_{cw} and γ' in Eqs. (10.13) or (10.16) are unchanged.

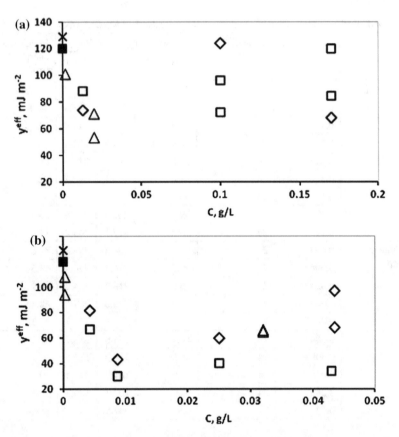

Fig. 10.10 The effective interface free energy, γ^{eff}, for calcite with adsorbed polymers. **a** Alg and **b** pAsp (obtained by several methods) versus polymer concentration: triangles, from precipitation kinetics; squares, from vapour adsorption measurements (24 h polymer adsorption); rhombi, from vapour adsorption measurements (2 h polymer adsorption); crosses, from vapour adsorption measurements (pure calcite); filled squares, theoretical value, γ_{cw}, for a pure calcite–water interface (Söhnel and Mullin 1978)

However, water or alcohol (reference vapor) is also adsorbed on the polymer; therefore, it is necessary to introduce an additional term related to the energy of the polymer–water interface. During polymer adsorption on calcite, carboxyl groups interact with calcium ions on calcite surface. At low polymer concentration and, therefore, with a low surface coverage, the polymer is adsorbed in the form of fully elongated chains, most of whose segments are attached to the surface and cannot rotate freely (Fernandez et al. 1992). Polymer molecules are located in flat position on the surface, and only a small fraction of the functional groups is free.

With increasing polymer concentration, the surface coverage increases, and some parts of the chains are still firmly attached to the surface, while others are turned into the solution in the form of loops or clusters on which water is adsorbed (Tsortos and

Nancollas 2002). In addition, some parts of the macromolecules can be attached to the surface by their carboxylic groups, leaving more hydrophobic parts facing the solution, thereby promoting the adsorption of other organic molecules (Matthiesen et al. 2017). For general reasons, the number of chain segments that are not attached to the surface is expected to increase nonlinearly with increase of surface coverage, i.e. increase significantly when there is little free space on the surface. Thus, it can be assumed that with an increase in the surface coverage, the number of chain segments extending into the solution is expected to increase nonlinearly i.e., significantly increase, with a decrease in free space on the surface. At relatively low surface coverage, the hydrophilic carboxyl groups of the polymers react with the calcite surface, and therefore, more hydrophobic parts of the polymer molecule remain in the solution. Thus, in the method of vapor adsorption, with increasing adsorption time and polymer concentration, the influence of the polymer–water interface increases and consequently, the effective free interface energy, γ^{eff}, increases and tends to the value of γ_{pw} in the limit. This dependence can be seen in Fig. 10.10. As polymer concentration increases γ^{eff} first decreases according to Eq. (10.16) and then increases because of the dominance of polymer–water interface. Therefore, the values of free surface energy differ from the data obtained using the precipitation kinetics, where the effective free interface energy, γ^{eff}, varies from γ_{cw} for pure calcite to γ' for full surface coverage, $\theta = 1$. Based on published data (Lakshtanov et al. 2011; Tsortos and Nancollas 1999), we can assume that for Alg this condition is satisfied at C > 0.005 g/L, while for pAsp—at C > 0.01 g/L. Therefore, data obtained by method of vapor adsorption at higher concentrations of Alg or pAsp correspond to the exclusive free energy for the polymer–water interface γ_{pw}. Taking this into consideration, it can be confirmed that the method of vapor adsorption provides reasonable values for the free surface energy, γ, which are very close to those obtained by the analysis of the calcite precipitation kinetics in the presence of biopolymers.

10.10 Concluding Remarks

A distinctive feature of biogeochemical systems is the interaction of biopolymers with minerals. In the case of calcite, the presence of biopolymers leads to a change in the mechanism of its precipitation and inhibition of growth. Pure calcite grows by the second-order surface reaction mechanism. The presence of biopolymers initiates growth due to surface nucleation. Both alginate (Alg) and polyaspartate (pAsp) slow down the growth of calcite, the stronger the higher their concentration and the lower the supersaturation of the solution.

Rate of calcite precipitation on the surface of chalk particles also does not obey the parabolic law. Instead, the dependence of calcite deposition rate on supersaturation is exponential, as is the case of calcite growth with adsorbed biopolymers, which indicates growth due to surface nucleation.

The values of free surface energy, the most important parameter for controlling nucleation and growth, estimated from the analysis of precipitation kinetics, as well as from the data on vapor adsorption, are practically the same for alginate and polyaspartate. This is confirmed by electrokinetic measurements, which show the same ζ-potential for calcite with each of the adsorbed polymers. An increase in polymer concentration and adsorption time leads to a gradual decrease in γ^{eff}, which significantly reduces the critical supersaturation required for surface nucleation.

The values of the free surface energy γ obtained from the analysis of the kinetics of precipitation are comparable for all chalk samples. Non-polar organic compounds, such as biomarkers, reduce surface energy, but do not alter the calcite precipitation mechanism. The presence of polysaccharides on the surface of chalk particles determines the mechanism of calcite precipitation on chalk. An analysis of the kinetics of calcite precipitation on the surface of chalk particles in accordance with the growth mechanism due to surface nucleation gives realistic physical parameters.

Thus, at moderate surface coverage due to adsorption of the biopolymer, the effective free energy of the interface decreases significantly compared to that for pure calcite–water interface. This means a significant decrease in the energy barrier to surface nucleation, which, in turn, leads to a very weak dependence of the calcite precipitation rate on supersaturation. That is, at very low supersaturations characteristic for recrystallization, the precipitation rate may not be too low, remaining not too high at large supersaturations. This can explain why calcite recrystallization in the presence of alginate is significantly faster than expected (Lakshtanov et al. 2015). With a significant increase of surface coverage, the effective interface free energy again increases, increasing the energy barrier of nucleation, which for very low supersaturations means a complete suppression of recrystallization.

Acknowledgements Financial support of the IEM RAS project № AAAA-A18-118020590152-0.

References

Bathurst RGC (1975) Carbonate sediments and their diagenesis. Development in sedimentology. Elsevier, Amsterdam, 658 p

Belova DA, Johnsson A, Bovet N, Lakshtanov LZ, Stipp SLS (2012) The effect on chalk recrystallization after treatment with oxidizing agents. Chem Geol 291:217–223

Borman AH, de Jong EW, Huizinga M, Kok DJ, Westbroek P, Bosch L (1982) The role in CaCO$_3$ crystallization of an acid Ca^{2+}-binding polysaccharide associated with coccoliths of *Emiliania huxleyi*. Eur J Biochem 129:179–183

Buol SW, Hole FD, McCracken RJ (1973) Soil genesis and classification. Iowa State University Press, Ames, 360 p

Burton WK, Cabrera N, Frank FC (1951) The growth of crystals and the equilibrium structure of their surfaces. Philos Trans R Soc Lond 243:299–358

Cabrera N, Vermilyea DA (1958) The growth of crystals from solution. In: Doremus RH, Roberts BW, Turnbull D (eds) Growth and perfection of crystals. Chapman & Hall, London, pp 393–410

Cicerone DS, Regazzoni AE, Blesa MA (1992) Electrokinetic properties of the calcite/water interface in the presence of magnesium and organic matter. J Colloid Interface Sci 154:423–433

Crabtree M, Eslinger D, Fletcher P, Johnson A, King G (1999) Fighting scale—removal and prevention. Oilfield Rev 11:30–45

Cubillas P, Anderson MW (2010) Synthesis mechanism: crystal growth and nucleation. In: Čejka J, Corma A, Zones S (eds) Zeolites and catalysis: synthesis, reactions and applications, vol 1. Wiley-VCH Verlag GmbH & Co KgaA, Weinheim, Germany, pp 1–55

Dove P, Hochella MF (1993) Calcite precipitation mechanisms and inhibition by orthophosphate: in situ observations by scanning force microscopy. Geochim Cosmochim Acta 57:705–714

Fernandez VL, Reimer JA, Denn MM (1992) Magnetic resonance studies of polypeptides adsorbed on silica and hydroxyapatite surfaces. J Am Chem Soc 114:9634–9642

Gauthier G, Chao Y, Horner O, Alos-Ramos O, Hui F, Lédion J, Perrot H (2012) Application of the fast controlled precipitation method to assess the scale-forming ability of raw river waters. Desalination 299:89–95

Giuffre AJ, Hamm LM, Han N, De Yoreo JJ, Dove PM (2013) Polysaccharide chemistry regulates kinetics of calcite nucleation through competition of interfacial energies. Proc Natl Acad Sci USA 23:9261–9266

Gomez-Morales J, Torrent-Burgues J, Rodriguez-Clemente R (1996) Nucleation of calcium carbonate at different initial pH conditions. J Cryst Growth 169:331–338

Hamdi R, Khawari M, Hui F, Tlili M (2016) Thermodynamic and kinetic study of $CaCO_3$ precipitation threshold. Desalin Water Treat 57:6001–6006

Janczuk B, Bialopiotrowicz T (1986) Spreading of a water drop on a marble surface. J Mater Sci 21:1151–1154

Karaseva ON, Lakshtanov LZ, Okhrimenko DV, Belova DA, Generosi J, Stipp SLS (2018) Biopolymer control on calcite precipitation. Cryst Growth Des 18:2972–2985

Lakshtanov LZ, Bovet N, Stipp SLS (2011) Inhibition of calcite growth by alginate. Geochim Cosmochim Acta 75:3945–3955

Lakshtanov LZ, Belova DA, Okhrimenko DV, Stipp SLS (2015) Role of alginate in calcite recrystallization. Cryst Growth Des 15:419–427

Lattemann S, Höpner T (2008) Environmental impact and impact assessment of sea-water desalination. Desalination 220:1–15

Manoli F, Dalas EJ (2002) The effect of sodium alginate on the crystal growth of calcium carbonate. J Mater Sci Mater Med 13:155–158

Marsh ME (2003) Regulation of CaCO3 formation in coccolithophores. Comp Biochem Physiol, Part B: Biochem Mol Biol 136:743–754

Matthiesen J, Hassenkam T, Bovet N, Dalby KN, Stipp SLS (2017) Adsorbed organic material and its control on wettability. Energy Fuels 31:55–64

Meng H, Zhang X, Chen Q, Wei J, Wang Y, Dong A, Yang H, Tan T, Cao H (2015) Preparation of poly(aspartic acid) superabsorbent hydrogels by solvent-free processes. J Polym Eng 35:647–655

Milchev A, Stoyanov S (1976) Classical and atomistic models of electrolytic nucleation: comparison with experimental data. J Electroanal Chem 72:33–43

Montanari G, Lakshtanov LZ, Tobler DJ, Dideriksen K, Dalby KN, Bovet N, Stipp SLS (2016) Effect of aspartic acid and glycine on calcite growth. Cryst Growth Des 16:4813–4821

Nielsen AE (1964) Kinetics of precipitation. Pergamon, 153 p

Nielsen AE, Söhnel O (1971) Interfacial tension electrolyte crystal-aqueous solution, from nucleation data. J Cryst Growth 11:233–242

Nielsen AE, Toft JM (1984) Electrolyte crystal growth kinetics. J Cryst Growth 67:278–288

Okhrimenko DV, Dalby KN, Skovbjerg LL, Bovet N, Christensen JH, Stipp SLS (2014) The surface reactivity of chalk (biogenic calcite) with hydrophilic and hydrophobic functional groups. Geochim Cosmochim Acta 128:212–224

Parkhurst DL (1995) User's guide to PHREEQC—a computer program for speciation, reaction-path, advective transport, and inverse geochemical calculations. In: US geological survey water-resources investigations report 95-4227, 143 p

Pokrovsky OS, Mielczarski JA, Barres O, Schott J (2000) Surface speciation models of calcite and dolomite/aqueous solution interfaces and their spectroscopic evaluation. Langmuir 16:2677–2688

Reddy MM (1978) Kinetic inhibition of calcium carbonate formation by wastewater constituents. In: Rubin AJ (ed) Chemistry of wastewater technology, vol 3, pp 1–58. Ann Arbor Science

Sand KK, Pedersen CS, Sjöberg S, Nielsen JW, Makovicky E, Stipp SLS (2014) Biomineralization: long-term effectiveness of polysaccharides on the growth and dissolution of calcite. Cryst Growth Des 14:5486–5494

Skovbjerg LL, Okhrimenko DV, Khoo J, Dalby KN, Hassenkam T, Makovicky E, Nielsen AE, Toft JM (1984) Electrolyte crystal growth kinectics. J Cryst Growth 67:278–288

Söhnel O, Mullin JW (1978) A method for the determination of crystallization induction periods. J Cryst Growth 44:377–382

Söhnel O, Mullin JW (1982) Precipitation of calcium carbonate. J Cryst Growth 60:239–250

Stipp SLS, Hochella MF (1991) Structure and bonding environments at the calcite surface as observed with X-ray photoelectron spectroscopy (XPS) and low energy electron diffraction (LEED). Geochim Cosmochim Acta 55:1723–1736

Torrent J, Rodriguez R, Sluiters JH (1993) The induction time in the electrocrystallization of lead chloride and lead sulphate on an amalgamated lead electrode. J Cryst Growth 131:115–123

Tsortos A, Nancollas GH (1999) The adsorption of polyelectrolytes on hydroxyapatite crystals. J Colloid Interface Sci 209:109–115

Tsortos A, Nancollas GH (2002) The role of polycarboxylic acids in calcium phosphate mineralization. J Colloid Interface Sci 250:159–167

Van Cappellen P, Charlet L, Stumm W, Wersin PA (1993) Surface complexation model of the carbonate mineral-aqueous solution interface. Geochim Cosmochim Acta 57:3505–3518

Van der Leeden MC, Kashchiev D, van Rosmalen GM (1993) Effect of additives on nucleation rate, crystal growth rate and induction time in precipitation. J Cryst Growth 130:221–232

Vdović N (2001) Electrokinetic behaviour of calcite—the relationship with other calcite properties. Chem Geol 177:241–248

Vdović N, Bišćan J (1998) Electrokinetics of natural and synthetic calcite suspensions. Colloids Surf A 137:7–14

Verdoes D, Kashchiev D, Rosmalen GM (1992) Determination of nucleation and growth rates from induction times in seeded and unseeded precipitation of calcium carbonate. J Cryst Growth 118:401–413

Wang Z, Neville A, Meredith AW (2005) How and why does scale stick—can the surface be engineered to decrease scale formation and adhesion? In: SPE inter-national symposium on oilfield scale, SPE 94993. Aberdeen, United Kingdom, pp 1–8

Zhao Q, Liu Y (2004) Investigation of graded Ni–Cu–P–PTFE composite coatings with anti-scaling properties. Appl Surf Sci 229:56–62

Chapter 11
The Evolutionary Types of Magmatic Complexes and Experimental Modeling Differentiation Trends

N. I. Bezmen and P. N. Gorbachev

Abstract Many minerals are associated with massifs of magmatic origin, in particular, sulfide copper-nickel ores, chromite and titanium-magnetite deposits, deposits of the platinum group elements (PGE), etc., so many aspects of petrology, in particular petrochemical analysis of the structure and differentiation of massifs are also gaining economic importance in addition to petrological. Despite the great success in the study of the magmatic formations of specific regions: the Baltic Crystal Shield, the Voronezh Crystalline Massif, Transbaikalia and Siberia, South Africa and other regions of the world, many geological, petrological and geochemical features of the evolution of magmatic massifs have not yet been fully studied, or performed at a low physical and chemical level. In the present article the petrochemical systematization of intrusive magmatic complexes has been developed, which became the basis for the petrologic and metallogenic classification of differentiated intrusions with the separation of massifs, promising for the content of ore mineralization.

Keywords Experiment · Intrusions · Petrology · Diagram · Norilsk · Stratified intrusions · Differentiation

11.1 Introduction

Petrochemical systematization of massifs was based on the types of intrachamber differentiation, which are clearly revealed on petrochemical diagrams by trends in the evolution of magmatic complexes of different formations. As it was shown experimentally, superliquidus nanocluster differentiation of fluid saturated melts occurs at the level of complexes (associates, cybotaxises) close to minerals in structure (Bezmen 1992; Bezmen and Gorbachev 2017). The existing petrochemical diagrams, for which a limited number of components are used, are not quite suitable for studying the processes of intrachamber differentiation of melts, as they are usually used to

N. I. Bezmen (✉) · P. N. Gorbachev
D.S. Korzhinskii Institute of Experimental Mineralogy, Russian Academy of Sciences, Academician Osipyan Street, 4, Chernogolovka, Moscow Region, Russia 1423432
e-mail: bezmen@chgnet.ru

© The Editor(s) (if applicable) and The Author(s), under exclusive license to Springer Nature Switzerland AG 2020
Y. Litvin and O. Safonov (eds.), *Advances in Experimental and Genetic Mineralogy*, Springer Mineralogy, https://doi.org/10.1007/978-3-030-42859-4_11

generalize the normative compositions of rock-forming minerals: olivine and pyroxene on the one hand, feldspartoids and quartz on the other. In this connection, we have developed a petrochemical diagram on which almost all normative rock-forming minerals (olivine, pyroxene, plagioclases, quartz, nepheline, leucite, spinel, chromite, titanomagnetite, etc.) are fixed by figurative points.

11.2 Methodological Features

11.2.1 Petrochemical Diagrams Construction Method

The whole spectrum of stratified rocks of the massifs (ultrabasic, basic, acidic, alkaline and even ores) is compared on the diagram. It is based on the separation of femical components, sodium-calcium feldspar and silica. Since the role of calcium is ambiguous: on the one hand, it is concentrated together with aluminum, bind in the main plagioclase, on the other—is accumulated together with iron in clinopyroxenes, the total calcium content was divided by calculation:

$$Ca^{Pl} = 0.5(Al + Na + K), \quad Ca^{Px} = Ca - 0.5(Al + Na + K).$$

On the diagram $Mg + Fe + Mn + Ca^{Px} + Cr + Ti + P(Na + K + Ca^{Pl} + Al) - Si$, in at.%, petrochemical differences of rocks are clearly distinguished, as all main rock-forming elements are used.

The some interesting trends of superliquidus nanocluster differentiation were confirmed experimentally.

11.2.2 Methodology of Experiment

As a rule, the fluid phase of massifs contains water, hydrogen, fluorine and other fluid components, so experiments were conducted in the installation of high gas pressure at the controlled fugacities of gases in the system H–O–C–F–Cl, the main components of the magmatogenic fluid. The peculiarity of the experimental method was the direct dosing of hydrogen in the composition of the fluid by means of its diffusion through the walls of platinum ampoules with the use of improved methods of the Show membrane (Bezmen et al. 2016) and the control of fugacities of other gases in the system H–O–C buffer reactions with carbon or Ni–Fe in the presence of hydrogen-containing vistite. The scheme of a hydrogen cell with a charged ampoule is described in detail in the works (Bezmen 2001; Bezmen et al. 2016). The glass powder made of oxides, corresponding to the weighted average composition of the trend rocks, was used as the initial silicate composition. In order to avoid iron diffusion into the platinum ampoule from the melt during the experiment, the experiments were carried out in vitreous carbon or Ni–Fe of the alloy crucibles, which were also indicators of carbon activity or oxygen fugacity. Fluorine and chlorine were injected as an acid solution. The crucible was inserted into a platinum ampoule with a diameter of 8 mm and a height of 50 mm with a wall thickness of 0.2 mm. Then water was poured into the ampoule, the weight of which was determined by temperature and pressure of the run. The presence of water in the reaction ampoule after the experiment was considered a prerequisite for the reliability of the experiment. The welded ampoule was inserted into the Re-reactor, which was filled under pressure of 100 atm with argon-hydrogen mixture with a given molar fraction of hydrogen.

The samples were analyzed with a wide probe (20–40 μm) on the Tescan Vega II XMU scanning electron microscope (SEM) with energy dispersive (INCAx-sight) and wave (INCA wave 700) X-ray spectrometers. The location of the analyses in the figures is shown by symbol.

11.3 Research Results

11.3.1 Differentiation Types of Magmatic Complexes

The majority of differentiated magmatic complexes is located within the limits of ancient shields among deeply metamorphosed formations of the Archean or Proterozoic and have pre-Cambrian age (Bushveld, Stillwater, etc.). Less frequently stratified massifs date back to the Early Paleozoic age, being located on the activated margins of ancient platforms or in the zones of tectonomagmatic activation of consolidated median massifs. Single layered intrusions of later age up to tertiary (Skaergaard, Ram Island intrusions) are known. Regardless of age, at the early stages of development of folded regions, magmatism is represented by the effusives of komatiite or picrite-basalt composition, which in time are replaced by ophiolite plutonic complexes. Stratified differentiated stratiform massifs are formed in the consolidated areas of the Earth's crust, i.e. they are typical postorogenic formations associated with the stages of tectonic-magmatic activation of rigid structures. A special formation consists of differentiated trap intrusions formed under shallow depths in the effusive-sedimentary cover of the platforms.

Thus, magmatic complexes are characterized by different tectonic mode of formation and, as a consequence, were formed in different physicochemical conditions, as well as under the pressure of the magmatic fluid of different compositions, which directly determines the evolution of intrachamber melts, i.e. the type of differentiation. That is why stratified massifs belonging to the same type of differentiation, but formed in different geological epochs and in different regions (and even on different continents) have close and stable characteristics of structure and ore specialization (for example, massifs of Bushveld, South Africa, Stillwater, USA and the Fedorovo-Panskie tundras, Kola Peninsula).

Ophiolite complexes. Geological formations of greenstone belts are the oldest rocks within the shields, having the features of ophiolite associations. Ore formations, which are spatially and genetically related to volcanosedimentary and intrusive complexes of komatiites, form in general a metallogenic specialization, typical for ophiolite complexes, of course, with their own peculiarities, probably due to the restorative regime of the processes of magmatism, metamorphism and ore formation, which took place at the early stages of the Earth's evolution. Components of this association are: bodies of ultrabasic composition containing chromite mineralization; intrusions of gabbroids with copper-nickel mineralization; formation and vein bodies of gold ore and copper-polymetallic (kies) deposits; in the main metavulcanoes of ferrous quartzite deposits (Watson 1980). In addition, the formational differences of the ancient greenstone belts are undoubtedly conditioned by the structures of the rift location (Grachev and Fedorovsky 1980; Moralev 1986).

In the full bottom-up ophiolite series the following types of rocks are presented: a complex of intrusive rocks consisting of harzburgites, lherzolites, wehrlites, dunites, peridotites, gabbroids and plagiogranites (Fig. 11.1); a complex of parallel dikes (sheeted complex) of the main composition, sometimes containing dikes and bodies of irregular shape of picrites and pyroxenites (perknites, Coish et al. 1982); volcanic complex, usually composed of pillow lava and more leucocratic ball-and-pillow rocks. The stratigraphic analogy of the ophiolite and oceanic crust has led many researchers to the concept of the oceanic origin of the ophiolite and the subsequent thrust to the continents in the subduction zones (Colman 1979). However, more detailed geological and petrological studies show their genetic differences. Still Miyashiro (1975) is tried to identify three types of ophiolite associations, relying on the petrochemical features of the volcanic series. Then such attempts were made

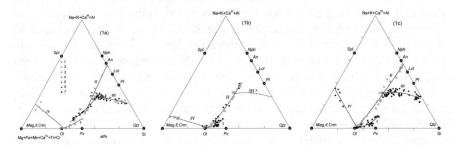

Fig. 11.1 Ophiolite complexes: (1a) Trends of differentiation of the Troodos massif (Cyprus). *I*—dunite-harzburgite basal zone; *II*—pyroxenite-anorthosite zone; *III*—gabbro-granitoids of the upper massif zone, sheeted dyke complex; *IV*—dunite-chromitite; Rocks of the massif: *1*—dunites, harzburgites, Pl-peridotites; *2*—Pl-pyroxenites, anorthosites; *3*—ol-gabbros, Qtz-diorites, trondhjemites, tonalites, plagiogranites; *4*—magnetite ores; *5*—chromitite ores; (1b) Kimpersaisky massif. *I*—basal zone (dunite-garzburgite-lherzolite); *II*—lherzolite-anorthosite; *III*—supposed gabbro-plagiogranite; *IV*—separation of chromitite ores from dunite liquid. (1c) Jil-Satanakhachsky massif (Caucasus). *I*—basal zone (dunite-garzburgite-wehrlite); *II*—the middle zone (olivinite-troctolite-anorthosite); *III*—upper zone (ol-gabbro—gabbro—diorites—plagiogranites); *IV*—trend of the chromitite ores separation from dunite liquid

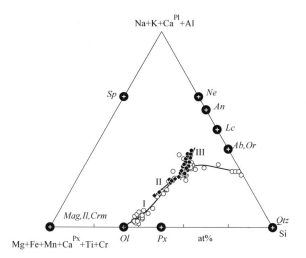

Fig. 11.2 Differentiation trends of oceanic rocks of the Atlantic Ocean. *I*—intrusive rocks of the mid-oceanic ridge and transform faults; *II*—picritic rocks; *III*—olivinite-plagioclase and plagioclase basalts (MORB)

repeatedly (Serri 1982; Beccaluva et al. 1983; Boudier and Nicolos 1985; Pierce et al. 1987). Now it has become clear that the set of rocks of mid-ocean ridges is one of the types of magmatic complexes, with its specific differences (Fig. 11.2). As the volcanic rocks of the mid-ocean ridges evolve, the picrite magmatism turns into a basaltic one, in which the anorthositic tendency of differentiation (MORB basalts, Fig. 11.2) manifests itself more typical for intrusive rocks. As the oceanic ridges consolidate, magmatism acquires a ferrous character, with the formation of Icelandic series that turn into liparites and complementary alkaline rocks. Intrusive rocks of mid-ocean ridges, as can be seen from the few samples associated with basalts, form a primitive olivine-pyroxene-plagioclase trend (*I–II*, Fig. 11.2).

A classic example of the plutonic phases of ophiolite complexes is the well-studied Troodos ophiolite complex in Cyprus, whose plutonic phase consists of three series of rocks that develop as if independently (Fig. 11.1a). Lower, dunite-harzburgite (*I*), up the section is replaced by pyroxenite-anorthozite rocks (*III*), which in turn are covered by gabbro, tonalites, trondjemites, plagiogranites (*IV*). All three series represent the ways of autonomous stratification of melts, probably related to each other by the primary trend of latent superliquidus nanocluster stratification. Petrochemically, diorites, diabases, gabbro-dolerites, and quartz-albite porphyries of the dyke complex coincide with both the upper gabbro-plagiogranitic series of the plutonic phase and with basalts, andesites, and dacites of the overlapping effusions. The dikes penetrate the upper part of the differentiated massif and intrude in volcanic rocks. From this position, intrusive formations can be considered as deep-seated chamber of volcanoes, and the study of the peculiarities of the formation and consolidation of massifs is directly related to the processes of magma evolution in volcanic series. If this is the case, it is the regularities of melts evolution in the upper part of magmatic chambers that determine the petrochemical characteristics of volcanic formations in combination with the intensity of tectonic movements that determine the periodicity of eruptions in time, as seen on petrochemical diagrams of plutonic complexes. As the magmatic chamber develops, eruption processes are likely to capture the melts of pyroxene—anorthosite trend, especially in the early stages of development of these rocks, forming dykes of appropriate composition, such as in the Betts Cove Ophiolite Complex, Canada (Coish et al. 1982).

As spreading evolves, dunite-garzburgite ophiolite complexes (Fig. 11.1a) occur from the oceanic side, dunite-garzburgite-lherzolite (Fig. 11.1b) and dunite-garzburgite-wehrlite (Fig. 11.1c) from the continental, and dunite-wehrlite-clinopyroxenite massifs (Urals type of differentiation, Fig. 11.3) in folded belts transient to platform-type formations (Marakushev and Bezmen 1983). In more consolidated conditions, dyke belts degenerate and volcanic intensity decreases. In the plutonic phase, due to the more contrasting development of differentiation, the role of clinopyroxene-containing rocks increases, while the share of acid rocks (plagiogranites and tonalits) increases, the thickness of which is sometimes commensurate with the capacity of gabbroid rocks (for example, in the ophiolite complex Bail Mountain, in Appalachians, Coish et al. 1982). In this evolution, there is an increase in the accumulation of calcium, iron, titanium, and vanadium in magma, which is associated with the development of the alkaline trend of magmatism, which is peculiar to platforms.

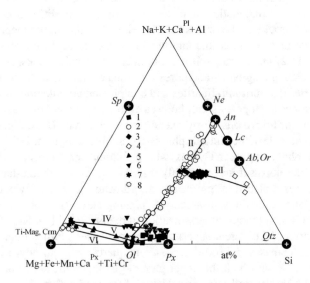

Fig. 11.3 Urals type of differentiation. The Kytlymsky massif (Middle Urals). Trends of differentiation: *I*—dunite-clinopyroxenite; *II*—olivinite-troctolite-anorthosite; *III*—gabbro—granitoid; *IV*—*Ti-mag*-clinopyroxenites; *V*—*Ti-mag*-olivinites; *VI*—dunite-chromitite. Rocks of massif: *1*—dunites, wehrlites; *2*—olivinites, troctolites, gabbro-troctolites, anorthosites; *3*—gabbro-norites; *4*—granitoids; *5*—*Ti*-magnetite clinopyroxenites; *6*—*Ti*-magnetite olivinites; *7*—*Ti*-magnetite ores; *8*—chromitite veins with nugget platinum

Urals type of differentiation. This type of differentiation (Fig. 11.3) shows a shift in rock compositions towards the development of anorthosites, troctolites, clinopyroxenites and associated magnetite ores. An example of dunite—clinopyroxenite formation is an ultrabasic belt in the Urals, in which intensive accumulation of iron in the troctolite (olivine-anorthozite) differentiation trend is traced petrochemically (*II*, Fig. 11.3) with the following decomposition of rocks into iron-rich and comparatively titanium-poor (up to 4% TiO_2) ores, Ti-magnetite bearing clinopyroxenites (*V*) and olivinites (*VI*, Fig. 11.3).

For experimental simulation of the gabbro-plagiogranite trend (*III*, Fig. 11.3), quartz gabbro glass was used. In order to avoid diffusion of iron into the platinum ampoule, the charge was pressed into a crucible made of Ni, Cr, Fe alloy. The experiments were maintained for 4 days at the pressure of $P_{total} = 4$ MPa, $X_{H2} = 0.5$, $X_{H2O} = 0.5$ and the temperature of 1150 °C, which is significantly higher than the liquidus temperature, which, as shown by our experiments, is below 950 °C at the water pressure equal to the partial pressure in the experiments with water-hydrogen fluids ($P_{total} = P_{H2O} = 200$ MPa).

The results of one of the experiments are shown in Fig. 11.4. During the experiment, three layers were formed (Fig. 11.4a): the upper, very narrow layer (1) of acidic composition (*Qtz*-diorite), the middle (2)—diorite and lower layer (3) of allivalite composition as fibrous-tangled clusters (Fig. 11.4b). The analysed compositions on the petrochemical diagram form a trend similar to the trend of differentiation of the upper series of Urals type massifs as well as the upper zones of ophiolite complexes and, as it will be shown below, of stratiform massifs.

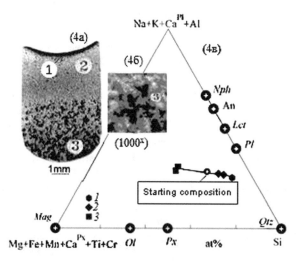

Fig. 11.4 Experimental study of the troctolite-granitoid trend of evolution (Qtz-gabbro—starting composition). (4a)—photo of the sample after the experiment; (4b)—cluster aggregates (increase 1000^X); (4c)—petrochemical diagram of the results: 1—dacite layer; 2—matrix; 3—aggregates of clusters of allivalite composition

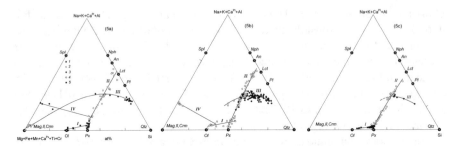

Fig. 11.5 Trends in differentiation of stratiform massifs. (5a) Bushveld Complex (South Africa). *I*—dunite—bronzitite; *II*—pyroxenite—anorthosite; *III*—granophyre—*Fe*-gabbro-magnetitite; *IV*—bronzitite—chromitite; Rocks of massif: *1*—dunites, harzburgites, bronzitities of the Basal zone; *2*—Pl-bronzitities, norites, anorthosites of the Critical and Main zones; *3*—felsites, granites, diorites, ferrodiorites, gabbro, norites, melanogabbro of the Upper zone; *4*—magnetite mineralization of the Upper zone; *5*—chromitities of the Critical zone. (5b) Massifs: (white symbols) Stillwater (USA); (black symbols) Sudbury (Canada). *I*—dunite—bronzitite trend of the Stillwater ultrabasic series; *II*—bronzitite-anorthosite trend of the Stillwater Banded series; *III*—norite-granite trend of the Upper Sudbury zone; *IV*—chromitite ores in peridotite melts. (5c) Munni-Munni massif (Australia). *I*—dunite—wehrlite trend of the lower ultrabasic zone; *II*—pyroxenite-anorthozite trend of the Critical zone; *III*—Fe-gabbro-granitoid trend of the Upper zone

Stratiform massifs. From the point of view of completeness of the presented sections, the stratiform massifs located within the continental platforms and were developing in the absence of large tectonic movements. This led to the formation of intrusive complexes with a far gone differentiation from dunites to granites with the separation of ore magmas: sulfide, chromite, titanomagnetite. The most fully studied is the Bushveld complex, occupying about 29,000 km^2, timed to the most ancient deeply eroded structure of the South Africa shield type, formed in the Achean and then subjected to activation for a long time. The petrochemical diagram of this complex is shown in Fig. 11.5a.

The Basal zone (*I*) is composed of dunites, harzburgites, and bronzitites. The Critical series is characterized by the clearest stratification in comparison with other zones of the massif. In general, the Critical zone is characterized by a rhythmical alternation of bronzitites, norites, gabbronorites, anorthosites and chromites. The Main zone is the largest unit of the massif. In contrast to the previous zones, a thin rhythmic stratification was weakly manifested here. The rocks in general are represented by gabbro-norites and anorthosites. The petrochemical diagram of rocks, Critical and main zones form a pyroxene-anorthite trend (*II*). At the base of the Upper zone lies a horizon, which for the first time appears a significant amount of titanomagnetite, forming layers of monominal rocks, with a thickness of 0.3–1.8 m. Ferro-gabbro (ferro-diorites) prevail, which are replaced by thick layer of felsites and red Buschweld granites up the section. In general, the petrochemical diagram of the zone shows the granophyre—*Fe*-gabbro—*T*i-magnetite trend (*III*).

The calm tectonic situation contributed to the separation of chromite (*IV*, Fig. 11.5a), titanomagnetite (V, Fig. 11.5a), and sulfide composition ore magmas, which are associated with PGE deposits (Naldrett 2003). The complex multi-trend structure of stratified basite-ultrabasite plutons with a far-gone differentiation is characterized by a complex formation. Chromite and copper-nickel sulfide mineralization are confined to the lower part (ultrabasic zones), sulfide PGE mineralization to the middle (pyroxene-anorthite) zone, and iron-titanium mineralization to the upper, ferrogabbro-anorthozite or ferrogabbro-granite zone, where rocks of the final stages of differentiation are concentrated.

The Stillwater and Sudbury stratiform massifs, which are well studied, are characterized by a similar type of differentiation (Fig. 11.5b). The first one in the modern erosion section is represented by its ultrabasic and main part, the second one—by rocks of the upper, granophyr-norite series.

Similar to the plutonic zones of the ophiolite complexes, which by their mineralogical and pentrochemical peculiarities form a series from dunite-garzburgite to dunite-garzburgite-wehrlite rocks, stratiform complexes are also characterized by formational diversity. In contrast to the Stillwater and Bushweld massifs, in which the ultrabasic rocks are represented by dunites, harzburgites, bronzitites and anorthosites, there are massifs with the composition corresponding to labradorites, in the structure of which the main role is played by lherzolites, wehrlite and more acidic anorthosites (andesinites). The representative of the latter is the Munni-Munni complex of Australia (Fig. 11.5c). This is a fairly large massif of about 225 km^2 with a capacity of 5500 m and lies among the Archean (3600–2650 million years) granite-green shale rocks of the Pilbara block of Eastern Australia and is 2925 million years old. Low-sulphide (2% wt% sulphides) disseminated PGE-like mineralization (up to 2.5 ppm) is confined to porphyrite plagioclase websterites and gabbro-norites and traced in the form of a reef with a thickness of several tens of meters over 12 km.

Compared to Bushveld massif, granophyrea rocks are almost completely absent in the Munni-Munni complex. Unfortunately, we do not have analyses of acid rocks confined to the tops of the section in the form of low-power layers, lenses, and schlieren accumulation in order to compare gabbro-granophyre differentiation trends (*III*, Fig. 11.5a).

The observed differences are probably determined not only by the initial gross composition of magma, but also by the peculiarities of differentiation. In the Bushveld massif in the process of evolution there was a contrasting separation of granophyre magmas with the formation of a powerful zone of upper granites, as well as the separation of anorthositic rocks enriched with calcium from magmas of the basite composition. In the Munni-Munni massif, differentiation was less contrasting, especially at the early stages of development (at the stage of latent stratification), which led eventually to the concentration of calcium in the form of clinopyroxene in the ultrabasic rocks, and sodium in anorthosites.

Skaergaard type. Among the stratified complexes, the Skaergaard massif in East Greenland is an excellent type of intra-chamber diffraction of magmatic melts. His petrochemical diagram is shown in the Fig. 11.6, which shows two trends: olivine-plagioclase (*I*) binding rocks of the main and ultrabasic composition and granophyr-ferrogabbro (*II*), reflecting the differentiation of detached magmas of acidic composition with complementary iron enriched melts.

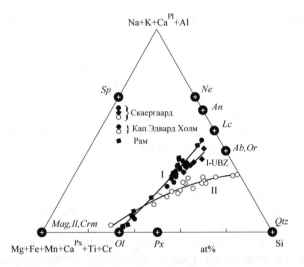

Fig. 11.6 Scaergaard type of stratiform differentiation of massifs. Skaergaard (Greenland), Cap Edward Hill (Greenland), Rum Massif (island of Rum). Differentiation trends: *I*—dunite—anortosite stratified series; *I-UBZ*-Upper Border zone of the Skaergaard massif; *II*—Fe-pyroxenite-granophyre of the Upper zone

The Skaergaard massif is an example of layered plutons with a well exposed top of the section. In the majority of other intrusions of a similar type in the British-Arctic province (Belkheevsk, Ram, Sky, etc.), the lower parts of the sections are preserved, formed by the rhythmic interlayering of dunites and troctolites with gabbroids, which supplement the ultrabasic part of the olivine-plagioclase trend in the diagram (Fig. 11.6). Industrial concentrations of platinum and low-sulphide type gold have been found in the Scaergaard-type massifs (Nielson and Garmicott 1993). This type of differentiation is characteristic of the following massifs: Tsipring, Karelia, Talnakh, Siberia, Ioko-Dovyrensky, Transbaikalia, Volkovskoye, Ural, Kiglapait, Labrador.

Gabbro—anorthosite, gabbro—granite and gabbro—syenite massifs. These massifs are characterized by similar trends in the differentiation (Fig. 11.7). The trend of the Tzaginsky massif (*1*) form the following rocks: labradorites—troctolites—gabbro-norites—gabbro—olivinites—pyroxenites—*Ti*-magnetite ores. The trend of the Chineisky massif (*2*) form the following rocks: anorthozites—gabbro—gabbro-norites—*Ti-Mag*-pyroxenites—*Ti*-magnetite ores. The trend of the NovoMirgorodsky massif (*3*) form the following rocks: anorthozites—gabbro-anorthozites—gabbro—*Fe*-peridotites—*Ti*-magnetite ores. The trend of the Kalarsky massif (*4*) form the rocks: anorthozites—*Ilm* and *Ti-mag*-ores. The trend of the Northern Timan massif (*5*) form the rocks: nepheline syenites—syenites—gabbro—*Fe*-gabbro. The

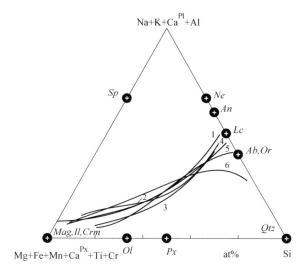

Fig. 11.7 Layered gabbro—anorthosite, gabbro—syenite, gabbro—granite intrusion with ilmenite and *Ti*-magnetite ores. Massifs: Gabbro—anorthsite: (1–4): *1*—Tzaginsky, Kola Peninsula; *2*—Chineisky, Transbaikal; *3*—Novomirgorodsky, Ukraina; *4*—Kalarsky, Siberia. Gabbro—syenite: *5*—Northern Timan, Siberia. Gabbro—granite: Malyi Kuibas, Urals

trend of the Malyi Kuibas (*6*) form the rocks: granites—diorites—gabbro—gabbro-pyroxenites—*Ti-Mag-* ores. They were formed in the activation zones of the platforms in a relatively calm tectonic environment and are therefore contrasted with the separation of rich *Ti*-magnetite and ilmenite ores (up to 15% Ti).

As an example, we studied the superliquidus nanocluster differentiation of the melt of anorthosite gabbro (33 wt%) and ilmenite (67 wt%) composition of the Novomirgorodsky massif in Ukraine. Pyroxenites of this massif contain phenocrystals of atomic carbon, which testifies to the reducing environment of formation of the chosen massif rock association. The experiments were carried out at 1250 °C and at 400 MPa in a high gas pressure vessel. Homogeneous glass (210 mg) was pressed into a vitreous carbon crucible, which was placed into a 7 mm platinum ampoule. Then a solution (150 mg) of acids (HF—0.25 wt%, HCl—0.25 wt%, H_3PO_4—0.15 wt% and H_3BO_3—0.15 wt%) was poured into the ampoule and paraffin C_nH_{n+2} (50 mg) was added. The fluid phase composition was controlled by the molar fraction of hydrogen ($X_{H2} = 0.18$) in the argon—hydrogen cell of the reactor (Bezmen and Gorbachev 2017). Within 3 days the melt is layered into three compositions: ilmenite, rutile and anorthosite gabbro (Fig. 11.8).

Differentiated sills and traps. As well as for stratiform massifs, the formation of differentiated traps and sills occurred in a relatively calm tectonic environment, but the small size of intrusions does not allow to reach a high degree of differentiation, leading to the separation of anorthosite or granite melts.

The gabbro-dolerite traps are characterized by several types of differentiation. In layered subsurface gabbro-diorite-granophyre sills: Polysides, New Jersey, USA

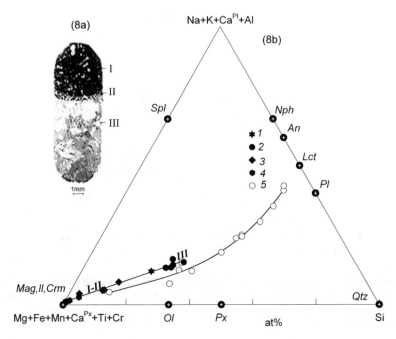

Fig. 11.8 The result of experimental modeling of ilmenite ore formation in the Novomirgorodsky gabbro-anorthosite massif (Ukraine). (7a)—micro-photography of a stratified sample obtained as a result of melting of the primary homogeneous melt of the ilmenite-anorthosite composition (T = 1250 °C, P = 400 MPa). *I*—ilmenite melt; *II*—rutile melt layer; *III*—*Fe*-gabbro—anorthosite melt; (7b)—petrochemical diagram of experimental and natural data: 1—initial glass sample; 2—lower layer of *Fe*-gabbro—anorthosite composition (*III*); 3-upper zone (ilmenite, *I*); 4—rutile layer (*II*); 5—rocks of the Novomirgorodsky massif

(Pearce 1970; Shirley 1987); Anakinskiy sill, Lower Tunguska (Marakushev and Kaminsky 1990) and other olivine-containing gabbro-norites, gabbro, diorite and granophyres are united by olivine—plagioclase komatite-like differentiation trend, passing into enriched quartz rocks.

Layered Pechenga massif have a more complex two-zone structure. In the lower zone, olivinites, wehrlites, clinopyroxenites are developed, with which sulfide copper-nickel ores are associated. The upper part of the massif is represented by rocks of the basic and subalkaline composition. There are layers of *Ti*-magnetite clinopyroxenites and *Ti*-*Mag*-monomineral ores on the border of the two zones. Accordingly, the diagram (Fig. 11.9) shows two differentiation trends reflecting the complex intra-chamber evolution of the melts, first nano-clucters separation of the basic and ultrabasic zones, then their autonomous development with the concentration of sulfide ores in dunites and wehrlites melts of the lower zone and *Ti*-magnetite in plagioclase clinopyroxenite melt of the upper zone. It has been confirmed by experimental researches: influence of hydrogen-containing fluids ($P_{total} = 100$ MPa)

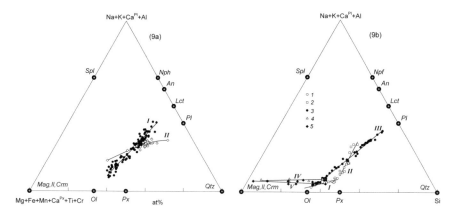

Fig. 11.9 Differentiation trends in stratified sills and traps. Comparative characteristics of differentiation trends of the Talnakh intrusion of the Norilsk district (Siberia) and Pilgujarvi massif (Kola Peninsula). (8a) Talnakh massif. I—peridotite—anorthosite trend; II—peridotite—diorite—granitoid (Skaergaard type of differentiation, see Fig. 11.6). (8b) Pilgujarvi massif. I—olivinite—peridotite trend of the Lower zone; II—pyroxenite—leucogabbro trend of the Middle zone of the massif; III—*Ti-Mag*-pyroxenite—essexite trend of the Upper zone; IV—*Ti-Mag*-pyroxenite trend of ferrous mineralization; V—Ni-Cu ores—peridotite trend of sulfide immiscibility. Rocks of the massif: 1—olivinites, peridotites, olivine pyroxenites; 2—plagioclase pyroxenites, gabbro, leukogabbro; 3—*Ti-Mag*-pyroxenites, gabbro, essexite; 4—*Ti*-magnetite ores, *Ti-Mag*-pyroxenites; 5—pentlandite—chalcopyrite ores in the olivinite and peridotite

on the melts corresponding on the bulk composition (Marakushev et al. 1984) and the upper zone of Pilgujarvi massif, Pechenga.

The traps of Pechenga and the Norilsk Region differ sharply in the type of differentiation (Fig. 11.9). Picrite dolerites, troctolites, olivine gabbrodolerites, gabbrodolerites of the most fully studied ore-bearing Talnakh massif form an olivine-plagioclase trend complicated by the appearance of diorites and quartz diorites. In general, the trend of differentiation of the Norilsk massifs is similar to that of the Scaergaard type differentiation (Fig. 11.6), which is probably connected with the proximity of physicochemical conditions of formation, especially with respect to the oxidation-reduction regime.

11.4 Discussion of Results

The need to classify differentiated complexes is not in doubt, as the classification can serve as a basis for forecasting and searching for magmatic mineral deposits. Therefore, it is not by chance that the systematization of massifs was carried out repeatedly (Kuznetsov 1964; Naldrett and Cabri 1976; Besson et al. 1979; Belousov et al. 1982). However, either classification features external to magmatic bodies, such as connection with effusives, belonging to different tectonic structures, sizes of intrusive bodies, etc., were chosen as the basis (Besson et al. 1979), or compositions of magmatic rocks in respect of individual oxides or groups of oxides. As a rule, the regularities of crystallization of fluid-free silicate petrologically important systems based on the physicochemical analysis of simple state diagrams based on experimental data in dry conditions were taken as the genetic basis of intrachamber differentiation. The petrochemical material presented in the article shows that, relying only on crystallization processes, it is impossible to deduce the whole set of rocks of magmatic complexes from granophyres to dunites, so it is no coincidence that hypotheses of multiple introduction and mixing of magmas in the chamber appeared (Irvine 1979; Campbell and Turner 1986; Naldrett et al. 1990, etc.). The multiphase magmatic introduction for most massifs is not supported by geological data. This process is contradicted by the following: in general, the natural change upward along the section of ultrabasic rocks of the main rocks, complicated by the rhythmic arrangement of the layers, parallel arrangement of the layers, if they crystallized before the manifestation of tectonic movements, the absence of numerous feeders, which we observe in the form of a layer of parallel dikes in the ophiolite complexes (sheeted dyke complex), with volcanic eruptions, as well as similarity in the structure of numerous massifs of one formation type. Besides, the difference in the composition of magma portions should be explained by some peculiarities of deep differentiation, which we do not see and have a weak hypothetical idea of, in addition, many of the phases of introduction should correspond to the compositions that are absent among effusive rocks.

At present, it is unlikely that anyone will deny the essential role of fluids in geological processes, including in magmatism. In terms of weight, the total number of fluid components dissolved in magma probably does not exceed 10%. Thus, for example, 5 wt% of water in basalt melt is 50–60 mol%, 0.5 wt% of hydrogen—60–70 mol%. Thus, magmatic melts in the majority, obviously, represent the fluid-silicate substance in which the silicate fraction makes a smaller part.

As our experimental studies (Bezmen 1992; Bezmen and Gorbachev 2014) have shown, nanoclusters can be separated in the Earth's gravitational field at superliquidus temperatures, forming initially cryptic stratification and then layers of different compositions in fluid magma. As a result of our experimental data it follows that the fluid melts of magmatic chambers are completely differentiated in the liquid state. Thus, the gravitational migration of different densities of nanoclusters in the magmatic chamber forms flotation, sedimentation and rhythmic types of melt stratification in the absence of temperature gradients in magma. Due to the migration of fluid and fluid-enriched clusters to the top of the massif, the crystallization processes are activated from the bottom to the roof of the magmatic chamber.

The developed diagram, which formed the basis for comparative petrochemical analysis of magmatic complexes, in an accessible and understandable form reflects the genetic features of the formation of massifs in the form of differentiation trends, the set and location of which in the diagram and were the main features of classification by type of differentiation.

It should be emphasized that massifs with a similar bulk composition are often characterized by different types of differentiation, which is difficult to explain from the standpoint of crystallization fractionation, as they are formed in close physico-chemical conditions. As the rock varieties are determined by the dry crystallization with cotectics and eutectics, the melts similar in composition would have a single trend of differentiation, and the massifs would differ only in the quantitative ratios of rocks. It is necessary to emphasize that magmatic complexes were formed in the Earth's crust in a relatively narrow pressure range (100–500 MPa), so the fluids could not affect to the order of crystallization of minerals. In this connection, without involving the mechanism of pre-crystallization differentiation, it is impossible to explain the detected regularities of evolution of intrachamber differentiation of initial melts.

It should be noted that among the basic and ultrabasic rocks in the intrusive complexes in the regular interlaying there are also more acid rocks, up to silicites (Anhaeusser 1971), which are genetically related to the evolution of magmatic chambers. If this is the case, we are convinced once again that crystallization differentiation cannot be the only and universal mechanism of magmatic differentiation. It is the liquid separation of clusters that leads, as experimental studies show, to the liquid formation of acidic and ultrabasic melts in the differentiation of basite magma. Problems of relations between rocks of homodrome and antidrome sequences, as well as gradual transitions between them, often observed in sections, are easily explained from the standpoint of liquid differentiation. In general, more basic magmas can be formed in any part of magmatic chambers: everything depends on the ratio and content of fluid components in the fluid shells of nanoclusters (Bezmen 2001).

According to the location of trends on the diagram and the totality of the rocks composing them, the ophiolite intrusive complexes form subtypes, the family—from dunite-garzburgite to dunite-garzburgite-wehrlite. Common for all opolite massifs is the occurrence at the base of the section of interlaying dunites and harzburgites, sometimes with lenses or layers of chromites. It is believed that these are the restite, remaining from melting melts from the mantle and their subsequent transfer to the overlying horizons. Maybe it is. However, their layered structure, absence of high-bar minerals, and absence of direct signs of melting such as reverse zoning in minerals, and often gradual transitions with overlying rocks, testify in favor of their origin by magmatic differentiation.

Depending on the tectonic situation, as the Earth's crust consolidates, the role of clinopyroxene-containing rocks as well as lherzolites and wehrlites increases in the massifs. At the same time, the associations of overlying rocks are also naturally changing. In dunite-garzburgite complexes they are represented by pyroxenite-anorthozite rocks in dunite-garzburgite-wehrlite rocks—olivinites, troctolites and anorthosites. The corresponding trend of differentiation of the dunite-garzburgite-lherzolite massifs occupies an intermediate position. In the upper parts of the section differentiation leads to the formation of tonalities and plagiogranites, the thickness of which is determined by the tectonic regime of massif formation. In a quieter environment, dyke belts degenerate, the thickness of effusive rocks decreases and the number of plagiogranite layers increases.

Ophiolite complexes are notable for the fact that they clearly show the genetic links between effusive and intrusive magmatism. The geological structure of ophiolite belts, the relationship between intrusive, subvolcanic and effusive rocks, and the comparison of their petrochemical characteristics (Fig. 11.1) show that the processes of differentiation in magmatic chambers in many cases determine the structure and composition of associated effusive differentiated complexes.

The dunite-pyroxenite massifs, as well as the ophiolite complexes, form belts shifted towards the continent in relation to the ophiolites (for example, the Urals platinum-bearing belt). They are characterized by a subplatform mode of formation, so they are often combined with alkaline rocks (syenites, granosyenites), which may be genetically related. Despite the absence of harzburgites, the dunite-clinopyroxenite massifs correspond to the ophiolite-like type of differentiation in terms of the set of trends, but occupy a kind of extreme position in relation to them (Figs. 11.5 and 11.6). Compared to the ophiolite massifs, there is an intensive accumulation of iron in the troctolite (olivinite-anorthosite) differentiation trend with the subsequent formation of magmatic iron ores. The geochemical behavior of chromium is exactly the opposite. Large chromite deposits associated with dunite-garzburgite massifs (Troodos, Cyprus) and especially dunite-garzburgite-lherzolite deposits (Kempirsaysky massif, South Ural) are replaced by poor chromite-hercynite mineralization in dunite-garzburgite-wehrlite complexes, while in dunite-clinopyroxenite massifs there is a vein chromite mineralization, confined to dunites. With regard to chromium, it has no industrial significance, but chromite-silicate separation of melts, as shown experimentally (Bezmen 1992), leads to the extraction of PGE into a chromite liquid, which crystallizes to form veins with native platinum. Later on, they served as a source of large placer deposits in the Urals, Siberia (Kondersky massif), Alaska and other regions of the world.

By their petrochemical peculiarities, stratiform complexes are divided into two types of differentiation: Bushveld (South Africa) and Skaergaard (Greenland). The first one combines three trends, the second one—two (Figs. 11.5 and 11.6).

Like ophiolite massifs, the Bushveldt type stratified complexes have a three-zone structure: the lower zone is Basal, composed of ultrabasic rocks of dunite (peridotite)-pyroxenite series (dunites, peridotite, bronzitite, websterite), the middle, Critical zone is composed of pyroxenite-norite and anorthosite rocks and the Upper zone is composed of melanocratic gabbro and granophyres. They also form a family, which is reflected in the diagrams of shifting trends in differentiation. In ophiolite complexes, as the role of clinopyroxene increases, trends shift towards troctolite. Here the picture is reverse, as soon as bronzitises are replaced by lherzolites, wehrlites, websterites, rocks are saturated with silica, thus the composition of anorthosites changes from labradorites in the Bushveld massif, to andesinites in the complex Munni-Munni (Australia). It should be noted that in contrast to the ophiolite massifs, differentiation in stratiform complexes is more contrasting, which leads to the appearance of almost monominal pyroxenites and anorthosites. It is explained by the quieter tectonic situation and deeper conditions of formation. Long-term differentiation, first cluster formation, then crystallization, promotes the accumulation of fluids in certain parts of the section—as a result of which the following concentrations can be achieved when ore magmas are separated: chromite, sulfide with PGE, titanomagnetite (Bezmen 2001).

Stratiform complexes are remarkable in comparison with other types of stratified massifs, high concentrations of platinum and gold, genetically related to poor sulfide copper-nickel mineralization and chromitite, but their content is naturally decreasing with the increasing role of clinopyroxene.

Stratiform massifs of the Skaergaard type have a simpler structure. They are characterized by the troctolite (olivine-plagioclase) differentiation trend, which is complicated by the splitting off from granophyre magmas (Fig. 11.6).

The rocks composing trap formations also form several differentiation trends, but the small size of magmatic chambers and their rapid loss of volatiles do not lead to a contrasting separation of melts as in large stratified plutons.

Thus, the considered extensive petrological material testifies to numerous ways of melts evolution in layered magmatic complexes, the formation belonging of which is characterized by a set of differentiation trends associated with different parts of the section of magmatic chambers.

In connection with the consideration of the peculiarities of the evolution of magmatic complexes, the main question arises:—What is the reason for the diversity of types of differentiation? The line-up? Of course, this is an important factor, but the bulk compositions of many massifs, calculated even without taking into account the granophyres, which are often not taken into account, because they do not fit into the processes of crystallization differentiation, occupy a rather narrow field on the petrochemical diagram, corresponding to the main rocks. Indeed, if one does not take into account plagioclase basalts of mid-ocean ridges, which are widespread in young geological structures, picrite and tholeitic basalts, as well as basalts of island arcs, are limited to a narrow field coinciding with the compositions of differentiated magmatic complexes. It remains to be assumed that the fluid component mode is also the main factor of differentiation.

It has been shown experimentally (Bezmen 1992) that the composition of the fluid phase has a significant influence on the differentiation of the melts. Since petrogenic components have different chemical affinity to volatile elements, nanoclusters contain fluids of different compositions.

The presence of carbon-containing gases in the fluid phase, especially CH_4, stimulates the separation of ultramafic melts in the upper parts of the sample. As an example, we present the results of Ol–Pl stratification of the basal zone of the Bushveld complex in South Africa. The main role in this part of the section belongs to dunites and olivine bronzitites containing a small amount of plagioclase. The average composition of this zone is close to plagioclase-bearing olivine bronzitite. As the initial composition for the experimental simulation, the glass of the following composition was synthesized from oxides in the vacuum high-temperature furnace at 1700 °C: SiO_2-49.1; TiO_2-0.2; Al_2O_3-7.56; Fe_2O_3-9.48; FeO-5.36; MnO-0.16; MgO-22.79; CaO-4.52; K_2O-0.11; Na_2O-0.71. The experiments were carried out under the pressure of complex gas mixtures of H–O–C–S system at P = 400 MPa and T = 1350 °C. The molar fraction of hydrogen in the experiment was controlled by the argon-hydrogen mixture and amounted to $X_{H2} = 0.2$. The fugacities of other gases was controlled by the presence of elemental carbon from which the crucible was made and the melt of pyrrhotite. The fluid phase at experiment parameters had the following composition: $X_{H2O} = 0.31$; $X_{H2} = 0.2$; $X_{H2S} = 0.003$; $X_{CO2} = 0.077$; $X_{CO} = 0.19$; $X_{CH4} = 0.218$; $\log f_{O2} = -9.9$; $\log f_{S2} = -3.71$. It is assumed that the given parameters are as close as possible to the conditions of formation of the Bushveldsky complex, in rocks of which elemental carbon is present (Touysinhthiphonexay et al. 1984), in gas-liquid inclusions of minerals reduced gases, including CH_4 (Ballhaus and Stushpfl 1985), in addition, oxidation-reduction conditions of formation of rocks are also close to natural (Elliott et al. 1982).

Under the pressure of hydrogen-containing fluids in the absence of a thermal gradient, the initial melt is stratified with accumulation of liquid close to the dunite in the upper zone of the crucible (Fig. 11.10). Super-liquid temperatures were confirmed by experiments under the pressure of pure water ($P_{total} = P_{H2O} = 120$ MPa), equal to the partial pressure in the mixtures ($X_{H2O} = 0.31$). After 3 days of experiment the upper peridotite zone is represented by large (up to 1 mm) crystallized olivine idiomorphic crystals, between which there is almost no intergranular liquid. This zone has the following composition: SiO_2-42.76; TiO_2-0.2; Al_2O_3-4.2; FeO-17.41; MnO-32.69; CaO-2.14; K_2O-0.007; Na_2O-0.44. The bottom zone, transparent glass,

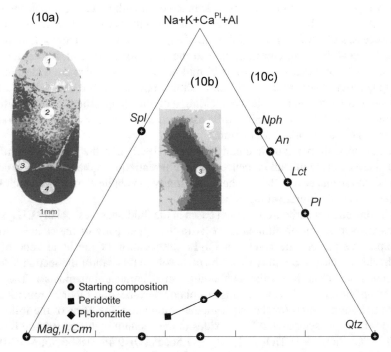

Fig. 11.10 Super-liquidus nano-cluster differentiation of *Ol-Pl*-pyroxenite (example of antidromic stratification). (9a)—microphotography of the layered sample: upper zone, 1—peridotite, 2—*Pl*-bronzitite matrix, 3—bronzitite clusters aggregates, 4—pyrrhotite melt (sulphur fugacity indicator). (9b)—linear structure of nanocluster aggregates with large magnification (1000^x). (9c)—Petrochemical diagram of experimental data

has composition close to plagioclase bronzitite: SiO_2-50.80; TiO_2-O.2; Al_2O_3-9.99; FeO-14.34; MnO-0.15; MgO-17.61; CaO-5.78; K_2O-0.18; Na_2O-0.95.

In the lower zone there is a new liquid stratification (Fig. 11.10) in the form of packets or spheroids of accumulation of macromolecules, which are collected at the bottom of the sample, where they form a liquid enriched with bronzite: SiO_2-51.09; TiO_2-O.2; Al_2O_3-9.37; FeO-15.13; MnO-0.15; MgO-18.36; CaO-4.71; K_2O-0.17; Na_2O-0.84. After quenching, it has a tangled fibrous structure, typical for hardened, colloidal, macromolecular liquids (Strepikheev and Derevitskaya 1976). In general, the initial composition of the fluid in the experiment determined the antidromic type of differentiation. Antidromic differentiation is characteristic of the Ural-Alaska, Elan-Vyazovsky and Podkolodnovsky massifs of the Voronezh crystalline shield and the Burpalinsky massif in Siberia. As a result of migration of volatile components to the upper part of the massif, the fluid composition changes, which affects the content and specialization of fluid components in the melts of cluster shells. As a result, in some parts of the section of stratified complexes, an antidromic type of differentiation appears (for example, in the Khibiny alkaline massif on the Kola Peninsula).

Unlike crystals or droplets of immiscible liquids that have phase interface boundaries, clusters or macromolecules are constantly exchanging individual molecules or parts of molecular groups with the matrix. These are fluctuating systems that do not have an interface and are permanently in dynamic equilibrium with the melt. This reduces the free energy of cluster formation, which prevents their unlimited increase in size, as opposed to crystal differentiation, which according to the laws of thermodynamics must constantly increase in size in magma by dissolving the smaller crystals and the growth of larger ones.

The interaction of clusters with the melt, the composition of which changes at the gravitational separation, can lead to a corresponding change at a constant temperature of the cluster composition (macromolecules), which we observe in the studied series of experiments.

The crystallization of ultrabasic melts raises separating fluids to higher horizons and interacts with overlying magma or partially or completely crystallized rocks, causing their melting and subsequent liquid differentiation (Fig. 11.11). Petrogenetically, this process is fixed by zones, often ore-bearing, with low temperatures of formation of finely stratified contrasting rocks, which are located in poorly differentiated strata.

The evolution of magmatism is associated with a restorative regime of magma formation, especially at the early stages of formation of magmatic chambers (Kadik et al. 1990; Bezmen 1992). At low chemical oxygen potential in the fluid phase, the share of hydrogen increases, which, interacting with magmatic melts, causes their intensive depolymerization (Bezmen 1992). In previous studies (Bezmen et al. 1991, 2005) it was shown that the presence of hydrogen leads to the breaking of the Si–O–Si bridge bonds in silica oxygen anions and the replacement of part of the oxygen atoms by hydrogen ions (fluorine, chlorine, phosphorus, boron) according to the following scheme:

$$-Si-O-Si + H = Si-O + Si-H.$$

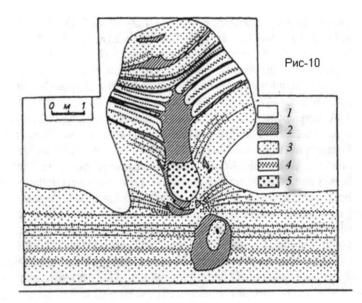

Fig. 11.11 Photograph of outcrop in the upper stratified zone of the Panskiye tundra massif. Melt remelting and liquid differentiation of melts under the influence of ascending fluids. 1—Quaternary sediments, 2—spotted anorthosites, 3—gabbro-norites, 4—norites, 5—pyroxenites

According to the data of photoelectron spectroscopy in silicate melts even at low molar fractions of hydrogen ($X_{H2} \geq 0.1$) elementary silicon, which is probably formed during quenching, was found (Bezmen et al. 1991). Fluid-saturated magmatic melts produce fluctuating molecular groups—nanoclusters or nanomolecular groups, similar in structure to minerals, but containing hydrogen, water and other components, including ore ones. In the Earth's gravitational field, clusters move to form a cryptic and then contrasting stratification in the magmatic melts. Fluorine has a strong influence on the separation of melts enriched with anorthosite molecule, i.e. depending on its calcium potential it is concentrated in anorthosites or pyroxenites. Apparently, the composition of volatile components adequately influences the distribution of alkalis and silica in the cross-section of the massif, however, for the final conclusions about the influence of the fluid phase composition on the peculiarities of liquid differentiation of magmatic melts we need more targeted experimental data covering the natural range of fluid magmas, but in all cases it is necessary to have the presence of hydrogen in experiments or hydrogen-containing gases, CH_4, NH_3, HCl, H_2S, etc.

In general, massifs are characterized by simultaneous manifestation of the processes of liquid differentiation and crystallization in the recumbent of the massifs as the volatile components migrate to the upper part of the magmatic chamber. The wide development of evenly grained or fine-grained rocks, especially monominal rocks, in the sections of stratified massifs is almost unambiguous evidence of the predominance of the processes of superliquidus differentiation, while the origin of rocks with unevenly grained or taxitic structures may be due to the crystallization of magmas saturated with fluids.

11.5 Conclusions

The paper presents the petrochemical classification of differentiated magmatic complexes. For this purpose, a petrochemical diagram has been developed: Mg, Fe, Mn, Ca^{Px}, Cr, Ti–Na, K, Ca^{Pl}, Al–Si on which the main normative minerals are fixed by figurative points. As the main classification feature of magmatic complexes, the peculiarities of intrachamber evolution of magmatic melts were accepted.

The rocks of layered complexes are combined into differentiation trends, the totality and location of which determines the type of differentiation in the diagram. Within a single type, there is a shift in trends due to the composition of rocks of different massifs forming subtypes. Each differentiation trend has its own ore specialization. The type of differentiation determines the complex of ore mineralization of the massif.

The number and location of massif differentiation trends depends not only on the composition of the initial magmatic melts, but also on the physicochemical conditions of formation, with a special role to be played by the mode of fluid components.

Superliquidus differentiation of magmatic melts, crystallization of magmas at different levels of the section, separation of volatiles enriched with ore components, and their interaction with overlying magmas or rocks represent a complex petrogenetic system of evolution of magmatic chambers. At various stages of this process, fluid concentrations occur, leading to the separation of ore magmas, which are naturally confined to certain zones of the section of the massifs. Since the fluid components migrate both by themselves and in the fluid-containing magmatic shells of migrating clusters, stratification effects further develop under the influence of fluid inflow into the upper part of the chamber. In the lower parts of the magmatic chamber from bottom to top, the crystallization processes caused by the formation of solidus temperatures during the dissipation of fluid components from the melt are activated.

Acknowledgements Financial support by the IEM RAS project № AAAA-A18-118020590141-4.

References

Anhaeusser CR (1971) Evolution of Arhaean greenstone belts. Geol Soc Aust Spec Publ 3:178–182

Ballhaus CG, Stumpfl EF (1985) Fluid inclusions in the Merensky and Bastard reefs western Bushveld complex (abstract). Can Mineral 23:294

Beccaluva L, Girolamo PD, Macciotta G, Morra V (1983) Magma affinities and fractionation trends, in ophiolites. Ofioliti 8(3):307–324

Belousov AF, Krivenko AP, Polyakova ZP (1982) Volcanic formations. Science. Sib. part, Novosibirsk, 280p (in Russian)

Besson M, Boud R, Czamanske G, Foose M (1979) IGCP project no 161 and a proposed classification of Ni–Cu–PGE sulfide deposits. Can Mineral 17:143–144

Bezmen NI (1992) Hydrogen in magmatic systems. Exp Geosci 1(2):1–33

Bezmen NI (2001) Superliquidus differentiation of fluid-bearing magmatic melts under reduced conditions as a possible mechanism of formation of layered massifs: experimental investigations. J Petrol 9(4):345–361 (English translation Petpologiya from Russian)

Bezmen NI, Gorbachev PN (2014) Experimental investigations of superliquidus phase separation in phosphorus rich melts of Li–F granite cupolas. J Petrol 22(6):620–634

Bezmen NI, Gorbachev PN (2017) Experimental study of gabbro-syenite melt differentiation in superliquidus conditions on the example of Northern Timan massife. Exp Geosci 23(1):114–117

Bezmen NI, Zharikov VA, Epelbaum MB et al (1991) The system $NaAISi_3O_8$–H_2O–H_2 (1200 °C, 2 kbar): the solubility and interaction mechanism of fluid species with melt. Contrib Min Petrol 109:89–97

Bezmen NI, Zharikov VA, Zavel'sky VO, Kalinichev AG (2005) Melting of Alkali aluminosilicate systems under hydrogen-water fluid pressure, P = 2 kbar. J Petrol 13(5):407–426

Bezmen NI, Gorbachev PN, Martynenko VM (2016) Experimental study of the influence of water on the Buffer equilibrium of magnetite-Wüstite and Wüstite-metallic iron. J Petrol 24(1):93–109

Boudier F, Nicolas A (1985) Harzburgite and lherzolite subtype in ophiolitic and ocenic environments. Earth Planet Sci Lett 76:84–92

Campbell IH, Turner JS (1986) The influence of viscosity on fountains in magma chamber. J Petrol 27:1–30

Coish RA, Hickey R, Frey FA (1982) Rare earth element geochemistry of the Betts Cove ophiolite formation. Geochim Cosmochim Acta 46:2117–2134

Colman RG (1979) Ophiolites. Mir, Moscow, 262p (in Russian)

Elliott WC, Grandstaff DE, Ulmer GC, Buntin T, Gold DP (1982) An intrinsic oxygen fugacity study of platinum-carbon associations in layered intrusions. Econ Geol 77:1439–1510

Grachev AF, Fedorovsky VS (1980) Greenstone precambrian belts: riffle zones or island arcs? Geotectonics 5:3–24 (in Russian)

Irvine TN (1979) Convection and mixing in layered liquids. Carnegie Institution Washington. Year book, pp 257–262

Kadik AA, Lukanin OA, Lapin IV (1990) Physico-chemical conditions for the evolution of basaltic magmas in near-surface foci. Science, Moscow, 346p (in Russian)

Kuznetsov YA (1964) Main types of magmatic formations. Nedra, Moscow, 387p (in Russian)

Marakushev AA, Bezmen NI (1983) The evolution of meteoritic matter, planets and magmatic series. Science, Moscow, 185p (in Russian)

Marakushev AA, Kaminsky AD (1990) The nature of the stratification of the sills of ferrous dolerites. Dokl USSR Sc 314(4):935–939 (in Russian)

Marakushev AA, Bezmen NI, Skufin PK, Smolkin VF (1984) Stratified nickel-bearing intrusions and Pechengi volcanic series. Essays on physico-chemical petrology. Science, Moscow XI:39–63 (in Russian)

Miyashiro A (1975) Classification, characteristics and origin of ophiolites. J Geol 83:249–281

Moralev VM (1986) The early stages of the evolution of the continental lithosphere. Science, Moscow, 165p (in Russian)

Naldrett AJ (2003) Magmatic sulfide deposits of Nickel–Copper and Platinum-metal ores. St. Petersburg State University, St. Petersburg, 487p

Naldrett AJ, Gabri LJ (1976) Ultramafic and related mafic rocks: their classification and genesis with special reference to the concentration of nickel sulfides and platinum-group elements. Econ Geol 76:1131–1158

Naldrett AJ, Brugmann GE, Wilson AH (1990) Models for the concentration of PGE in layering intrusions. Can Miner 28:389–408

Nielson TFD, Gannicott RA (1993) The Au-PGM deposit of Scaergaard intrusion. Abst IAGOD. Suppl 2. Terra Nova, East Greenland 5(3):38

Pearce TH (1970) Chemical variations in the Palisade sill. J Petrol 11(1):15–32

Pierce J, Lippard SJ, Roberts S (1987) Features of the composition and tectonic significance of ophiolites over the subduction zone. Geology of marginal basins. Mir, Moscow, 134–165p (in Russian)

Serri G (1982) The petrochemistry of ophiolite gabbroic complexes: a key the classification of ophiolites in low-Ti and high-Ti types. Earth Planet Sci Lett 52:203–212

Shirley DN (1987) Differentiation and compaction in the Palisades sill New Jersey. J Petrol 28(5):835–865

Strepikheev AA, Derevitskaya VA (1976) Fundamentals of the chemistry of high-molecular compounds. Chemistry, Moscow, 437p (in Russian)

Touysinhthiphonexay Y, Gold DP, Deines P (1984) Same properties of grafite from the Stillwater complex, Montana and the Bushveld igneous complex, South Africa. Geol Soc Am Abst Progr 16:677

Watson J (1980) Ore mineralization in the Archean provinces. The early history of the Earth. Mir, Moscow, 115–122p (in Russian)

Chapter 12
Influence of C–O–H–Cl-Fluids on Melting Phase Relations of the System Peridotite-Basalt: Experiments at 4.0 GPa

N. S. Gorbachev, A. V. Kostyuk, P. N. Gorbachev, A. N. Nekrasov, and D. M. Soultanov

Abstract The article presents the results of experiments at pressure 4.0 GPa and temperature 1400 °C on the influence of fluids (H_2O, $H_2O + CO_2$, $H_2O + HCl$) on the composition of restite and magma formed during the melting of the peridotite-basalt-$(K, Na)_2CO_3$ system as a model analogue of the mantle reservoir contaminated with protoliths of subducted oceanic crust. The composition of the fluid has a significant impact on the phase relations. In "dry" conditions and with H_2O fluid alkaline melts of phonolite type are formed, at $H_2O + CO_2$ fluid composition—trachiandezybasalts, with $H_2O + HCl$ fluid—riodacite melts. The alkaline melts coexist with an olivine-free restite pyroxene-phlogopite composition. Critical relations between fluid and silicate melt are observed in the water-bearing system. Interaction of supercritical fluid melts with peridotite restite leads to the formation of clinopyroxene, K-amphibole, phlogopite, carbonate, quenching silicate globules. Newly formed clinopyroxene and K-amphibole are in reactionary relations with olivine, orthopyroxene and clinopyroxene of peridotite restite. The revealed effects testify to instability of olivine at melting of peridotite-basalt mixture in the presence of fluid, effective influence of fluid composition on phase composition and critical ratios.

Keywords Experiment · Mantle · Crust · Fluid · Melt · Interaction · Melting · Critical relations

12.1 Introduction

The most important mechanism of large-scale exchange of matter between the crust and mantle is the subduction of the oceanic crust, leading to the eclogitization of basalts and the formation of reservoir in the mantle with protoliths of the subducted crust (Taylor and Neal 1989). The volatile enriched fluid has an effective effect on the phase composition, melting point of the mantle, the composition of the formed

N. S. Gorbachev (✉) · A. V. Kostyuk · P. N. Gorbachev · A. N. Nekrasov · D. M. Soultanov
D.S. Korzhinskii Institute of Experimental Mineralogy, Russian Academy of Science,
Academician Osipyan Street 4, Chernogolovka, Moscow Region, Russia 142432
e-mail: gor@iem.ac.ru

© The Editor(s) (if applicable) and The Author(s), under exclusive license to Springer Nature Switzerland AG 2020
Y. Litvin and O. Safonov (eds.), *Advances in Experimental and Genetic Mineralogy*, Springer Mineralogy, https://doi.org/10.1007/978-3-030-42859-4_12

fluid-containing silicate and salt (carbonate, chloride, sulfide) melts. High extracting and transport properties of fluids at high pressures lead to mobilization and transport of main and impurity elements. The processes of mantle metasomatism are related to their interaction with the mantle substrate.

The experimental model of the mantle reservoir with protoliths of the subducted oceanic crust is the system of peridotite-basalt-fluid. This type of system has been experimentally studied under "dry" conditions (Yaxley 2000; Tumiati et al. 2013; Mallik and Dasgupta 2012). Fluid-containing systems have been studied to a lesser extent. The system of peridotite-basalt in the presence of water and water-carbonate fluids was studied experimentally in the interval of 1250–1400 °C, 1.5–2.5 GPa (Gorbachev 1990). Instability of olivine in the presence of water-containing fluid was revealed. It is shown that the formation of magnesia magmas of the picritobasalt type does not require overheating of the mantle, their formation occurs at temperatures close to the mantle adiabatic point (Gorbachev 2000).

An important feature of fluid-containing silicate systems is the existence of critical relations between silicate melt and fluid at high pressures and temperatures, caused by their high mutual solubility. At critical $P_C T_C$, there is a complete mixing between them with the formation of supercritical fluid melt, and in the 2nd end critical point $(2P_C T_C)$ with the equality of P-T silicate solidus and $P_C T_C$ equilibrium melt fluid—a complete mixing between the liquidus phases, melt and fluid, which makes it difficult to determine P-T silicate solidus (Keppler and Audetat 2005; Litasov and Ohtani 2007).

The existence in the upper mantle of critical relations and composition of super-critical fluids-melt is proved by fluid-melt inclusions in diamonds, in minerals of mantle xenoliths of peridotite and eclogites from kimberlites and alkaline basalts, as well as in tectonically embedded carbonate-containing peridotite and eclogite massifs. Three main final types of such fluids are distinguished: carbonate enriched with Ca, Mg, Fe; salt enriched with Na, K, H_2O and silicate enriched with Si, Al (Klein-Bendavid et al. 2007; Navon et al. 1988; Weiss et al. 2009; Bogatikov et al. 2010). There is a complete mixture between alkaline and carbonate, carbonate and silicate end compositions of mantle fluids, and at pressure and temperature of the second critical point—between silicates and fluid (Wyllie and Ryabchikov 2000; Kessel et al. 2005).

Critical ratios were experimentally studied in water-containing silicate systems of monomineral (SiO_2, albite, nepheline, jadeite) and granite compositions. In these systems the critical pressure and temperature of $P_C T_C$ and $2P_C T_C$ lie in the range of 0.7–2.3 GPa, 550–1050 °C, increasing in the sequence of quartz–nepheline–albite–jadeite–granite. Critical pressures decrease when components (e.g. fluorine) are added to the fluid, increasing the mutual solubility of the melt and fluid, and increase when a component is added to the fluid, reducing the mutual solubility of the melt and fluid (e.g. CO_2) (Bureau and Keppler 1999). It has been established that at $H_2O + CO_2$ composition of the fluid in the system of peridotite-basalt critical ratios of silicate melt—fluid do not occur up to pressures of 4 GPa, but are observed at the water composition of the fluid. It was shown that 4 GPa and 1400 °C in the system of peridotite-basalt-H_2O characterize parameters of the 2 end critical points (Gorbachev 2000).

In the main and ultrabasic systems, much attention was paid to the $P_C T_C$ of the second end critical point of $2P_C T_C$. In the Fo–En–H_2O system, $2P_C T_C$ pressure is estimated at 12–13 GPa (Stalder et al. 2001), eclogite-H_2O at 5–6 GPa (Kessel et al. 2005), basalt-peridotite-H_2O at 3.8–4.0 GPa (Gorbachev 2000), and peridotite-H_2O at 3.8 GPa (Mibe et al. 2007). The Fo + En + H_2O + CO_2 $2P_C T_C$ system is estimated at 12–15 GPa (Willy and Rhyabchikov 2000). Critical ratios also exist between silicate and carbonate melts (Kisseva et al. 2012; Dasgupta et al. 2005; Gorbachev et al. 2015a, b).

In the majority of experimental works the transition of the system to the supercritical state and critical equilibria of melt and fluid were determined in situ at extreme (supra-liquidus) temperatures. The complete melt and fluid miscibility has been recorded either visually or by the homogenization of the 2-phase melt fluid association on diamond anvil equipment (Bureau and Keppler 1999) or by X-ray radiography (X-ray radiography) (Mibe et al. 2007). In some studies, the quenching method was used, where the transition from subcritical to supercritical state in the eclogite-H_2O system was fixed by the composition of quenching water-bearing glass and fluid in diamond traps (Kessel et al. 2005). However, these methods do not allow us to study the peculiarities of the phase composition in the interaction of supercritical fluid melts with restite. Moreover, most of the experiments were carried out at extreme temperatures. Taking into account the large temperature interval between solidus and liquidus in the main and ultrabasic systems, as well as the formation of mantle magmas in the partial melting of the mantle substance, it is of great interest to study the critical relations between partial (subliquidus) melts and fluid in the presence of refractory restite. In the works (Gorbachev 2000; Gorbachev et al. 2015a, b), the texture and phase composition of experimental samples can be used as a critical ratio test in such experiments. The composition of supercritical fluid melts and the nature of their interaction with the silicate substrate depend on the ratio of silicate and fluid components. It is believed that at intermediate compositions in supercritical fluid melts there may exist clusters, typical for both melt and fluid (Wyllie and Rhyabchikov 2000). The dual nature of supercritical fluid melts may appear in the structure and phase composition of the material formed during quenching after the experiment.

To obtain new data on the influence of fluid composition on melting, phase composition and critical ratios in the fluid-containing upper mantle, the system of peridotite-basalt-$(Na, K)_2CO_3$ under "dry" conditions and in the presence of H_2O, $H_2O + CO_2$ and $H_2O + HCl$ fluids is studied at $P = 4$ GPa, $T = 1400$ °C as an experimental model of a mantle reservoir with protoliths of subducted oceanic crust.

12.2 Experiment

The experiments were carried out in IEM RAS on the anvil with hole apparatus (NL-40) using a multi-ampoule technique with a Pt-peridotite ampoule (Gorbachev 1990). A specially prepared peridotite ampoule was filled with a mixture of powders of tholeiite basalt and sodium and potassium carbonates. Start composition (wt%): peridotite—55 wt%, basalt—25 wt%, sodium and potassium carbonate ~10 wt%. Chemical compositions of peridotite and basalt are given in Table 12.1. The source of fluids (20 wt% versus silicate) was distilled H_2O, oxalic acid dihydrate $H_2C_2O_4 \times 2H_2O$, 1H solution HCl. Chromite was added as accessory minerals, and Ni–Cu–Pt containing pyrrhotite was used in experiments under "dry" conditions. The charged peridotite ampoule was placed in a "working" Pt ampoule (d = 5 mm), which was hermetically welded. This ampoule was placed in a larger Pt ampoule (d = 10 mm) containing a fO_2 buffer association close to the quartz-fayalite-magnetite buffer. Temperature was measured by Pt30Rh/Pt6Rh thermocouple, pressure at high temperatures was calibrated according to the quartz-coesite equilibrium curve. The accuracy of the temperature and pressure tests is estimated at ±5 °C and ±1 kbar (Litvin 1991). The duration of the experiment was 18–24 h. The products of the experiments—polished preparations of quenching samples—were studied on the electronic scanning microscope CamScan MV2300 with YAG detector of secondary and reflected electrons and energy dispersive X-ray microanalyzer with semiconductor Si(Li) detector Link INCA Energy.

12.3 Results

12.3.1 Peridotite-Basalt-Na₂CO₃–K₂CO₃ System

Microphotographs of quenching samples are given in Fig. 12.1, chemical compositions of coexisting phases are given in Table 12.2. In "dry" conditions the quenched samples inherit the structure of peridotite ampoule. The inner "basalt" part, the "reaction" zone at the contact basalt-peridotite and the outer "peridotite" part of the ampoule are separated. The texture is massive. Although there was no disintegration of the initial peridotite ampoule during the experiment, the mineral composition of the zones is similar as a result of the reactions occurring during the experiment. Their main resemblance is the absence of olivine. The samples consist of isolated secretions of pyroxenes, phlogopite, chromium-spinel, cemented with intergranular alkaline melt (Table 12.2, Fig. 12.1b, c). Pyroxene grains of 20–50 μ in size have a zonal structure. The reaction relations of clinopyroxene ← orthopyroxene in the peridotite part of the sample are observed. The central part of the crystal is characterized by a higher content of MgO (up to 33 wt%), low content of CaO (~3 wt%) is represented by orthopyroxene. The edges of the grain are lighter colored due to a decrease in the concentration of MgO (up to 20–22 wt%) and an increase in the

Table 12.1 Chemical compositions (in wt%) of the initial peridotite and basalt in experiments

	SiO$_2$	TiO$_2$	Al$_2$O$_3$	Cr$_2$O$_3$	FeO	MnO	MgO	CaO	Na$_2$O	K$_2$O	P$_2$O$_5$	H$_2$O$^-$	H$_2$O$^+$	LOI[a]	Total
Peridotite	42.84	0.04	1.74	0.56	7.19	0.13	40.52	3.39	0.12	0.03	0.02	0.41	2.00	1.00	100.0
Basalt	50.02	1.85	14.51	–	14.03	0.20	5.85	10.40	2.50	0.72	–	–	–	–	100.08

Note [a]LOI—loss on ignition

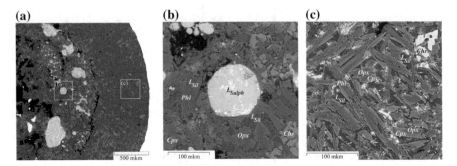

Fig. 12.1 *Peridotite-basalt-Na$_2$CO$_3$–K$_2$CO$_3$ system.* BSE images of experimental run products.
a General view of the sample consist of peridotite ampoule and basalt; **b** phase composition of
reaction zone consist of: orthopyroxene (*Opx*), clinopyroxene (*Cpx*), phlogopite (*Phl*), chromite
(*Chr*), with inclusions of sulphide globules (*L$_{Sulph}$*), cemented by alkaline silicate glass (*L$_{Sil}$*);
c peridotite ampoule consist of orthopyroxene (*Opx*), clinopyroxene (*Cpx*), phlogopite (*Phl*) and
chromite (*Chr*), cemented by alkaline silicate glass (*L$_{Sil}$*)

content of CaO (up to 12–14 wt%), belong to the pigeonites (Fig. 12.1c, Table 12.2).
In the reaction zone of the ampoule isolated orthopyroxene crystals coexist with
clinopyroxenes of avgite composition. Silicate melt has a phonolite composition.
SiO$_2$ content ~55 wt%, K$_2$O + Na$_2$O ~ 11–13 wt%, K$_2$O/Na$_2$O ~ 0.15. Sulfides
are concentrated in the reaction zone. The oval, molten form of sulfides indicates
partial melting of the original pyrrhotite. Sulfide globula matrix is represented by
Fe–Ni sulfide of pentlandite and Ni-pyrrhotite composition with inclusions of PtS
and Pt–Fe phase (Table 12.2, Fig. 12.1b).

12.3.2 Peridotite-Basalt-Na$_2$CO$_3$–K$_2$CO$_3$–(H$_2$O + CO$_2$) System

Microphotographs of the samples are shown in Fig. 12.2, chemical compositions of
the coexisting phases-in Table 12.3. The hardened sample has a solid texture. The
initial structure of the peridotite ampoule is not preserved. The texture and phase
composition of peridotite and its basaltic part are similar. The experimental sample
consists of clinopyroxene, K-clinopyroxene, phlogopite, single grains of orthopyroxene. Notice the absence of olivine among restite minerals. Klinopyroxene is moderately aluminous (up to 9 wt% Al$_2$O$_3$), concentration of micro impurities TiO$_2$, Cr$_2$O$_3$
does not exceed 0.5 wt%, Na$_2$O content is about 0.8 wt%. Potassium clinopyroxene
is more aluminous (up to 12 wt% Al$_2$O$_3$), with a reduced concentration of SiO$_2$
(46–47 wt%), concentration of micro impurities does not exceed 0.5 wt% for TiO$_2$
and 0.7 wt% for Cr$_2$O$_3$, enriched with Na$_2$O (up to 2.5 wt%), K$_2$O (up to 0.7 wt%),
ratio of K$_2$O/Na$_2$O ~ 0.3. Phlogopite is found in the form of thin, filiform secretions.

Table 12.2 Chemical compositions (in wt%) of coexisting phases in peridotite–basalt–$(Na, K)_2CO_3$ system

	SiO_2	TiO_2	Al_2O_3	FeO	MnO	MgO	CaO	Na_2O	K_2O	Cr_2O_3	SO_3	Total
Peridotite-basalt contact												
Opx	54.99	0.39	5.99	2.59	0.25	32.86	3.27	0.99	0.27	0.84	0.57	103.63
Cpx	51.05	0.66	5.61	2.09	0.25	16.46	17.57	1.87	0.00	0.67	0.31	96.54
Phl	40.77	0.50	13.11	1.24	0.00	25.54	0.13	1.25	8.03	1.94	0.48	93.85
Cht	0.30	0.32	16.61	5.90	0.57	20.03	0.00	0.26	0.15	54.95	0.04	99.99
L$_{Sil}$	55.02	0.40	19.18	0.46	0.07	0.00	0.63	9.92	1.52	0.02	0.29	87.50
Peridotite part												
Opx	54.23	0.47	5.06	2.61	0.03	30.40	3.93	0.53	0.04	1.35	0.00	98.66
Cpx	54.16	0.61	5.00	2.56	0.31	21.53	13.43	1.20	0.00	0.88	0.01	99.67
Phl	41.53	0.66	12.71	1.91	0.09	25.49	0.89	1.10	7.37	1.41	0.00	93.15
Cht	0.26	0.21	18.36	7.17	0.55	18.62	0.10	0.57	0.00	50.43	0.69	97.49
L$_{Sil}$	55.97	0.33	19.20	0.68	0.17	0.21	0.50	10.75	1.52	0.00	0.08	89.40

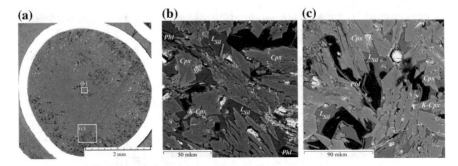

Fig. 12.2 *Peridotite-basalt-Na₂CO₃–K₂CO₃–(H₂O + CO₂) system.* BSE images of experimental run products. **a** General view of the sample. During melting, the composition averaged. Contact between the peridotite and basalt parts is almost not observed; **b** basalt, **c** peridotite part of the ampoule. Their texture and phase composition (Ca-clinopyroxene (*Cpx*), K-clinopyroxene (*K-Cpx*), phlogopite (*Phl*) and intergranular silicate glass (*L_Sil*)) are similar

Silicate melt of moderate alkalinity refers to trachiandezybasalts. The alkali content in the melt ($K_2O + Na_2O$) is about 6 wt%, $Na_2O/K_2O \sim 2.3$.

12.3.3 Peridotite-Basalt-Na₂CO₃–K₂CO₃–H₂O System

Microphotographs of the samples are shown in Fig. 12.3, chemical compositions of coexisting phases-in Table 12.4.

Disintegration of the peridotite ampoule was observed in the experimental products. As can be seen in Fig. 12.3a, the cross section of the sample only shows the components of the initial sample: disintegrated peridotite ampoule (Fig. 12.3c), reaction zone on the border of peridotite-basalt and "basalt" (Fig. 12.3b, c) the part of the initial sample formed during the quenching of supercritical fluid-melt consisting of a mixture of small crystals (up to 5 μ) of silicates, carbonates, sulfide and aluminosilicate microglobules. In contrast to subcritical conditions, massive silicate glass is not formed in the peridotite of the ampoule. The experimental sample consists of isolated relics or intergrowths of peridotite restite (olivine + orthopyroxene + clinopyroxene). Reaction relations between them and newly formed minerals with the replacement of type: orthopyroxene ← clinopyroxene ← K-amphibole (Fig. 12.3d, e), as well as the formation of newly formed and quenched phases—phlogopite, carbonate, Al–Si globules of phonolite glass (Fig. 12.3f). Figure 12.3(d–f) shows the types of reaction relations in peridotite restite, and Table 12.4 shows their chemical composition.

Table 12.3 Chemical compositions (in wt%) of coexisting phases in peridotite–basalt–(Na, K)$_2$CO$_3$–(H$_2$O + CO$_2$) system

	SiO$_2$	TiO$_2$	Al$_2$O$_3$	FeO	MnO	MgO	CaO	Na$_2$O	K$_2$O	Cr$_2$O$_3$	SO$_3$	Total
Basalt part												
L_{Sil}	53.00	0.14	17.94	0.56	0.00	0.23	3.92	3.38	1.51	0.00	0.00	80.68
K-Cpx	45.72	0.42	11.72	3.78	0.32	19.13	10.08	2.33	0.67	0.67	0.18	95.02
Cpx	50.95	0.47	8.84	4.83	0.17	16.05	16.51	0.84	0.00	0.47	0.00	99.12
Phl	41.35	0.36	14.44	2.94	0.20	22.29	0.06	1.33	6.76	0.38	0.82	91.2
Peridotite part												
L_{Sil}	52.55	0.10	17.32	1.03	0.03	0.25	3.54	3.38	1.45	0.11	0.43	80.2
K-Cpx	46.77	0.43	11.98	3.87	0.33	19.56	10.31	2.38	0.69	0.68	0.18	97.19
Cpx	50.84	0.46	8.82	4.82	0.17	16.02	16.48	0.84	0.00	0.46	0.00	98.92

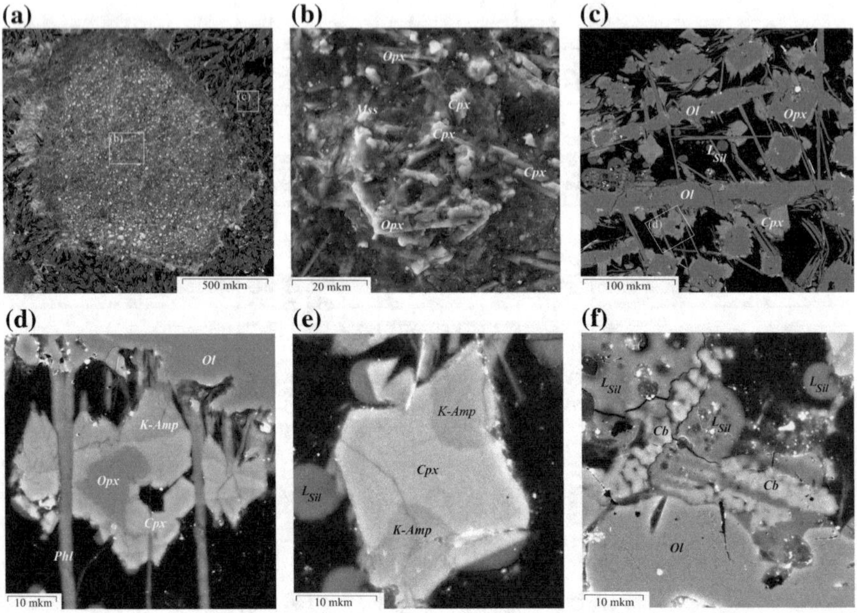

Fig. 12.3 *Peridotite-basalt-Na₂CO₃–K₂CO₃–H₂O system.* BSE images of experimental run products. **a** Peridotite-basalt contact; **b** basalt part, represented by a fine mixture of quenching silicate phases; **c** disintegrated peridotite part; **d–f** different areas of peridotite ampoules, characterizing the features of the phase composition, due to the reaction ratios in the peridotite restead with *Opx* ← *Cpx* ← K-*Amp* substitutions, newly formed *Phl*, *Cb* and quenched glass globules (*L_Sil*)

12.3.4 Peridotite-Basalt-Na₂CO₃–K₂CO₃–(H₂O + HCl) System

Microphotographs of quenching samples are shown in Fig. 12.4, chemical compositions of coexisting phases-in Table 12.5. The texture and phase composition of the quenching sample resembles that of the H₂O + CO₂ fluid experiment (Fig. 12.3). They differ in the presence of halite, riodacite composition of intergranular and quenched (in the form of Al–Si globules) melts. The matrix of experimental samples consists of relics and aggregates of restite minerals clinopyroxene, olivine, orthopyroxene, phlogopite.

Table 12.4 Chemical compositions (in wt%) of coexisting phases in peridotite–basalt–(Na, K)$_2$CO$_3$–H$_2$O system

	SiO$_2$	TiO$_2$	Al$_2$O$_3$	FeO	MnO	MgO	CaO	Na$_2$O	K$_2$O	Cr$_2$O$_3$	SO$_3$	Total
Ol	40.23	0.40	0.78	6.08	0.12	50.20	0.18	0.38	0.04	2.36	0.17	100.75
K-Amp	41.78	0.85	15.02	6.56	0.21	14.36	10.87	2.09	1.50	0.10	0.23	93.62
Cpx	49.11	0.61	8.34	4.16	0.23	16.00	19.69	1.07	0.00	0.00	0.00	99.20
Opx	54.29	0.28	4.88	3.54	0.13	33.33	1.69	0.00	0.00	1.03	0.06	99.23
Phl	40.67	0.35	14.20	2.89	0.20	21.92	0.06	1.31	6.65	0.37	0.81	89.70
L$_{Sil}$-glob	53.08	0.24	18.60	0.10	0.12	0.00	0.22	6.81	2.83	0.04	0.13	82.30
L$_{Cb}$	6.04	0.10	2.31	1.14	0.51	3.26	40.41	0.61	0.73	0.02	0.11	55.25
Coexisting phases on the Fig. 12.3d												
K-Amp	42.45	0.63	14.52	7.44	0.21	15.56	10.22	2.73	0.95	0.23	0.16	95.66
Cpx	47.71	1.57	8.70	3.00	0.35	14.08	21.33	0.95	0.02	0.52	0.08	98.41
Opx	53.50	0.14	5.23	3.31	0.19	33.20	1.45	0.09	0.06	1.68	0.02	98.91
Phl	41.82	0.29	13.09	2.62	0.00	24.52	0.00	0.84	7.85	0.55	0.00	91.70
Ol	40.23	0.40	0.78	6.08	0.12	50.20	0.18	0.38	0.04	2.36	0.17	100.9
Coexisting phases on the Fig. 12.3e												
K-Amp	41.56	1.15	15.54	6.80	0.15	15.41	10.91	2.91	1.42	0.00	0.20	96.08
Cpx	46.17	0.97	12.51	5.02	0.40	11.82	20.63	1.23	0.07	0.19	0.27	99.48
L$_{Sil}$-glob	52.31	0.01	18.11	0.07	0.00	0.04	0.18	5.76	2.71	0.00	0.08	79.27
Coexisting phases on the Fig. 12.3f												
Ol	40.23	0.40	0.48	6.08	0.12	50.20	0.18	0.38	0.04	2.36	0.17	100.75
L$_{Sil}$-glob	52.70	0.01	17.77	0.09	0.00	0.14	0.20	5.11	2.67	0.00	0.06	78.94
Cb	1.29	0.18	0.78	1.17	0.53	3.46	43.02	0.89	0.13	0.00	0.01	51.47
L$_{Sil}$	51.21	0.24	18.56	0.21	0.13	0.20	0.46	6.34	2.79	0.02	0.15	80.48

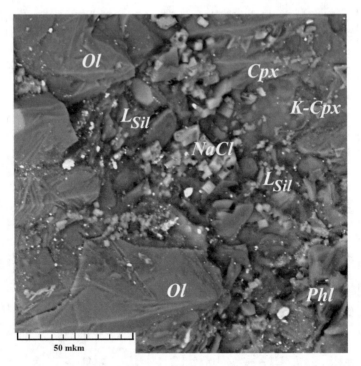

Fig. 12.4 *Peridotite-basalt-Na₂CO₃–K₂CO₃–(H₂O + HCl) system.* BSE image of experimental run products. The formation of NaCl crystals in a silicate matrix consisting of olivine (*Ol*), clinopyroxene (*Cpx*, K-*Cpx*), phlogopite (*Phl*), silicate melt (*L_Sil*)

12.4 Discussion

12.4.1 *Peridotite-Basalt-Na₂CO₃–K₂CO₃ System*

Partial melting of the original sample is observed. Although after the experiment the structure of peridotite ampoule is preserved, but as a result of reactions occurring during the experiment, the mineral composition of peridotite and basalt is similar. Their main feature is the absence of olivine, formation of alkaline melt of phonolite composition, concentration of sulfide melt in the reaction zone. Reaction ratios in pyroxenes of the type of pigeonite ← orthopyroxene is observed in the peridotite part of the ampoule enriched with MgO.

Table 12.5 Chemical compositions (in wt%) of coexisting phases in peridotite–basalt–(Na, K)$_2$CO$_3$–(H$_2$O + HCl) system

	SiO$_2$	TiO$_2$	Al$_2$O$_3$	FeO	MnO	MgO	CaO	Na$_2$O	K$_2$O	Cr$_2$O$_3$	SO$_3$	Total
L_{Sil}	71.45	0.20	18.76	0.59	0.21	2.08	2.82	1.95	1.65	0.00	0.00	99.71
Ol	40.57	0.90	0.59	8.20	0.37	46.82	0.00	0.16	0.05	2.24	0.07	99.97
Cpx	50.72	0.70	9.98	3.85	0.44	15.69	18.27	0.89	0.04	0.86	0.06	101.5
K-Cpx	46.75	0.49	12.69	3.63	0.00	19.72	10.32	2.38	0.68	0.66	0.05	97.42
Phl	41.55	0.31	12.88	2.35	0.41	23.46	0.18	1.56	7.01	0.88	0.16	90.78

12.4.2 Peridotite-Basalt-Na₂CO₃–K₂CO₃–(H₂O + CO₂) System

12.4.2 Peridotite-Basalt-Na$_2$CO$_3$–K$_2$CO$_3$–(H$_2$O + CO$_2$) System

In the case of partial melting of the initial sample, the structure of the initial peridotite ampoule is not preserved. The phase relations were determined by reactions between the fluid, alkaline fluid-containing melt and peridotite. As a result of reactions, olivine is dissolved, clinopyroxene, K-amphibole, and phlogopite are formed that coexist with the trachiandezybasalt melt.

12.4.3 Peridotite-Basalt-Na$_2$CO$_3$–K$_2$CO$_3$–H$_2$O System

At 4 GPa, 1400 °C, critical relations between the carbonated silicate melt and the fluid were achieved. This is evidenced by the peculiarities of texture and phase composition of the samples: disintegration of peridotite of the ampoule, absence of intergranular silicate glass in peridotite. Peridotite ampoule, which was filled with a mixture of basalt and carbonates during the assembly of the initial sample, after the experiment consisted of a compressed mixture of micron-sized particles of silicate and carbonate composition, formed during the quenching of fluid-melt. Reaction relations in restite with substitutions of olivine ← orthopyroxene ← clinopyroxene ← K-amphibole are determined by the interaction with supercritical fluid-melt. Wide development of reaction relations among minerals of peridotite restite testifies to high chemical activity of supercritical fluid-melt. The presence of quenching phases—phlogopite, Al–Si globules, carbonate—characterize the composition of supercritical fluid-melt. At melting of the supercritical mantle reservoir with protoliths of the subducted oceanic crust at *P-T*, exceeding the critical values, there should also be observed decompaction of the substrate as a result of disintegration of peridotite.

12.4.4 Peridotite-Basalt-Na$_2$CO$_3$–K$_2$CO$_3$–(H$_2$O + HCl) System

The quenching samples do not inherit the structure of the peridotite ampoule. At partial melting as a result of reactions between peridotite restite, fluid and fluid-containing melt there was an averaging of phase composition of peridotite and basalt components of the initial sample, formation without olivine association coexisting with riodacite melt. The texture and phase composition of the quenched samples are similar to the samples from the experiments with H$_2$O + CO$_2$ fluid. The formation of halite is observed.

12.4.5   K-Containing Phases

In experiments with $H_2O + CO_2$, $H_2O + HCl$ fluids, potassium phases are represented by K-containing clinopyroxenes with a similar chemical composition. In experiments with aqueous potassium-rich fluids, the K-amphibole phase is represented. Compared to K-clinopyroxene, K-amphibole is depleted in Na, Mg, Si and contains higher concentrations of K, Fe, Al, Ti (Table 12.6).

12.5   Conclusions

The composition of the fluid has a significant influence on the phase composition and critical ratios in mantle reservoirs with protoliths of the oceanic crust. In "dry" conditions and at acid composition of fluid alkaline melts of Na series are formed. The effect of acid-base interaction is clearly manifested, acidification of the fluid shifts the composition of the melt from trachiandezybasalt at $H_2O + CO_2$ composition of the fluid to riodacites with $H_2O + HCl$ fluid. Alkaline melts coexist with non-olivine metasomatically altered restite of pyroxene-phlogopite composition.

Instability of olivine at partial melting of peridotite-basalt mixture in the presence of acid fluid is established. The interaction of fluid and fluid-containing melts with restite in experiments with acid fluid ($H_2O + CO_2$, $H_2O + HCl$) was accompanied by pyroxenization and phlogopitization of peridotite:

$$Mg_2SiO_4(Ol) + MgSiO_3(Opx) + (3CaCO_3 + 4SiO_2)(L_{Sil} + Fl)$$
$$= 3CaMgSi_2O_6(Cpx) + 3CO_2(Fl);$$

$$2Mg_2SiO_4(Ol) + 2MgSiO_3(Opx) + (K_2O + Al_2O_3 + 2SiO_2 + 2H_2O)(L_{Sil} + Fl)$$
$$= 3CaMgSi_2O_6(Cpx) + 3CO_2(Fl).$$

The acid fluid produces K-containing clinopyroxene, the aqueous fluid produces K-containing amphibole, and the $H_2O + HCl$ fluid produces magmatic halite.

Critical relations between fluid and silicate melt are observed in the water-bearing system. Interaction of supercritical fluid melts with peridotite restitis leads to the formation of clinopyroxene, K-amphibole, phlogopite, carbonate, quenching silicate globules. Newly formed clinopyroxene and K-amphibole are in reactionary relations with olivine, orthopyroxene and clinopyroxene of peridotite restitis. Modal metasomatism of the upper mantle under the influence of water-containing supercritical fluid-melt leads to secondary enrichment—"refertilization" of depleted restite of the harzburgite composition, replacement of peridotite olivine-orthopyroxene association of newly formed phlogopite-clinopyroxene association. The texture and phase composition of the samples under supercritical conditions allow us to draw a conclusion about the zonal structure of reservoirs with protoliths of the subducted oceanic

Table 12.6 Chemical compositions (in wt%) of K-containing phases in systems with different fluid ($H_2O + CO_2$, $H_2O + HCl$, H_2O)

Phase	Fluid	SiO_2	TiO_2	Al_2O_3	FeO	MnO	MgO	CaO	Na_2O	K_2O	Cr_2O_3	SO_3	Total
K-*Cpx*	$H_2O + CO_2$	46.77	0.43	11.98	3.87	0.33	19.56	10.31	2.38	0.69	0.68	0.18	97.19
K-*Cpx*	$H_2O + HCl$	46.75	0.49	12.69	3.63	0.00	19.72	10.32	2.38	0.68	0.66	0.05	97.42
K-*Amp*	H_2O	41.78	0.85	15.02	6.56	0.21	14.36	10.87	2.09	1.50	0.10	0.23	93.6

crust: the outer zone—metasomatically altered under the influence of supercritical fluid-melt disintegrated restite of dunite or harzburgite composition, underwent "refertilization" as a result of the formation of new minerals in it, the inner zone—isolated lenses of supercritical fluid-melt. Disintegration of fluid-containing mantle peridotite substrate at supercritical pressures can lead to the formation of tectonically weakened zones, fluid paths and upper mantle plumes.

Acknowledgements Acknowledgement. The study was funded by the projects of RFBR №17-05-00930a and IEM RAS №AAAA18-118020590140

References

Bogatikov OA, Kovalenko VI, Sharkov EV (2010) Magmatism, tectonics, geodynamics of the Earth. Nauka, Moscow, p 615

Bureau H, Keppler H (1999) Complete miscibility between silicate melts and hydrous fluids in the upper mantle: experimental evidence and geochemical implications. Earth Planet Sci Lett 165:187–196

Dasgupta R, Hirschmann MM, Dellas N (2005) The effect of bulk composition on the solidus of carbonated eclogite from partial melting experiments at 3 GPa. Contrib Mineral Petrol 149:288–305

Gorbachev NS (1990) Fluid-magma interaction in sulfide-silicate systems. Int Geol Rev 32(8):749–831

Gorbachev NS (2000) Supercritical state in the hydrous mantle: evidence from experimental study of fluid-bearing peridotite at $P = 40$ kbar and $T = 1400$ °C. Dokl Akad Nauk SSSR 370:147–150

Gorbachev NS, Kostyuk AV, Shapovalov YB, (2015a) Experimental study of the peridotite-H_2O system at $P = 3.8$–4 GPa, $T = 1000$–1400 °C: critical relations and vertical zoning of the upper mantle. Dokl Earth Sci 461(2):360–363

Gorbachev NS, Kostyuk AV, Shapovalov YB (2015b) Experimental study of the basalt–carbonate–H_2O system at 4 GPa and 1100–1300 °C: origin of carbonatitic and high-K silicate magmas. Dokl Earth Sci 464(2):1018–1022

Keppler H, Audetat A (2005) Fluid-mineral interaction at high pressure. Mineral behavior at extreme conditions. EMU Notes Mineral 7:225–251

Kessel R, Ulmer P, Pettke T, Schmidt MW, Thompson AB (2005) The water-basalt system at 4–6 GPa: Phase relations and second critical endpoint in a K-free eclogite at 700–1400 °C. Earth Planet Sci Lett 237:873–892

Kiseeva E, Yaksley GM, Hermann J, Litasov KD, Rosenthal A, Kamenetsky VS (2012) An experimental study of carbonated eclogite at 3.5–5.5 GPa—implications for silicate and carbonate metasomatism in the cratonic mantle. Journal of Petrol 53(4):727–759

Klein-BenDavid O, Izraeli ES, Hauri E, Navon O (2007) Fluid inclusions in diamonds from the Diavik mine, Canada and the evolution of diamond-forming fluids. Geochim Cosmochim Acta 71:723–744

Litasov KD, Ohtani E (2007) Effect of mater on the phase relations in Earth's mantle and deep water cycle. Special paper Geol Soc of Amer 421:115–156

Litvin YuA (1991) Physical and chemical studies of the melting of the Earth's deep matter. Nauka, Moscow, p 312

Mallik A, Dasgupta R (2012) Reaction between MORB-eclogite derived melts and fertile peridotite and generation of ocean island basalts. Earth Planet Sci Lett 329(330):97–108

Mibe K, Kanzaki M, Kawamoto T, Matsukage KN, Fei Y, Ono S (2007) Second critical endpoint in the peridotite–H_2O system. J Geophys Res 112:B03201

Navon O, Hutcheon ID, Rossman GR, Wasserburg GJ (1988) Mantle–derived fluids in diamond microinclusions. Nature 335:784–789

Stalder R, Ulmer P, Thompson AB, Gunther D (2001) High pressure fluids in the system MgO–SiO_2–H_2O under upper mantle conditions. Contrib Miner Petrol 140:607–618

Taylor LA, Neal CR (1989) Eclogites with oceanic crustal and mantle signatures from the Bellsbank kimberlite, South Africa, part 1: mineralogy, petrography, and whole rock chemistry. J Geol 97:551–567

Tumiati S, Fumagalli P, Tiraboschi C, Poli S (2013) An experimental study on COH-bearing peridotite up to 3.2 GPa and implications for crust-mantle recycling. J Petrol 54:453–479

Weiss Y, Kessel R, Griffin WL, Kiflawi I, Klein–BenDavid O, Bell DR, Harris JW, Navon O (2009) A new model for the evolution of diamond–forming fluids: Evidence from microinclusion–bearing diamonds from Kankan, Guinea. Lithos 112(2):660–674

Wyllie PJ, Rhyabchikov ID (2000) Volatile components, magmas, and critical fluids in upwelling mantle. J Petrol 41:1195–1206

Yaxley GM (2000) Experimental study of the phase and melting relations of homogeneous basalt plus peridotite mixtures and implications for the petrogenesis of flood basalts. Contrib Mineral Petrol 139:326–338

Chapter 13
Inter-phase Partitioning of Pb and Zn in Granitoid Fluid-Magmatic Systems: Experimental Study

V. Yu. Chevychelov

Abstract Experimental studies of the partitioning of Pb and Zn between the fluid and the granite melt have shown that relatively low pressure of 100 MPa and temperature of 750 °C and acidic chloride composition of the (0.1N HCl + 1N NaCl) fluid will promote the best extraction these metals from the melt. Herein the $^{\text{fluid/melt}}D_{\text{Pb,Zn}}$ partition coefficients are ≥ 10. In systems with acidic chloride fluids these coefficients increase proportionally to the square of the fluid chloride concentration: $^{\text{fluid/melt}}D_{\text{Pb}}$ $\approx (7 \pm 1.6) \times {}^{\text{fluid}}C_{\text{Cl}}^2$ and $^{\text{fluid/melt}}D_{\text{Zn}} \approx (5.7 \pm 2.0) \times {}^{\text{fluid}}C_{\text{Cl}}^2$, which allows us to assume pure chloride complexes in these systems for Pb^{+2} and Zn^{+2}. Increases in temperature and, in particular, pressure significantly reduce the partition coefficients of these metals. At the conditions of two-phase fluid of H_2O–NaCl composition there is a significant (~5–7 times) accumulation of Pb and Zn in the dense brine (water–salt) fluid phase relatively low-density vapor (essentially aqueous) fluid phase. The addition of CO_2 to the fluid increases the concentration of these metals in the brine phase, and for Pb it is much higher than for Zn. In the process of crystallization of granite melt at pressure of about 500 MPa polymetals do not pass into the separated (nascent) fluid. Herein Pb is concentrated in crystallizing feldspars (*Olg* and *Kfs*), and Zn remains in the residual melt. With decreasing pressure (up to about 250 MPa) the fluid is visibly enriched with polymetals, and the content of Pb and Zn become almost the same in the fluid and in feldspars, except for the sharply lower content of Zn in *Ab* and *Ancl*. At change of composition of melt from granite to andesite and basalt the $^{\text{fluid/melt}}D_{\text{Pb,Zn}}$ partition coefficients sharply decrease (on 1.5–2 orders of magnitude). In this case positive linear correlation dependences between the $^{\text{fluid/melt}}D_{\text{Pb,Zn}}$ coefficients and silica content in melt are established ($R_{\text{Pb}} = 0.95$, and $R_{\text{Zn}} = 0.93$).

Keywords Geochemistry · Solubility · Partitioning · Lead · Zinc · Fluids · Magmatic melts · Granites · Experiment

V. Y. Chevychelov (✉)
D.S. Korzhinskii Institute of Experimental Mineralogy, Russian Academy of Sciences, Academician Osipyan Street 4, Chernogolovka, Moscow Region 142432, Russia
e-mail: chev@iem.ac.ru

© The Editor(s) (if applicable) and The Author(s), under exclusive license to Springer Nature Switzerland AG 2020
Y. Litvin and O. Safonov (eds.), *Advances in Experimental and Genetic Mineralogy*, Springer Mineralogy, https://doi.org/10.1007/978-3-030-42859-4_13

13.1 Introduction

Non-ferrous heavy metals or polymetals Pb and Zn according to Kigai and Tagirov (2010), Kigai (2011) are basitophilic metals of mantle genesis, as opposed to granitophilic metals of crustal origin (Ta, Nb, Sn, W, Mo). Lead and zinc in the Earth's crust are typical chalcophile elements, which are mainly found in the form of sulfides. Pb and Zn in the conditions of post magmatic processes usually are found together (Tugarinov 1973). However, in magmatic bodies they behave differently (ion radius $Pb^{2+} = 0.119$ nm, much more than $Zn^{2+} = 0.074$ nm), so Pb mainly replaces K, and Zn replaces Mg or Fe. As a result, lead is mainly concentrated in feldspar, while zinc is concentrated in biotite. The partitioning of Zn and Pb in the series of magmatic differentiates from diorites to granites shows that Zn is contained in these rocks in more or less equal quantities, and the content of Pb increases markedly in the most acidic differentiates enriched with potassium feldspar.

This paper is a generalizing review of the author's works performed in the IEM RAS on the experimental study of some geochemical features of the behavior of polymetals and potential ore-bearing of high-temperature magmatic fluid at various stages of fluid-magmatic (granitoid) evolution. The paper presents the results of experimental study of the influence of (NaCl, HCl, HF, NaOH) composition, (0.1–1.1N) concentration and acidity-alkalinity of aqueous salt fluid on the extraction of polymetals from the granite melt into the fluid in hypabyssal conditions imitating the formation of intrusion ($T = 750$ °C and $P = 100$ MPa). The obtained results confirm the possibility of formation of a high-temperature ore-bearing fluid as a result of extraction of ore substance from the magmatic melt by a upward intratelluric fluid (Chevychelov and Epelbaum 1985; Chevychelov 1987). The influence of pressure (100–500 MPa) and temperature (700–900 °C) on the evolution of the fluid-magmatic (granitoid) system was also studied. In these experiments, the conditions for the formation of granite magma at depths of about 20 km (abyssal conditions near the Conrad boundary) and hypabyssal conditions simulating the formation of intrusion at a depth of 4–5 km were simulated (Chevychelov 1993; Chevychelov et al. 1994). The results of the study of partitioning of polymetals in the fluid-magmatic system at heterogenization of aqueous-chloride and aqueous-chloride-carbon-acid fluids at hypabyssal conditions of intrusion formation (at $P = 100$ MPa and $T = 800$ °C) are presented (Chevychelov 1992, 1993; Chevychelov et al. 1994). The resulting aqueous salt fluids play an important role in processes of extracting from the melt and transport of ore metals. Besides, the partitioning of polymetals in the process of separation of aqueous-chloride fluid from granite melt at melt crystallization in subsolidus (near-solidus) conditions has been experimentally investigated (Chevychelov 1993; Chevychelov et al. 1994). Finally, at $T = 800$–1000 °C and $P = 100$ and 500 MPa, the partitioning of Pb and Zn between acidic chloride fluid and model leucogranite, granite and granodiorite melts was studied (Chevychelov 1987, 1998; Chevychelov and Chevychelova 1997).

13.2   Review of the Results of Previous Experimental Studies

13.2.1   Partitioning of Pb Between Aqueous Fluid and Aluminosilicate Melt

In the Earth's crust, lead is found actually only in the form of Pb^{2+}. High (n $\times$ 10 wt%) solubility of lead in silicate glasses (crystal glass, SiO_2–Na_2O and other compositions) is known. At magmatic P-T conditions, the solubility of Pb in granitoid melts is not experimentally studied. The strong influence of oxygen fugacity (f_{O2}) on lead solubility in SiO_2–CaO–FeO slags of mine melting is established: at $T = 1220$ °C in oxidizing conditions ($f_{O2} = 10^{-4}$ Pa) the melt can contain up to 12 wt% Pb, and in reducing conditions ($f_{O2} = 10^{-8}$ Pa) the maximum content of lead decreases sharply to 0.5 wt% (Zaitsev et al. 1982; Kukoev et al. 1982).

There are relatively not many the experimental works on studying the partitioning of Pb in the fluid-granitoid melt system. They were carried out only at the increased pressure of 160–500 MPa. The $^{fluid/melt}D_{Pb}$ partition coefficient at $1m$ NaCl of the initial fluid, 800 °C and 160 MPa is 1.2–2.1 according to the data (Urabe 1987a), but it can increase with the increase in alumina content and decrease with the increase in the alkalinity of the melt (Urabe 1985), i.e. this coefficient can be both greater and smaller than unity with a change in the melt composition.

With increasing temperature (700–800–900 °C), the $^{fluid/melt}D_{Pb}$ coefficient slightly increases, for example, at 1N NaCl—0.14–0.20–0.26 (Khitarov et al. 1982). Negative baric dependence of lead partitioning is established (Urabe 1987a). Thus, with increasing pressure from 160 to 450–500 MPa, the Pb partition coefficient decreases at the initial $2m$ NaCl solution from 7–11 to 0.5–2.2, and at $1m$ HCl from 8–20 to 2–5. For andesite melts in the range of 200–800 MPa, the extreme baric dependence of this coefficient with a maximum near by 400 MPa was obtained (Khodorevskaya and Gorbachev 1993b), i.e. in the beginning the dependence is positive: with the pressure increase from 200 to 400 MPa, the coefficient increases and only with a further pressure increase from 500 to 800 MPa it decreases.

The composition of the high-temperature fluid has a significant influence on the partitioning of Pb. The maximum coefficients of $^{fluid/melt}D_{Pb}$ were obtained in the presence of chloride solutions: up to 6.5–19.5 (Khitarov et al. 1982; Urabe 1985, 1987a). T. Urabe also is shown the increase of Pb partition coefficient by 1.5 order of magnitude with the increase of chloride fluid concentration from 0.5–1m to 2–3m NaCl. In the case of initial HCl solutions, the $^{fluid/melt}D_{Pb}$ coefficient can be several times higher compared to NaCl solutions of the same concentration. The $^{fluid/melt}D$ coefficient for Pb^{+2} as a divalent ion increases in proportion to the square of the chloride concentration of the aqueous fluid (Urabe 1985, 1987a). In the presence of 1N $NaHCO_3$ and 1N NaOH solutions, this coefficient does not exceed 0.05–0.10, and the extraction capacity of pure water is below the detection limit (Khitarov et al. 1982). The main forms of Pb in the fluid are probably chloride complexes, most likely hydroxychloride or oxychloride. In concentrated chloride solutions, the formation of more chlorine-rich complexes can be expected. Adding sulfur has little or no effect on the extraction of Pb from the melt.

A very strong influence of granite melt composition on the partitioning of Pb is established (Urabe 1985). So, at change of the melt composition from slightly alkaline ($Wol_{1.9}Ab_{66}An_{5.4}Qtz_{27}$) to weakly alumina ($Crn_{2.4}Ab_{56}An_{9.1}Qtz_{32}$), the partition coefficient $^{fluid/melt}D_{Pb}$ increases almost on two orders (from $0.0474 \times (^{fluid}C_{Cl})^2$ to $2.26 \times (^{fluid}C_{Cl})^2$). Such a sharp increase in this coefficient and the Pb content in the fluid are caused by a strong increase in the acidity of the chloride fluid as a result of its interaction with the melt. The threshold value of the molecular $(Na + 2Ca)/Al$ ratio in quenching glasses equal (0.85 ± 0.03) has been calculated, above which low $^{fluid/melt}D_{Pb}$ coefficients and near-neutral pH of solutions have been obtained in experiments with slightly alkaline melts, and below which high $^{fluid/melt}D_{Pb}$ coefficients and acid solutions have been obtained in experiments with weakly alumina melts.

At the melt composition changes from granite to andesite and basalt, the Pb partition coefficient decreases by 1.5–2 of the order from 8–20 at $1m$ HCl and 160 MPa to 0.25 (andesite melt) and 0.14 (basalt melt) at the same fluid and 200–100 MPa (Urabe 1987a; Khodorevskaya and Gorbachev 1993a, b).

The partitioning of Pb between fluid phases and granite melt at conditions of heterogenization of chloride fluid at $P \leq 100$–150 MPa has not been experimentally studied.

13.2.2 Partitioning of Zn Between Aqueous Fluid and Aluminosilicate Melt

Zinc has only one constant valence of $+2$ in its compounds. By means of diffusion technique (Chekhmir et al. 1991; Epelbaum 1986) it was established that the melt contains 3.74 and 1.96 wt% Zn at dissolution of willemite Zn_2SiO_4 in the melt of rare-metal akchatau granite at pressure 100 MPa and temperature 1000 and 800 °C, respectively. Extrapolation of the obtained data to the temperature of granite melt formation of $T_{eutectic} = 750$–770 °C gives the value of effective solubility of zinc in the melt about 1.77 wt%.

Experimental studies of Zn partitioning in the fluid–granitoid melt system are very few in number, especially at P below 200 MPa. The partition coefficients of $^{fluid/melt}D_{Zn}$ at $1m$ NaCl initial solution, 800–850 °C and 160–200 MPa are from 1–2 according to data (Holland 1972) to 9.5 according to (Urabe 1987a) and can significantly varies at the change of alkalinity-alumina content in the melt composition (Urabe 1985), i.e. this coefficient can be both larger and smaller than unity depending on the melt composition.

At increasing temperature (700–800–900 °C), the $^{fluid/melt}D_{Zn}$ coefficient remains practically constant (NaCl and NaHCO$_3$ solutions) or slightly increases, for example, in experiments with 0.1N NaCl and elementary sulfur—0.21–0.26–0.42 (Khitarov et al. 1982). The negative baric dependence of zinc partitioning was obtained Urabe (1987a) for the granite system. So, at increasing pressure from 160 to 450–500 MPa, the Zn partition coefficient decreases at the initial 2m NaCl solution from 11–13.5 to 5–8, and at 0.5m NaCl solution from 5 to 0.01–0.05. According to Kravchuk et al. (1991), the $^{fluid/melt}D_{Zn}$ coefficient between 1M NaCl fluid and granodiorite melt decreases by 3–5 times at increasing pressure from 200 to 400 MPa. The extreme baric dependence of the Zn partition coefficient with a maximum in the region of about 400 MPa was obtained for the andesite melts in the range of 300–800 MPa (Khodorevskaya and Gorbachev 1993b). Namely at first at the increase in pressure from 300 to 400 MPa, the coefficient increases and only at a further increase in pressure from 500 to 800 MPa it begins to decrease.

The composition of the high-temperature fluid has a significant influence on the partitioning of Zn. The maximum coefficients of $^{fluid/melt}D_{Zn}$ are obtained in the presence of chloride solutions; they can reach 24–30 (Holland 1972; Urabe 1985, 1987a). Positive dependence of Zn partition coefficient on concentration of chloride solution was established (Khitarov et al. 1982; Kravchuk et al. 1991). According to the results of the studies (Holland 1972; Urabe 1985, 1987a), the $^{fluid/melt}D_{Zn}$ coefficient, as a divalent ion, increases in proportion to the square of the chloride concentration of the aqueous fluid. Herein the maximum values of $^{fluid/melt}D_{Zn}$ were obtained in the presence of acid fluids (pH solution after experiment about 1–2). In the presence of 1N $NaHCO_3$ and 1N NaOH solutions, this coefficient is very small and amounts to 0.02–0.29, and the extraction capacity of pure water is below the detection limit (Khitarov et al. 1982). The main forms of Zn in the fluid are probably chloride complexes, most likely hydroxychloride or oxychloride. Adding sulfur doubles the Zn content of the fluid. This is due to the instability of ZnS in experimental conditions, as well as the fact that sulfur, oxidizing, can significantly acidify the fluid composition, increasing the extraction of Zn from the melt (Khitarov et al. 1982).

A very strong influence of granite melt composition on the Zn partitioning is established (Urabe 1985). So, at change of the melt composition from slightly alkaline ($Wol_{1.9}Ab_{66}An_{5.4}Qtz_{27}$) to weakly alumina ($Crn_{2.4}Ab_{56}An_{9.1}Qtz_{32}$), the $^{fluid/melt}D_{Zn}$ partition coefficient increases on two orders of magnitude (from $0.0833 \times (^{fluid}C_{Cl})^2$ to $9.40 \times (^{fluid}C_{Cl})^2$). Such a sharp increase in this coefficient and the Zn content in the fluid are caused by a strong increase in the acidity of the chloride fluid as a result of its interaction with the melt. The threshold value of the molecular $(Na + 2Ca)/Al$ ratio in quenching glasses equal (0.85 ± 0.03) has been calculated, above which low $^{fluid/melt}D_{Zn}$ coefficients and near-neutral pH of solutions have been obtained in experiments with slightly alkaline melts, and below which high $^{fluid/melt}D_{Zn}$ coefficients and acid solutions have been obtained in experiments with weakly alumina melts. It can be assumed that the increase in the Zn content in the slightly alkaline melt in comparison with the weakly alumina melt is explained by the formation of compounds such as $(Na,K)_4Zn_2(SiO_4)_2$, similar to alkaline zirconium silicates, the existence of which in the melt was assumed by Watson (1979).

At the melt composition changes from granite to andesite and basalt, the Zn partition coefficient decreases by about 1.5–2 and more on the order from 13–18 at

$1m$ HCl and 160 MPa to 0.27–0.68 (andesite melt) and 0.06 (basalt melt) at the same fluid and 300–100 MPa (Urabe 1987a; Khodorevskaya and Gorbachev 1993a, b).

13.2.3   Partitioning of Zn Between Brine (Water–Salt) and Low-Density Vapor (Essentially Aqueous) Fluids Phases in the H_2O–NaCl System

Processes of fluid-magmatic interaction in the Earth's crust occur within wide limits of P-T conditions and compositions. At these conditions, fluids can often be in heterogeneous (two, three or more phases) states. Available information on physical and physical-chemical parameters of relatively simple composition of fluids testifies to the existence in natural conditions of a number of areas of immiscibility of fluid phases, which significantly affect the partitioning of components. Experimental data on the partitioning of zinc between brine and low-density vapor phases of heterogeneous fluid of H_2O–NaCl (interesting for geology and the most investigated system) are shown in Table 13.1.

In experimental paper (Kravchuk et al 1992) the partitioning of zinc between brine and vapor fluids phases in the H_2O–NaCl system was investigated. Artificial fluid inclusions formed in the experiments at healing cracks in quartz were obtained by the technique described in the papers (Sterner and Bodnar 1984; Ishkov and Reyf 1990). According to data (Chou 1987; Anderko and Pitzer 1993) at $T = 800$ °C, $P = 100$ MPa and 30 wt% NaCl initial solution, the H_2O–NaCl system decomposes into a low-density vapor and brine phases, differing from each other in composition and density. Herein the low-density fluid phase contains from 1.9 to 6.9 wt% NaCl, and the brine (water-salt) phase contains 60–62 wt% NaCl. As normal inclusions

Table 13.1 Literature data on the partitioning of Zn between brine (water–salt) and low-density vapor (essentially aqueous) fluids phases in the system H_2O–NaCl

$^{brine/vapor}D_{Zn}$[a]	Experimental method[b]	Composition of initial fluid	Experimental conditions	Analytical procedure[c]	Literature source
>1	The method of synthesis of synthetic fluid inclusions in quartz	30 wt% NaCl + $ZnCl_2$ + CuCl	800 °C, 100 MPa, 3–3.3 days	Laser spectral analysis	Kravchuk et al. (1992)
≈8–85 (independent of S)	The same	10–30 wt% NaCl + $ZnCl_2$ + CuCl ± S	500–650 °C, 35–100 MPa	X-ray fluorescence synchrotron radiation	Nagaseki and Hayashi (2008)

[a] The partitioning coefficient of Zn
[b] The method of the experiment
[c] The method of analysis of the phases compositions after the experiment

that captured only the gas or only liquid phase, and combined inclusions, in that the mixture was preserved, were formed in quartz in the process of experiments. Since the content of the combined inclusions during them isolation was various, they are not homogenized at the formation temperature and therefore they are easily rejected by preheating. The densities and masses of water-salt and low-density fluids phases in separate inclusions were determined, that allowed to present the results of laser-spectral analysis in units of concentration. As a result for the first time, the Zn partition coefficient between water-salt and low-density fluids phases ($^{\text{brine/vapor}}D_{Zn} > 1$) was estimated and it was shown that zinc enriches the brine fluid phase relatively low-density vapor fluid phase. Due to the relatively high detection limit, 2×10^{-10} g, and the small size of the inclusions, less than 10^{-8} ml, only the upper limit of Zn content in the inclusions with the low-density vapor fluid phase was estimated.

In experimental paper (Nagaseki and Hayashi 2008) partition coefficients of Zn between brine and low-density vapor fluids phases are determined at 500–650 °C, 35–100 MPa, 10–30 wt% NaCl fluid, both containing sulfur (1–2 mol/kg H_2O), and without sulfur. The experimental method of artificial fluid inclusions in quartz was used. The composition of inclusions was analyzed by X-ray fluorescence synchrotron radiation. It is established that Zn is mainly concentrated in the brine phase regardless of the sulfur content in the system. Zinc partition coefficients ($^{\text{brine/vapor}}D_{Zn}$) in the system vary from 8 to 85. These values are well consistent with the analyses of natural fluid inclusions from ore deposits (Audetat et al. 2000; Audetat and Pettke 2003; Ulrich et al. 2001).

Thus, it was established that zinc enriches the brine phase relatively low-density vapor phase at conditions of heterogenization of the model chloride H_2O–NaCl fluid at $T = 500$–800 °C and $P = 35$–100 MPa. The $^{\text{brine/vapor}}D_{Zn}$ partition coefficient between water-salt and low-density fluids phases is ~8–85 and it does not depend on the S content in the system.

13.3   Used Experimental and Analytical Techniques

The initial glasses enriched in polymetals, that were used to study the influence of *P-T* conditions, concentration and composition of the fluid, contained 0.5 wt% PbO + 0.5 wt% ZnO (0.46 wt% Pb + 0.40 wt% Zn). They were prepared by a two-fold long melting of powdered akchatau granite at $T = 1200$–1350 °C and $P = 0.1$ MPa. The chemical compositions of individual glass grains according to the electron microprobe analyses varied in the range of ±8 relative%. Also in the experiments were used initial aqueous solutions of NaCl, HCl, HF, NaOH, (HCl + NaCl), (NaOH + NaCl) of different concentration (from 0.1N to 1–1.5N), not containing Pb and Zn, and pure H_2O.

For the experiments carried out at conditions of fluid heterogenization, the model granite sub-aluminous melt (in wt%: 72.7 SiO_2, 0.4 TiO_2, 13.2 Al_2O_3, ≈2.6 Fe_2O_3, ≈1.6 FeO, 0.3 MgO, 0.9 CaO, 2.8 Na_2O, 5.9 K_2O) and rich-aluminous (plumazite) melt (in wt%: 75.8 SiO_2, 0.2 TiO_2, 15.3 Al_2O_3, ≈0.6 Fe_2O_3, ≈0.4 FeO, 0.4 CaO, 3.0

Na$_2$O, 4.2 K$_2$O) were previously are prepared. Pb and Zn were introduced both into
granite melts and into the initial solution of H$_2$O–NaCl and H$_2$O–CO$_2$–NaCl. CO$_2$
was injected in the form of oxalic acid H$_2$C$_2$O$_4$ × 2H$_2$O plus hydrogen peroxide
H$_2$O$_2$, and polymetals—in the form of salts Pb(NO$_3$)$_2$ and ZnCl$_2$. Chlorine was
injected into solutions both in the form of NaCl and in the form of well soluble salt
ZnCl$_2$. It was supposed that adding ZnCl$_2$ does not significantly change the H$_2$O–
NaCl diagram in conditions of our experiments. The equilibrium of granite melt with
brine and vapor fluids phases were studied in turn in different experiments. A similar
experimental technique has been proposed by (Kotelnikov et al. 1990).

For the experiments to study the influence of the magmatic melt composition
on the partitioning of polymetals, the water-saturated granodiorite melt (in wt%:
68 SiO$_2$, 17–19 Al$_2$O$_3$, 4.0–5.2 CaO, 5.4–6.6 Na$_2$O, 1.7–3.3 K$_2$O), granite melt
(in wt%: 73 SiO$_2$, 15–17 Al$_2$O$_3$, 1.4–2.4 CaO, 5.1–6.7 Na$_2$O, 2.2–3.2 K$_2$O) and
leucogranite melt (in wt%: 76–76.5 SiO$_2$, 13.5–14.5 Al$_2$O$_3$, 1.1–1.7 CaO, 4.8–6.3
Na$_2$O, 1.9–2.7 K$_2$O) were previously are prepared from gel mixtures. Experiments
to study the partitioning of Pb and Zn were carried out by the method of "approach
to the equilibrium from both sides" that is to say from the melt and from the fluid.
A modified combined technique was used in accordance with which aluminosilicate
glasses with the same polymetallic content were used, and the metal contents in
the initial solutions were different in different experiments. Each of the three initial
glasses, in addition to the main rock-forming components, contained 0.19 wt% Pb +
0.16 wt% Zn. The initial 0.1N HCl + 1N NaCl solution was used in the experiments,
in which polymetals were injected in the form of ZnCl$_2$ and PbCl$_2$. In most of the
experiments the initial solution contained from 0.3 to 0.9 wt% Pb and 0.6 (0.9) wt%
Zn, some experiments were carried out with the solution initially containing only
Zn. The amounts of polymetals added into the initial glasses and solutions were
minimum sufficient for electron microprobe analysis of experimental products. One
or two pieces of water-saturated Pb,Zn-containing granitoid glass wrapped in gold
foil were hung over the bottom of the gold capsule. The initial solution was loaded
inside the capsule, and then capsule was blown for two minutes with argon to remove
air and then welded shut. The high temperature (800–1000 °C) of the experiments
was related to the high liquidus (T_L) temperature of the melts, which contained
appreciable amounts of CaO. At these conditions the first crystallizing phase from
the melt was plagioclase.

All experiments were carried out in gold and platinum capsules in conventional
hydrothermal vertical externally heated cold seal pressure vessels (CSPV) made of
a nickel-based alloy at the value of f_{O2}, close to the buffer association Ni–NiO, as
well as in gas internally heated pressure vessels (IHPV) at log f_{O2} ≈ NNO + 3.5
(Berndt et al. 2005). Herein Pb and Zn were divalent. Initial ratio of solution: glass
was (6–3):1. The experiments duration ranged from 2.5 to 10 days depending on the
P-T conditions.

After the experiments, the chemical composition of glass (about 15 μm from
the edge and in the center) and fluid were determined by the method of electron
microprobe analysis, the Pb and Zn analyzed using a wave dispersion spectrometer
(the conditions of the analysis are given in the notes to Tables 13.2 and 13.3). The

Table 13.2 The results of experimental study of the Pb partitioning between various fluids and granitic melt at $T = 750\,°C$ and $P = 100$ MPa. The method of electron microprobe analysis[a]

Composition of initial fluid	$^{melt}C^b_{center,}$ wt%	$^{melt}C^b_{rim,}$ wt%	$^{fluid}C,^b$ wt%	$^{fluid/melt}D_{Pb}^b$
0.1N HCl	0.22	0.10	0.03	0.3
0.5N HCl	0.15	0.04	0.13	3.3
0.1N HF	0.32	0.23	0.002	0.009
0.55N HF	0.46	0.40	0.004	0.01
H_2O	0.46	0.40	0.002	0.005
0.5N NaCl	0.05	<0.04[c]	0.11	>3.0
1N NaCl	0.10	<0.03[c]	0.16	>5.8
0.1N NaOH	0.39	0.08	0.001	0.013
0.5N NaOH	0.36	0.03	0.002	0.055
0.1N HCl + 1N NaCl	0.06	<0.02[c]	0.25	>10.8
0.1N NaOH + 1N NaCl	0.08	<0.02[c]	0.09	>3.8
0.5N NaOH + 1N NaCl	0.19	0.08	≤0.01	≤0.13

[a]Analysis conditions: wave dispersion spectrometer, Pb M_α analyzing crystal PET, $U = 25$ kV, $I \approx 15–25$ nA, $d_{probe} \approx 5\,\mu m$, standard—PbS

[b] $^{melt}C_{center}$ and $^{melt}C_{rim}$—content of lead, respectively, in the central and rim (about 15 μm from the boundary) parts of the glass sample. $^{fluid}C$—Pb content in fluid. $^{fluid/melt}D_{Pb} = {}^{fluid}C/^{melt}C_{rim}$—the partition coefficient of lead between the fluid and the melt

[c]Component detection limit. It was calculated individually for each experiment

detection limit for Pb and Zn was 0.02–0.03 wt%. It was calculated for each series of analyses by formula (Romanenko 1986):

$$C_{min} = \left(3\sqrt{2} \times \sqrt{N_o/(N - N_o)}\right) \times C_{sample}, \qquad (1)$$

where N_o and N are the number of pulses registered for a reference sample for time of measurements of the background and analytical signals, respectively. C_{sample} is concentration of PbO or ZnO in the reference sample equal to 0.5 wt% for each.

The solution after experiment was prepared for analysis by the method developed by Salova et al. (1983). For this purpose, the solution was evaporated to a dry residue, which was alloyed together with $LiBO_2$. The melting was carried out in graphite crucibles at $T = 1000–1100\,°C$. In the case of a significant NaCl content, SiO_2 or GeO_2 was added to the fluid dry residue and lithium metaborate to ensure glass transition so that the ratio was about 1:(3–4):12. To control the results obtained by the method of electron microprobe analysis were carried out parallel analyses of solutions after experiments by atomic absorption and spectrophotometric methods. The differences between the individual analyses for each sample were usually less than 20–25 relatively%.

The effect of the duration of the experiments on the partitioning of Pb and Zn was studied in a series of kinetic experiments lasting about 1, 3, 7 (5.8–7.8) and 17 days. At the initial 0.5N HCl solution, two parallel experiments of 1 day duration, two

Table 13.3 The results of experimental study of the Zn partitioning between various fluids and granitic melt at $T = 750\ °C$ and $P = 100$ MPa. The method of electron microprobe analysis[a]

Composition of initial fluid	$^{melt}C_{center}^{b}$, wt%	$^{melt}C_{rim}^{b}$, wt%[a]	$^{fluid}C,^{b}$ wt%	$^{fluid/melt}D_{Zn}^{b}$
0.1N HCl	0.19	$\leq 0.04^{c}$	0.013	≥ 0.3
0.5N HCl	0.18	0.04	0.09	2.0
0.1N HF	0.26	0.25	0.008	0.03
0.55N HF	0.30	0.21	0.016	0.08
H_2O	0.32	0.20	0.004	0.02
0.5N NaCl	0.21	0.03	0.04	1.3
1N NaCl	0.13	$<0.02^{c}$	0.05	>2.2
0.1N NaOH	0.32	0.13	0.0004	0.003
0.5N NaOH	0.34	0.17	0.003	0.02
0.1N HCl + 1N NaCl	0.15	$<0.02^{c}$	0.13	>8.8
0.1N NaOH + 1N NaCl	0.19	0.03	0.04	1.4
0.5N NaOH + 1N NaCl	0.30	0.10	≤ 0.006	≤ 0.06

[a]Analysis conditions: wave dispersion spectrometer, Zn K_α analyzing crystal LiF, $U = 25$ kV, $I \approx 15$–20 nA, $d_{probe} \approx 5$ μm, standard—Zn_{metal} and ZnO

[b] $^{melt}C_{center}$ and $^{melt}C_{rim}$—content of zinc, respectively, in the central and rim (about 15 μm from the boundary) parts of the glass sample. $^{fluid}C$—Zn content in fluid. $^{fluid/melt}D_{Zn} = {}^{fluid}C/^{melt}C_{rim}$—the partition coefficient of zinc between the fluid and the melt

[c]Component detection limit. It was calculated individually for each experiment

experiments of 6 days duration and one seventeen-day experiment were carried out (Table 13.4 and Fig. 13.1). At pure H_2O, two one-day, one three-day, and three seven-day experiments were carried out. At initial 1N NaCl solution, one three-day and one seven-day experiments were carried out. And finally at the solution of 0.1N HCl + 1N NaCl, there were two three-day and one seven-day experiments. The results of the kinetic experiments show that, due to the diffusion control of the process of extraction from the melt, the partition coefficients of Pb and Zn with an increase in the duration of the experiments 1–3–6 (7) and even up to 17 days significantly increase (see Table 13.4 and Fig. 13.1), and it can be assumed that the equilibrium in the partitioning of Pb and Zn in the studied system has not been achieved. At the same time, in the same kinetic experiments on the majority of rock-forming components (SiO_2, Al_2O_3, Na_2O, CaO, TiO_2) the equilibrium in the studied system is established within 1–7 days. Unfortunately, it was not possible to carry out a mass series of experiments at duration up to 17 days or more, and our main experiments to study the partitioning of polymetals had duration of about 7 (6–8) days. At the same time, it is possible that, in the course of more prolonged experiments, higher partition coefficients of Pb and Zn would be obtained. To reduce the equilibrium time, the ore components, if possible, were introduced into the melt and fluid, and these "contaminated of ore substance" phases were used both alternately in different experiments and together in the same experiments (combined and reference techniques described below). It

Table 13.4 The influence of the experiments duration on the partitioning of Pb and Zn in a series of kinetic experiments with durations of 1, 5.8, and 17 days

Experiment duration, days (number of experiments)	Components	$^{melt}C^a_{center,}$ wt%	$^{melt}C^a_{rim,}$ wt%	$^{fluid}C,^a$ wt%	$^{fluid/melt}D^a$
0.75–1.1 (2)	Pb	0.21	≤0.06	0.06	≥1.0
	Zn	0.24	0.06	0.06	1.15
5.8 (2)	Pb	0.15	0.04	0.12	3.2
	Zn	0.18	0.05	0.09	2.0
17.0 (1)	Pb	0.12	≤0.02[b]	0.22	≥11.4
	Zn	0.12	≤0.02[b]	0.18	≥9.3

The 0.5N HCl initial solution

[a] $^{melt}C_{center}$ and $^{melt}C_{rim}$—content of the polymetal, respectively, in the central and rim (about 15 μm from the boundary) parts of the glass sample. $^{fluid}C$—polymetal content in fluid. $^{fluid/melt}D$ = $^{fluid}C/^{melt}C_{rim}$—the partition coefficient of polymetal between the fluid and the melt in the rim part

[b] Component detection limit. It was calculated individually for each experiment

Fig. 13.1 Influence of experiment duration on the partitioning of Pb and Zn between fluid and granite melt (0.5N HCl, T = 750 °C, P = 100 MPa)

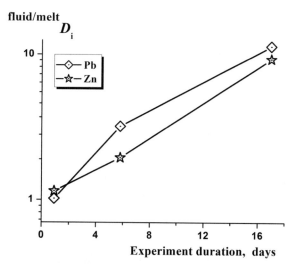

should be noted that experimental studies of polymetals partitioning performed by other researchers at similar conditions, with which we compare our results, were mostly performed in shorter (0.7–3 days) experiments (Khitarov et al. 1982; Urabe 1985, 1987a; Kravchuk et al. 1991; Khodorevskaya and Gorbachev 1993b) and only in research (Holland 1972) in 7 days duration.

The component partition coefficients between the coexisting phases are the ratios of concentrations in equilibrium (close to equilibrium) phases. However, they can be obtained in dynamic experiments, so the problem of achieving of equilibrium

Table 13.5 Comparison of the results of lead partitioning obtained using three different techniques ($T = 750\,°C$, $P = 100$ MPa, 0.1N HCl + 1N NaCl, duration 5.5–7 days)

Technique	The content of Pb in the fluid, wt%		The content of Pb in the granite melt, wt%		$^{\text{fluid/melt}}D_{Pb}$	
	Before the experiment	After the experiment	Before the experiment	After the experiment		
Extraction	0	0.23	0.46	≤0.022[a]	≥10.5	
Reference	0.93 Pb + 1.61 Zn	By charge 0.93 Pb	0.09	>0.09	<10	
			0.19	<0.19	>5	
Combined	0.93 Pb + 1.61 Zn	0.89 Pb	0.09	0.14	6.4	$^{\text{fluid/melt}}D_{\text{average}}$ = 7.6 ± 1.2
			0.19	0.10	8.8	
			0.09	0.10	8.8	
			0.19	0.14	6.4	

[a]Pb concentration close to detection limit

acquires fundamental importance. Three series of experiments were carried out using extraction, reference and combined techniques to study the partition of Pb between acidic chloride fluid and granite melt ($T = 750\,°C$, $P = 100$ MPa, 0.1N HCl + 1N NaCl solution, 5.5–7 days duration) (Table 13.5). The first technique is based on the extraction of components from the granite melt into the fluid. Ore components were introduced only into the granite melt. Such a system is slowly approaching the equilibrium, especially in the case of slow-moving W and Mo. The reference technique, described in detail in the work (Chekhmir 1988), is that the several pieces of granite glass (references) with different content of ore components, and also the solution containing ore components together placed in the capsule. After the experiment, the content of metals in the boundary parts of granite glasses was determined, and the excess solution was not analyzed. Those experiments, in which there were two pieces of glass, in one of which the metal content increased in comparison with the initial one, and in the other—decreased (the approach to equilibrium from both sides), were considered as correct. Assuming that the metal content in the fluid does not change significantly, it was considered that the correct partition coefficient is in the range between the two partition coefficients calculated for the initial fluid and the initial metal content in these two reference pieces of glass. The reference method has its disadvantages and advantages. The disadvantage is the presentation of the results in the form of a range of values, and the advantage is the rapidity (1 day experiment duration is enough to determine the direction of change in the metal content at the melt). The third combined technique as respects preparation and carrying out of experiment is the same reference technique, but in this case partition coefficients can been calculated, using analyses of fluid and reference pieces of glass after experiment that has allowed to raise appreciably accuracy of definition. In our experiments, the coefficients obtained using the combined technique ($^{\text{fluid/melt}}D_{Pb} = 7.6 ± 1.2$, see Table 13.5) fall into the middle of "the fork" calculated using the reference method ($10 > {}^{\text{fluid/melt}}D_{Pb} > 5$), while the partition coefficient obtained using the extraction method was slightly overestimated ($^{\text{fluid/melt}}D_{\text{Pb}} \geq 10.5$). Thus, the comparison of the results of lead partitioning from experiments using three different techniques shows the similarity of the obtained values, which may indirectly indicate their correctness. In our experiments, mainly, extraction and combined techniques were used. In addition, a modified combined technique was used, in which the granitoid melts initially contained identical amounts of ore components, and in the fluid the initial contents of these components were different in different experiments. The convergence of the results was estimated by calculating the relative deviations from the mean value of the partition coefficient from several parallel experiments. These deviations, estimated over the entire data set, were ±20–40% for Pb and ±30–50% for Zn, respectively.

The partition of components between three and more phases (separated (nascent) fluid, formed crystalline phases, and residual melt) was studied at the investigation of the behavior of Pb and Zn in the process of crystallization of granite melt. It turned out that at the temperature above solidus the crystallization of granitoid melt is extremely slow, due to its high viscosity, especially at low pressure (100–200 MPa). Therefore, the experiments were carried out in the sub-solidus (but close to solidus) region (according to our estimates for the akchatau granite at $P = 250$ MPa, $T_{\text{sol.}} = 630$–680 °C, and at $P = 500$ MPa, $T_{\text{sol.}} = 600$–650 °C) and at pressure not lower than 250 MPa. The sufficiently high degree of crystallization (above 50 vol.%) was possible to obtain at these conditions. The experiments were conducted as follows. A pre-prepared rod of granite glass enriched in Pb, Zn and saturated at a given pressure of 0.1N HCl + 1N NaCl solution was placed in Pt capsule. The crystallization was carried out by means of slow temperature decrease at two pressures ($P \approx 270$ and ≈ 510 MPa): $T = 630 \rightarrow 580$ °C, $P = 285 \rightarrow 255$ MPa and $T = 600 \rightarrow 550$ °C, $P = 520 \rightarrow 500$ MPa, during 7–14 days. Isochoric quenching method (as opposed to the isobaric quenching method) was used to extract the separated fluid from intercrystalline pores. For this purpose the pressure was released at such a rate as to maintain the maximum allowable free volume of the capsule as the temperature was reduced. After the experiment, the main part of the solution was in the sample-free part of the capsule and was easily extracted under the influence of internal pressure by gentle piercing of the capsule wall by a thin needle.

13.4 Experimental Results and Their Discussion

13.4.1 Influence of Fluid Concentration and Composition on Pb and Zn Partitioning Between Fluid and Granite Melt

The results of the experimental study of the influence of composition (NaCl, HCl, HF, NaOH), concentration (0.1–1.1N) and acidity-alkalinity of aqueous-chloride fluid on the partitioning of polymetals between the fluid and the melt of akchatau granite are presented at $T = 750$ °C and $P = 100$ MPa (Chevychelov and Epelbaum

1985; Chevychelov 1987). Experimental data have been obtained for the entire range of acidity-alkalinity of fluid realized in natural conditions (pH from <1 to 9). The main results on the partitioning of Pb and Zn are summarized in Tables 13.2, 13.3 and Figs. 13.2, 13.3, 13.4 and 13.5. Minimum partition coefficients were obtained in systems with pure H_2O, as well as with NaOH and HF fluids (0.003–0.08), and moreover these coefficients increase slightly with increasing fluid concentration from 0.1N to 0.5N (Figs. 13.2 and 13.3). In systems with NaCl and HCl, the partition coefficients increase significantly, making up at 0.5N solution >3 and 3.3 for Pb and 1.3–2 for Zn. In systems with more complex solutions, the partition coefficients

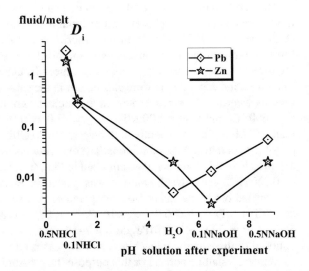

Fig. 13.2 Influence of fluid acidity (HCl, NaOH, pure H_2O) on the partitioning of Pb and Zn in the fluid—granite melt system at $T = 750\ °C$ and $P = 100\ MPa$

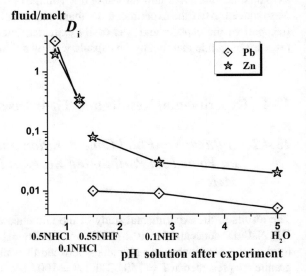

Fig. 13.3 Influence of acidic fluid composition (HCl, HF, pure H_2O) on the partitioning of Pb and Zn in the fluid—granite melt system at $T = 750\ °C$ and $P = 100\ MPa$

Fig. 13.4 Influence of composition and acidity of (0.1(0.5)N HCl,NaOH + 1N NaCl) complex fluid on the Pb and Zn partitioning in the fluid–granite melt system at $T = 750\,°C$ and $P = 100$ MPa

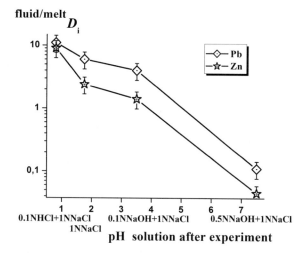

Fig. 13.5 Influence of concentration and composition of (HCl, NaCl, pure H_2O) chloride fluid on the partitioning of Pb and Zn in the fluid–granite melt system at $T = 750\,°C$ and $P = 100$ MPa

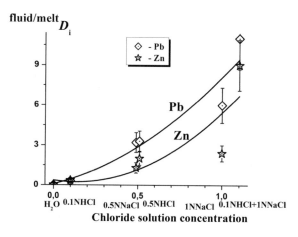

are small at a slightly alkaline 0.5N NaOH + 1N NaCl solution ($^{fluid/melt}D_{Pb\ and\ Zn} \leq$ 0.13–0.06, $pH_{quench} = 8$–7) and increase by two orders of magnitude to the maximum ($^{fluid/melt}D_{Pb} > 10.8$ and $^{fluid/melt}D_{Zn} > 8.8$) at an acidified 0.1N HCl + 1N NaCl solution ($pH_{quench} = 1$) (Fig. 13.4). Thus, the partitioning of Pb and Zn between about 1N chloride fluid and granite melt is determined by the fluid acidity. The obtained results are described by the equation: $\log\ ^{fluid/melt}D_{Pb,Zn} \approx -1/3 \times pH_{quench}$ (Chevychelov 1987). Similar results at higher pressure 370 MPa ($T = 800\,°C$) were obtained by Urabe (1984, 1985). In these studies the pH value of the solution after experiment also changes from near-neutral (6.6–8.0) to acid (1.0–1.4), and the increase in the partition coefficient is mainly due to the increase in the content of Pb and Zn in the fluid.

In systems with acidic chloride fluids, the $^{fluid/melt}D_{Pb,Zn}$ coefficients also increase significantly in proportion to the square of the chloride concentration of the fluid

(Fig. 13.5), which allows us to assume purely chloride complexes in these systems for Pb and Zn as divalent ions. These our results correspond to the results of other experimental studies (Holland 1972; Urabe 1985, 1987a). Our data in systems with chloride fluids can be described using the following equations: $^{\text{fluid/melt}}D_{\text{Pb}} \approx (7 \pm 1.6) \times {}^{\text{fl}}C_{\text{Cl}}^2$ and $^{\text{fluid/melt}}D_{\text{Zn}} \approx (5.7 \pm 2.0) \times {}^{\text{fl}}C_{\text{Cl}}^2$ (Chevychelov 1987). As the acidity of the fluid increases, the partition coefficients of polymetals increase: for example, the partition coefficients are higher in similar concentrations of $\sum\text{Cl}$ but more acidic fluids, such as 0.5N HCl and 0.1N HCl + 1N NaCl, as against 0.5N NaCl and 1N NaCl. These differences, as a rule, are not large, because initially different fluids of HCl and NaCl approach each other in their chemical composition and, consequently, in acidity in the process of fluid-magmatic interaction due to the relatively long duration of our experiments.

Comparison of the obtained $^{\text{fluid/melt}}D$ coefficients for Pb and Zn with each other shows their proximity as for absolute values, and for the nature of dependencies on the composition and concentration of the fluid. Nevertheless, it can be noted that the coefficients for Pb are slightly higher in systems with chloride and alkaline fluids, and the coefficients for Zn are higher in case of fluoride fluid or pure H_2O.

In these experiments, the partitioning of volatile components (Cl and S) was also roughly assessed (Chevychelov and Epelbaum 1985). The $^{\text{fluid/melt}}D_{\text{Cl}}$ coefficients in systems with chloride fluids are ~10–30. When using alkaline solutions of NaOH and pure H_2O, initially not containing chlorine, significant content of Cl (1.8–3.3 wt%) were determined in the dry residue of the fluid (anhydrous component of the fluid). This chlorine was likely extracted from granite melt containing 0.0n wt% Cl. At the same time, no chlorine was detected in the dry residue of HF fluid ($\leq$0.2–0.4 wt%). The 1–1.1 wt% S was determined in the dry residue of 0.5N NaCl and 0.1(0.5)N NaOH + 1N NaCl fluids after the experiments. In these experiments, sulfur was probably also extracted into the fluid from granite melt containing 0.0n wt% S. Thus, it has been shown that the fluid that did not initially contain Cl and S, can be appreciably enriched by them due to extraction from the granite melt containing clarke amounts (bulk earth values) of these elements due to significant re-partitioning of these components in favor of the fluid.

Using the obtained partition coefficients for Pb and Zn, as well as such coefficients for rock-forming elements, we plotted the dependence of $^{\text{fluid/melt}}D_{\text{Me}}$ partition coefficients on the Gibbs free energy change ($\Delta_f G^{\circ}_{,\text{MeO}}$) of formation of oxides of these elements at $T = 1000$ K in accordance with (Robie and Hemingway 1995) (Fig. 13.6). The results from experiments with chloride fluids were used (0.5N HCl, 0.5N NaCl, and 1N NaCl). Negative correlation between $^{\text{fluid/melt}}D_{\text{Me}}$ coefficients and $\Delta_f G^{\circ}_{,\text{MeO}}$ of oxide formation of these elements was revealed in the papers (Malinin and Khitarov 1983, 1984). The values of the coefficients obtained by different experimenters and at different conditions were compared. Dependence, obtained from our data, generally confirms this negative correlation. Only the coefficients for Zn, as well as K at 0.5N HCl and Ca at 1N NaCl fall out of the total dashed dependence. The reason for this dependence in accordance with (Malinin and Khitarov 1983, 1984) is that the decrease in the partition coefficient of element reflects the increase in the strength of the bond of this element with oxygen in the structure of aluminosilicate

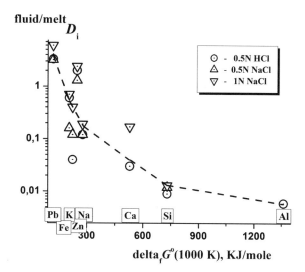

Fig. 13.6 Dependence of elements partition coefficients on the free energy of formation (ΔG^o) of their oxides taken from Robie and Hemingway (1995). Oxide Fe is taken as FeO. Experiments with HCl and NaCl fluids at $T = 750\,°C$ and $P = 100$ MPa

melt. Since metals in the melt are bound to aluminosilicate groups through oxygen atoms and the strength of these bonds is determined by the affinity of the metal to oxygen, we can assume that the strength (free energy of formation) of individual oxides of the corresponding metals must be in the first approximation a measure of this affinity. The obtained dependence, if necessary, can be used for a preliminary, of course very rough estimation of the unknown $^{fluid/melt}D_i$ coefficients by the known free energy of formation of oxides of these elements.

Based on the obtained experimental data, an attempt was made to estimate the scale of possible extraction of Pb and Zn by fluid from the granite melt. Similar assessments were made in the studies (Holland 1972; Khitarov et al. 1982; Malinin and Khitarov 1984) for the natural chamber of granite magma containing Clarke amounts of ore components and dissolved water. The mass of the ore substance that could be extracted by the fluid separating from the melt in the process of pressure reduction and crystallization of magma was calculated. Like Khitarov et al. (1982), assuming the amount of aqueous fluid of separated from the fluid-saturated granite magma was equal to ~ 3 wt%, a size of magmatic chamber of about 4 km³ (not large stock), a density of granite melt ~2.35 g/cm³ we obtain the results given in Table 13.6. According to calculations (Peretyazhko 2010), the density of granite melts enriched with volatile components at magmatic temperatures of 650–850 °C can be about 2.3–2.35 g/cm³. The amounts of Pb and Zn extracted by fluid in case of their subsequent compact deposition are quite sufficient to form a small industrial deposit (reserves about 200 thousand tons). At the same time, the mass and content of Pb and Zn in the granite massif should decrease quite noticeably (by about 1/3). Our data differ significantly from those presented in the paper (Khitarov et al. 1982), according to which the fluid extracted from the melt of relatively small amounts of ore substance, and the content of ore components in the granite massif after this extraction practically do not change. This fundamental difference is explained by

Table 13.6 Estimation of the scale of possible extraction of Pb and Zn by fluid from granite pluton about 4 km^3 in size

Ore component	Clarke[a] (according to A.P. Vinogradov), wt%	Mass of ore substance in granite melt, t	$^{fluid/melt}D_i$		The mass of ore substance extracted by fluid, t		The remainder of the ore substance in the massif, after the separation of the fluid, wt%	
			Our data[b]	Khitarov et al. (1982)[c]	Our data[b]	Khitarov et al. (1982)[c]	Our data[b]	(Khitarov et al. 1982)[c]
Pb	0.0020	1.88×10^5	>10.8	0.1	$>6.1 \times 10^4$	0.056×10^4	67.6	99.7
Zn	0.0060	5.64×10^5	>8.8	0.4	$>14.9 \times 10^4$	0.68×10^4	73.6	98.8

[a] Data taken from Voitkevich et al. (1977), Makrygina (2011)
[b] The initial solution of 0.1N HCl + 1N NaCl was taken as a model fluid
[c] The 1N NaCl initial solution

the difference in the results of the experiments: the obtained by us coefficients of $^{fluid/melt}D_{Pb}$ are about 100 times higher, and $^{fluid/melt}D_{Zn}$ is about 20 times higher than in the paper (Khitarov et al. 1982). At the same time, our data on zinc partitioning are close to those obtained by Holland (1972).

According to the data of Barnes (1982), based on the content of ore components in natural thermal springs and gas-liquid inclusions, the concentrations of Pb, Fe, Zn and Cu at hydrothermal ore formation are n × 10 − n × 1000 ppm, at a minimum productive ore-bearing concentration of about 10 ppm. According to our experimental data, the acidic chloride fluid in the equilibrium with granite melt can contain about 220 ppm Pb and about 530 ppm Zn (clarke concentration × $^{fluid/melt}D_{Pb(Zn)}$, Table 13.6). Therefore, the concentrations of lead and zinc in such a magmatic fluid will be quite sufficient for ore transport and subsequent ore deposition (Chevychelov 1987).

Thus, the experimental results obtained by us demonstrate the possibility of formation of a high-temperature ore-bearing magmatogenic fluid containing about 220 ppm Pb and about 530 ppm Zn, due to the mechanism of extraction of these metals by acidic chloride fluid from the granite melt at clarke concentrations of Pb and Zn. The degree of fluid extraction of these metals is sufficient for the formation of an ore body comparable in scale with a small industrial lead-zinc deposit and, consequently, the presence of metallogenic specialization of magmas for the formation of such deposits is not always necessary. At the same time, for rare metals such as W, Mo and Sn, in general, a similar mechanism of their extraction from granitoid melts by high-temperature fluids is less typical, due to lower partition coefficients between the fluid and the melt. The apparent contradiction with geological data, according to which the connection with granitoid complexes is more typical for rare-metal deposits, is explained by lower temperatures of polymetallic ore depositions and long-term evolution of ore-bearing fluid, during which the fluid can be significantly removed from the primary source. This is confirmed by the metallogenic zoning at many deposits, according to which polymetallic veins with sulfides Cu, Zn, Pb, sometimes Ag and Au, are manifested at the periphery of the ore field (e.g., Raguin 1979; Zaraisky et al. 1994).

The experimental data obtained by us are quite comparable with the fluid/melt partition coefficients determined according to the data of LA-ICP MS analysis of coexisting melt and fluid inclusions from quartz of different saturated by volatile components granite intrusions (from ultra-alkaline to ultra-aluminous, $\log f_{O2}$ from NNO -1.7 to NNO $+4.5$, the separating fluid contained from 1 to 14 mol/kg $\sum$Cl) (Zajacz et al. 2008). These authors have shown that Pb and Zn are soluble in the fluid mainly in the form of chlorine-containing complexes, since a close to linear positive correlation is in existence between the partition coefficients of these elements and the concentration of Cl in the fluid ($^{fluid/melt}D_{Pb} \approx 6 \times {}^{fluid}C_{Cl}$, $^{fluid/melt}D_{Zn} \approx 8 \times {}^{fluid}C_{Cl}$).

13.4.2 Influence of Pressure and Temperature on the Partitioning of Pb and Zn Between Acidic Chloride Fluid and Granite Melt

The main results on the partitioning of Pb and Zn at $T = 700–900$ °C and $P = 100$ and 500 MPa are presented in Table 13.7 and Figs. 13.7 and 13.8 (Chevychelov 1993; Chevychelov et al. 1994). It is shown above that at $T = 750$ °C and $P = 100$ MPa, the maximum partition coefficients Pb and Zn in favor of the fluid were obtained at the initial 0.1N HCl + 1N NaCl chloride acidified fluid. Therefore, the influence of temperature and pressure on the partitioning of polymetals was mainly studied in the presence of this fluid. It has been established that the temperature increase within the limits studied both at a pressure of 100 and 500 MPa in general decreases the values of the $^{fluid/melt}D_{Pb \text{ and } Zn}$ coefficients (see Fig. 13.7). In this case the content of these metals in the chloride fluid and the total mineralization of the fluid, and at $P = 100$ MPa the acidity of the fluid also decrease (see Table 13.7). This decrease in the $^{fluid/melt}D$ coefficients may be due to the retention of polymetals in the melt due to the increase in Cl concentration in the melt with the increase in temperature. In the paper (Hemley et al. 1986) the opposite (positive) temperature dependence of solubility of polymetals in a fluid with temperature growth from 400 to 600 °C is received. The change of a sign of this dependence allows to assume possible existence of a maximum of Pb and Zn solubility in chloride solutions in a range of temperature 600–750 °C at $P = 100$ MPa and about 1N $\sum$Cl in composition of the fluid. The negative temperature dependence of the polymetals partition coefficients obtained by us does not agree with the experimental data of Khitarov et al. (1982), which received a weak positive dependence of these coefficients in the same range $T = 700–900$ °C, but at $P = 200$ MPa. It can be assumed that the changes in the partition coefficients, in general, are weakly dependent on temperature, may be different sign depending on the composition of the granite melt.

Table 13.7 The partitioning of Pb and Zn between aqueous chloride fluids and granite melt at T = 700–900 °C, P = 100 and 500 MPa and the 6–8 days experiments duration

T, °C	P, MPa	The composition of the initial solution	pH solution after experiment	fluid/meltD_i Pb	fluid/meltD_i Zn	Mineralization[a] solution after experiment, wt%
750	100	0.1N HCl	1–1.5	0.3	0.3[b]	1.1
750	100	0.1N HCl + 1N NaCl	≥1	10.8[b]	8.8[b]	7.3
860	100	The same	2–1.5	6.5[b]	6.5[b]	5.3
750	100	1N NaCl	2–1.5	5.8[b]	2.2[b]	6.6
750	100	0.1N NaOH + 1N NaCl	3.5	3.8[b]	1.4	6.0
750	100	0.5N NaOH + 1N NaCl	8	0.13[c]	0.06[c]	6.8
700	500	0.1N HCl + 1N NaCl	4	0.64 ± 0.3	1.5 ± 0.4	9.9
800	500	The same	4	0.34	0.4	8.0
900	500	The same	4	0.3	0.25	6.2
700	500	1N NaCl	4.5	0.3	0.2	8.3
700	500	0.1N NaOH + 1N NaCl	6	0.4	0.2	7.8
700	500	0.5N NaOH + 1N NaCl	≥10	0.7	0.08	8.1

[a]The ratio of the weight of the dry residue to the total weight
[b]The concentration of metals in the glass after the experiment is close to the detection limit (0.02–0.03 wt%)
[c]The concentration of metals in the solution after the experiment is close to the detection limit (0.006–0.01 wt%)

Pressure increase from 100 to 500 MPa at T = 700(750)°C can significantly (up to 1–1.5 orders of magnitude) reduce the partition coefficients of Pb and Zn, except for systems with alkaline fluid (see Fig. 13.7). This regularity is mainly related to the reduction of polymetals content in the fluid. At the same time with increasing pressure, the solubility of alkaline components and, especially, silica increases in the chloride fluid (the total mineralization of the fluid after experiment increases), and acidity of the fluid significantly decreases or in the case of a initial near-neutral pH this fluid is alkalized (see Table 13.7). This is also confirmed by the experimental data of other authors. Negative influence of pressure on solubility of sulfides Pb and Zn in chloride fluid (about $1m$ Cl$^-$) at P = 50–200 MPa and T = 400–600 °C is shown in papers (Arutyunyan et al. 1984; Hemley et al. 1986; Malinin et al. 1989). The negative baric dependence of the partition coefficients obtained by us agrees well with the experimental data (e.g., Urabe 1987a; Kravchuk et al. 1991).

Fig. 13.7 Dependence the partition coefficients of Pb and Zn on temperature and pressure. The 0.1N HCl + 1N NaCl initial solution and the granite melt

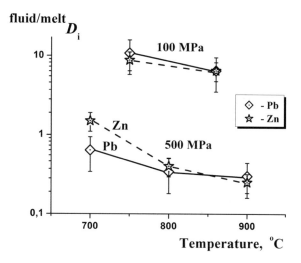

Fig. 13.8 Dependence the partition coefficients of Pb and Zn on the acidity-alkalinity of the fluid, temperature and pressure. The (0.1(0.5)N HCl,NaOH + 1N NaCl) initial complex solutions

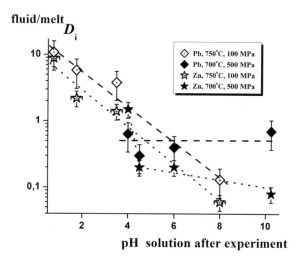

The influence of the acidity-alkalinity of the complex chloride fluid on the partitioning of Pb and Zn at different pressures differs (see Fig. 13.8). At 100 MPa the partition coefficients are inversely proportional to the pH_{quench} solution, and the obtained results can be represented as dependences: $\log \,^{fluid/melt}D_{Pb} \approx 1.2 - (0.2 \pm 0.05) \times pH$ and $\log \,^{fluid/melt}D_{Zn} \approx 1.1 - (0.3 \pm 0.1) \times pH$ (Chevychelov et al. 1994). At 500 MPa the lead partition coefficient practically does not depend on the acidity of the chloride fluid coexisting with the melt, and the zinc partition coefficient significantly (almost 4 times) decreases with the transition from 0.1N HCl + 1N NaCl to 1N NaCl fluid and slightly decreases with further increase in the alkalinity of the fluid (see Fig. 13.8). Possibly, the change in the slope of dependence in the latter case at Zn partitioning is connected with the change of the type of complexes Zn in the fluid. At both pressures the total mineralization of the solution after experiment decreases with decreasing acidity (pH growth) and slightly increases in the alkaline region with transition from 0.1N NaOH + 1N NaCl to 0.5N NaOH + 1N NaCl fluid (see Table 13.7).

Comparison of the obtained partition coefficients for Pb and Zn with each other shows their similarity both in absolute values and in the character of temperature and pressure dependences. It can be noted that, as a rule, these coefficients for Pb are slightly higher than for Zn, except for the results at 500 MPa and 0.1N HCl + 1N NaCl fluid.

Crystallization of granitoid massifs at different depths will produce magmatogenic fluids with different levels of potential ore-bearing in regard to polymetals all other things being equal. Hypabyssal (about 100 MPa) bodies will have higher potential productivity as compared to depth (about 500 MPa) bodies. Naturally, the ore-bearing of the intrusions due to the influence of many factors does not correlate directly with the differences in the depth of their occurrence. But the contents of the studied components in granitoids will correlate with the established regularity. Thus, depth intrusions, apparently, may contain large amounts of Pb and Zn in its composition in comparison with the hypabyssal ones. If in general the evolution of the magmatic system is associated with a decrease in temperature and pressure, it can be assumed that the portions of subsequent fluids should be increasingly enriched with Pb and Zn.

The paper by Urabe (1987b) shows that sulphide ores (kies) from hydrothermal deposits of the Kuroko type in Japan were formed with the active participation of polymetals enriched chloride magmatic fluids rising from depth magmatic chambers. The ore formation at the Kuroko deposits is spatially and genetically associated with the peak of granitoid magmatism that occurred 15 million years ago during the formation of the Sea of Japan. It is assumed that there are magmatic chambers of acid composition in the depth root zones of the deposits. The rhyolite lava outpourings and formation of lithoclastic tuff breccias preceded the ore formation. Simultaneously with the rhyolite lava outpourings, the magmatic fluids enriched with metals came from depth magmatic chambers to the upper crustal horizons. The hydrothermal system producing this type of deposits was formed by mixing of these magmatic fluids with sea water. The ore horizon was formed over a long period of time (several million years). Since the isotopic composition of lead is quite homogeneous, it is unlikely that the rocks of the foundation were its source. The analysis of data on the composition and zonality of wallrock-altered tertiary volcanogenic rocks has shown that these formations also could not serve as a source of ore elements. The increased chlorine contents and the meta-aluminous composition of rhyolite melts before deposition of Kuroko type ores confirm the enrichment with polymetals the magmatic fluid as a result of its interaction with granitoid melts. The only hypothesis consistent with the full spectrum of geological, geochemical, and experimental data is the concept of the formation of deposits of this type at the active participation of aqueous chloride fluids, gradually released from the depth chambers of acid magma.

According to our data as well as the investigations (Khodorevskaya and Gorbachev 1993a, b), the composition of the fluid interacting with the melt depends rather strongly on the pressure. At increasing pressure, the fluid is increasingly enriched with Cu relative to Zn, Pb and Fe. At the same time, the ratios of Zn, Pb and Fe do not depend on pressure. This means in respect to natural processes of magmatic hydrothermal ore formation, that the composition of the ore-bearing fluid and the inherited composition of ores depend on the depth of origin of the fluid during degassing of parental magmas. The concentration of Cu increases with increasing depth of fluid generation, and the concentrations of Zn and Pb in it decrease. The ratio of Zn to Pb must remain relatively constant.

From this point of view, it is interesting to consider sulphide ores (kies) deposits of the Kuroko type, Japan, which are characterized by wide variations in the Cu/Zn and Cu/Pb ratios. Khodorevskaya and Gorbachev (1993b) assume that variations in the Cu, Zn, Pb and Fe contents in the ores of these deposits may be associated with different depths of ore-bearing fluid formation. These deposits are associated with miocene submarine basalt-andesite-dacite-rhyolite (BADR) volcanic activity in the Japanese islands. The deposits are located in a belt of "green" tuffs, stretched over 1500 km long, 100 km wide, and parallel to the volcanic front. The magmatic nature of the ore-bearing fluids that form the Kuroko deposits is proved by the data on the isotopic composition of sulfur, lead, strontium, as well as data on the geochemistry of rare, rare-earth and ore elements in ores and effusives (Sato 1977). Data on the partitioning of Cu, Zn and Pb in ores from 9 Kuroko type sulphide deposits show a negative correlation between the Cu content in the ores relative to Zn and Pb (Khodorevskaya and Gorbachev 1993b). At the same time, the Zn/Pb ratio remains constant. Similar partitioning of Cu, Zn and Pb in ores can be explained by different depth of fluid generation during degassing of parental magmas. An attempt is made to estimate the generation depths of magmatic fluids for the deposits of this type, using as a geo-barometer the experimental baric dependences of exchange coefficients ($^{fluid/melt}K_D$) Cu/Zn and Cu/Pb and the calculated values of these coefficients for the Kuroko deposits. At calculating the latters, the mean values of 2 and 0.5 for andesites and dacites were taken as Cu/Zn and Cu/Pb ratios in the parental magma. According to the data (Bogatikov et al. 1987) the contents of Cu, Zn and Pb in the effusive rocks composing the series of "green" tuffs vary widely. It is made an assumption that the Cu/Zn and Cu/Pb ratios in ores are close to the value of these ratios in the ore-forming fluid, since the ores inherit the peculiarities of the fluid composition. The data obtained show that the formation of ore-bearing fluids at the studied deposits took place in a wide pressure range from less than 100 to 600 MPa (Khodorevskaya and Gorbachev 1993b).

Studies of melt and fluid inclusions in the quartz phenocryst from acid volcanites (rhyodacites and rhyolites) of the Uzelga ore field in the Southern Urals by the LA-ICP-MS method allowed us to determine a high saturation by ore metals of primary magmatic fluid and magmatic melt (estimated P-T parameters are about 850 °C and 510–680 MPa) (Vikentjev et al. 2012). Direct data on the potential ore-bearing of acid volcanic complexes, with which sulphide ores (kies) deposits of the Urals are spatially associated, have been obtained. Magmatic fluid contains elevated concentrations, g/t: Cu 300-3700, Zn 80-3400, Pb 14-1000, Sn 4-1600. Glass of melt inclusions is also significantly enriched with metals, g/t: Cu 1100, Zn 1400. Similar high-temperature high-baric ore-bearing aqueous fluids, appeared during degassing of magmatic melts of acid composition, have been established by a number of researchers (e.g., Zaw et al. 2003; Karpukhina et al. 2009; Prokofjev et al. 2011), which allows us to conclude that such magmatic fluids are actually involved in ore formation. Concentrations of metals in melt inclusions in porphyry phenocrysts of quartz are discussed in a number of papers devoted to the study of copper-porphyry deposits. High upper limits of copper content (from 450 to 7000 g/t) in the magmatic melts preceding the formation of such deposits have been established (Simonov et al. 2010).

13.4.3 Partitioning of Pb and Zn Between Brine (Water–Salt) and Low-Density Vapor (Essentially Aqueous) Phases of Water–Chloride (–Carbon Dioxide) Fluids and Granite Melt

Solutions of H_2O–NaCl composition and, especially, H_2O–CO_2–NaCl contain the most frequently occurring components of natural fluids. The possibility of the formation of coexisting immiscible fluids of a similar composition on natural objects is shown in a number of publications, for example, in the summary of Roedder (1987) based on fluid inclusions study data.

At $T = 800$ °C and $P = 100$ MPa in the two-phase region of the H_2O–NaCl system, phases containing about 62 wt% (33 mol%) and about 2 wt% (0.6 mol%) of NaCl coexist in equilibrium with each other, and at $T = 750$ °C, $P = 100$ MPa these phases contain about 57 wt% and about 3 wt% of NaCl, respectively (Fig. 13.9) (Bodnar et al. 1985). The triple system H_2O–CO_2–NaCl at magmatic P-T parameters is poorly studied. The schematic phase diagram of a part of this system (Fig. 13.10) is

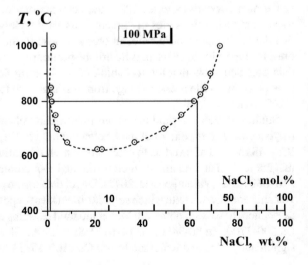

Fig. 13.9 Isobaric projection (T-X) of the H_2O–NaCl system at pressure of 100 MPa according to Bodnar et al. (1985)

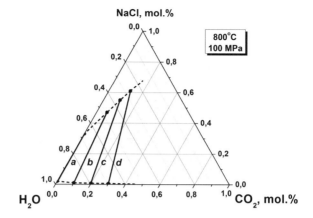

Fig. 13.10 Schematic phase diagram of a part of the H_2O–CO_2–NaCl system at $T = 800$ °C and $P = 100$ MPa; **a, b, c, d**—conodes

based on data of (Takenouchi and Kennedy 1968; Gehrig 1980; Bowers and Helgeson 1983; Bodnar et al. 1985; Korotaev and Kravchuk 1985; Popp and Frantz 1990). The diagram shows a wide field of fluid stratification into brine and vapor fluids phases. The conodes (a, b, c, d) connecting the compositions of the coexisting phases were represented conditionally taking into account the general trend and extrapolation of low-temperature experimental data.

The results of the study of the partitioning of Pb and Zn are presented in the form of partition coefficients between brine, low-density vapor fluids phases and granite melt at $P = 100$ MPa and $T = 800$ °C (Tables 13.8, 13.9 and Fig. 13.11a, b) (Chevychelov 1992, 1993; Chevychelov et al. 1994). The use of two different techniques (combined and original extraction) in the experiments yielded similar results, especially for lead, which indirectly confirms the closeness of the obtained results to the true ones. It should be noted that since the combined method for studying of partitioning of polymetals between fluids phases of different densities and granite melt was applied in separate experiments, the partition coefficients between brine and low-density vapor phases of the fluid ($^{brine/vapor}D$) were roughly calculated by dividing the $^{brine/melt}D$ coefficient from one experiment into $^{vapor/melt}D$ coefficient from another. At the same time, it was assumed that small differences in the total content of polymetals in these experiments do not have a large impact on the value of the calculated coefficient (see Tables 13.8 and 13.9). The second assumption is the mismatch between the CO_2 content in the experiments with the brine and vapor fluids phases used in calculating the $^{brine/vapor}D$ coefficients and the direction of the conodes in Fig. 13.10. However, as it is shown below the $^{brine/melt}D$ coefficients at these CO_2 content are practically constant.

The pH value of solutions after experiment for the H_2O–NaCl system was about 4.5 for the brine phase and about 1.5 for the low-density vapor phase, i.e. the latter, as expected, was more acidic. The addition of CO_2 to the fluid slightly reduces the pH of the solutions after experiment to about 4.0 and ≤ 1, respectively. The results of the experiments show the possibility of significant concentration of polymetals in the dense brine fluid phase. It has been established that at H_2O–NaCl fluid composition,

Table 13.8 The partitioning of Pb between the phases of two-phase H_2O–NaCl and H_2O–CO_2–NaCl fluids and granite melt at $T = 800\ °C$, $P = 100\ MPa$ and the combined technique

Composition of the initial solution				melt C_{Pb}, [a] wt%	brine C_{Pb}, [a] wt%	vapor C_{Pb}, [a] wt%	brine/melt D_{Pb} [b]	vapor/melt D_{Pb} [b]	brine/vapor D_{Pb} [b]
m Cl	NaCl equivalent		CO_2, mol%						
	mol%	wt%							
64	53	79	–	0.06–0.08	1.2	–	16 ± 2	–	6.7
0.5	0.9	2.9	–	0.08–0.11	–	0.23	–	2.4 ± 0.4	–
27[c]	32	61	–	0.03[d]	0.42	–	14 ± 3.5	–	4.7
0.4[c]	0.6	2.0	–	0.03[d]	–	0.10	–	3 ± 1	–
77	52	73	10	0.02–0.04	1.0	–	48 ± 22	–	41–45
0.5	0.8	2.3	10	0.28–0.29	–	0.23	–	0.8 ± 0.25	–
0.5–0.9	0.8–1.2	2.1–3.1	20	0.23–0.60	–	0.33–0.37	–	1.1 ± 0.5	–
70–79	46–48	65–67	18	0.02–0.025[d]	1.0–1.4	–	50 ± 19	–	42
0.6	0.8	1.9	29	0.14–0.18	–	0.19	–	1.2 ± 0.2	–

[a] The lead contents in the granite melt, the brine fluid phase, and the low-density vapor phase, respectively

[b] The partition coefficients of Pb between brine fluid and granite melt, vapor fluid and granite melt, and brine and vapor fluids phases, respectively

[c] Experiment by the extraction technique at $T = 750\ °C$, $P = 100\ MPa$ and duration of 7 days in the presence of two fluid phases and granite melt together in one experiment

[d] The concentration of Pb in the glass after the experiment is close to the detection limit (0.02–0.03 wt%)

Table 13.9 The partitioning of Zn between the phases of two-phase H_2O–NaCl and H_2O–CO_2–NaCl fluids and granite melt at $T = 800$ °C, $P = 100$ MPa and the combined technique

m Cl	NaCl equivalent mol%	NaCl equivalent wt%	CO_2, mol%	melt C_{Zn},[a] wt%	brine C_{Zn},[a] wt%	vapor C_{Zn},[a] wt%	brine/melt D_{Zn}[b]	vapor/melt D_{Zn}[b]	brine/vapor D_{Zn}[b]
64	53	79	–	0.10–0.14	1.6	–	13 ± 2.5	–	5.7
0.5	0.9	2.9	–	0.06–0.10	–	0.18	–	2.3 ± 0.4	–
27[c]	32	61	–	0.03[d]	0.25	–	8 ± 2	–	6.2
0.4[c]	0.6	2.0	–	0.03	–	0.04	–	1.3 ± 0.6	–
77	52	73	10	0.02–0.06	0.8	–	24 ± 11	–	14–16
0.5	0.8	2.3	10	0.18–0.27	–	0.29	–	1.4 ± 0.3	–
0.5–0.9	0.8–1.2	2.1–3.1	20	0.06–0.11	–	0.06–0.14	–	1.4 ± 0.9	–
70–79	46–48	65–67	18	0.03–0.09	1.3–1.4	–	22 ± 9	–	12
0.6	0.8	1.9	29	0.08–0.12	–	0.17	–	1.8 ± 0.4	–

[a]The zinc contents in the granite melt, the brine fluid phase, and the low-density vapor phase, respectively

[b]The partition coefficients of Zn between brine fluid and granite melt, vapor fluid and granite melt, and brine and vapor fluids phases, respectively

[c]Experiment by the extraction technique at $T = 750$ °C, $P = 100$ MPa and duration of 7 days in the presence of two fluid phases and granite melt together in one experiment

[d]The concentration of Zn in the glass after the experiment is close to the detection limit (0.02–0.03 wt%)

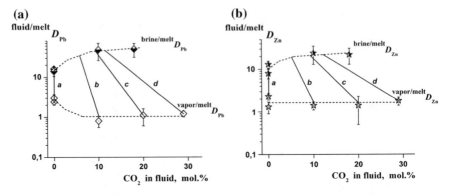

Fig. 13.11 Partitioning of Pb (**a**) and Zn (**b**) between fluid phases of different densities at hetero-genisation in H_2O–$NaCl$ and H_2O–CO_2–$NaCl$ systems and granite melt depending on CO_2 content at $T = 800(750)$ °C and $P = 100$ MPa. The a, b, c, d lines correspond to the conodes in Fig. 13.10

the partition coefficients Pb and Zn between the brine fluid and the granite melt are 5–7 times higher than those between the low-density vapor fluid and the melt (Fig. 13.11a, b).

Adding about 6–10 mol% CO_2 to the fluid composition significantly increases $^{brine/vapor}D_{Pb,Zn}$ coefficients (up to ~41–45 for Pb and up to ~12–16 for Zn), and with further increase in the system of CO_2 content $^{brine/vapor}D$ coefficients remain practically at this constant level. The difference in the partitioning of Pb and Zn at added to the CO_2 system is noteworthy: the enrichment of the brine fluid phase with lead is more significant than with zinc. It can be assumed that in the field of fluid immiscibility in natural fluid-magmatic systems the quantities dissolved in the brine fluid phase Pb and Zn will be close to each other, although the clarke content of lead in granites equal 20 g/t is three times less than that for zinc equal 60 g/t (according to A.P. Vinogradov). Data taken from Voitkevich et al. (1977), Makrygina (2011).

The fact that in the presence of H_2O–$NaCl$ fluid, $^{brine/melt}D_{Pb,Zn}$ partition coefficients significantly exceed $^{vapor/melt}D_{Pb,Zn}$ coefficients is not unexpected and is explained by the formation of stable chloride complexes in the brine fluid phase. The results obtained after adding CO_2 to the fluid are fundamentally new. Super-critical salting-out of CO_2 significantly expands the field of fluid separation on two phases (see Fig. 13.10). At the same time, the low-density vapor phase becomes depleted, and the brine phase essentially is enriched NaCl. But a significant increase in the $^{brine/vapor}D$ partition coefficient at the addition of about 6–10 mol% CO_2 cannot be explained only by these changes in the Cl content in fluid phases. Especially, taking into account the effect of approximate constancy of $^{brine/vapor}D_{Pb,Zn}$ coefficient at further increase of CO_2 content and changes of Cl content in separate phases (see Fig. 13.11a, b).

The obtained results confirm the hypothesis about the important role of salt fluid phases in processes of ore genesis. According to Marakushev (1979), Marakushev et al. (1983), Marakushev and Bezmen (1992), Marakushev and Shapovalov (1993,

1994) ore generating ability of granite systems arises in the evolution of immiscibility of fluids, expressed in the individualization of dense salt fluid phases in the silicate melt, which are effective concentrators of ore metals. Metallogenic specialization arises in relatively closed granite systems, in which the crystallization accumulation of ore metals is accompanied by a concentration of salt components. On the other hand, the reality of the existence of salt fluid phases containing up to 14–40m Cl and coexisting with granitoid melts at magmatic P-T conditions has been confirmed in a number of studies (e.g., Reynolds and Beane 1985; Chevychelov 1987; Reyf 1990; Solovova et al. 1991).

13.4.4   Partitioning of Pb and Zn Between Phases in the Process of Crystallization of Granite Melt

Herein the results of experiments at $T = 630 \rightarrow 580\,°C$, $P = 285 \rightarrow 255$ MPa and $T = 600 \rightarrow 550\,°C$, $P = 520 \rightarrow 500$ MPa, which simulate the partitioning of polymetals in the process of separation of the fluid from the granite melt during its crystallization in sub-solidus (but near-solidus) conditions, are presented (Chevychelov 1993; Chevychelov et al. 1994).

After the experiment in the central part of the rod of sample, the formation of large cavities was observed, the walls of which were folded by sufficiently large crystals of feldspar and quartz, up to 50–100 μm in size. The sizes of the crystals decrease significantly progressively as removal from the cavity into the sample. In our opinion, this indicates the recrystallization of crystals originally grown out of the melt by the action of the fluid, with the composition of the latter approaching the equilibrium with the recrystallized minerals.

The pH values of solutions after the experiment were about 4.0 at $P \approx 270$ MPa and about 6.0 at $P \approx 510$ MPa. These values are in good agreement with the previously revealed regularity of fluid acidity reduction with increasing pressure, despite the fact that the experiments considered here, in contrast to those described in the previous sections, were conducted at conditions of melt dominance over the fluid. The amount of fluid was about 5–7 wt% of the mass of the dry initial glass.

Solid phases crystallizing in the system are represented by quartz (Qtz) and feldspars: albite (Ab), oligoclase (Olg), anorthoclase ($Ancl$) and enriched in K_2O feldspar of undefined by us specie (Kfs). The chemical compositions of feldspars crystallizing in the system show in Table 13.10. No impurities of lead and zinc are found in quartz, so we do not consider the quartz in the sequel.

After the experiment the fields of uncrystalline (residual) melt (glass) are quite large. Their size reaches 50–1000 μm and more. In accordance with the composition, the uncrystalline melts are combined into four groups: two at $P \approx 270$ MPa and two at $P \approx 510$ MPa (Table 13.11). Despite some heterogeneity of these melts in terms of K, Na, Ca, and partly Fe and Al contents, the concentrations of polymetals in them are quite constant (about 0.2–0.3 wt%). They are lower than in the initial one

Table 13.10 The chemical composition (in wt%) feldspars crystallizing in the system

Components	P ≈ 270 MPa			P ≈ 510 MPa		
	Ab_1	$Ancl$	Kfs_1	Ab_2	Kfs_2	Olg
SiO_2	67.3[a] (67.0–67.6)[b]	66.4 (65.6–67.1)	63.9 (63.3–64.4)	67.3 (66.8–67.6)	62.9	62.5 (62.3–62.6)
TiO_2	–	0.1 (0–0.1)	0.4 (0.3–0.6)	–	0.1	–
Al_2O_3	19.2 (19.0–19.4)	18.7 (17.8–19.2)	17.6 (16.9–18.5)	19.5 (19.3–19.9)	19.1	21.0 (20.7–21.4)
FeO_{total}	0.3 (0.3–0.4)	0.3 (0.1–0.3)	3.1 (2.3–4.2)	0.1 (0–0.1)	0.2	0.9 (0.2–1.5)
MnO	–	–	0.1 (0–0.2)	–	–	–
ZnO	0.05 (0.04–0.07)	0.04 (0–0.09)	0.3 (0.2–0.4)	<0.02 (0–0.04)	0.1	0.3 (0.1–0.5)
PbO	0.3 (0.2–0.3)	0.3 (0.2–0.4)	0.3 (0.2–0.5)	0.4 (0.1–0.9)	1.9	1.9 (1.8–2.0)
CaO	1.8 (1.6–2.1)	0.8 (0.5–1.2)	0.9 (0.8–1.1)	0.8 (0.4–0.9)	0.5	2.8 (2.3–3.2)
Na_2O	8.8 (8.8–8.9)	6.5 (5.4–7.6)	6.1 (5.1–7.1)	10.1 (9.7–10.8)	3.4	8.1 (7.5–8.7)
K_2O	2.2 (2.0–2.5)	6.6 (4.9–8.2)	6.8 (5.7–8.7)	1.6 (1.1–2.3)	10.3	2.9 (2.1–3.7)
n[c]	3	3	3	8	1	2
CIPW norms	$Ab_{75}Or_{13}An_{6.4}$	$Ab_{55}Or_{39}An_{2.3}$	$Ab_{52}Or_{41}An_{0.5}$	$Ab_{86}Or_{9.5}An_{3.2}$	$Ab_{30}Or_{63}An_{2.6}$	$Ab_{67}Or_{17}An_{13}$

[a]The average content of the component
[b]The range of contents of this component
[c]The number of analyzed crystals

Table 13.11 The chemical composition (in wt%) uncrystalline (residual) melts (glasses)

Components	Initial glass	$P \approx 270$ MPa		$P \approx 510$ MPa	
		$Melt_1$	$Melt_2$	$Melt_3$	$Melt_4$
SiO_2	72.7	72.7[a] (72.3–73.1)[b]	72.6 (72.5–72.8)	72.7 (71.1–74.3)	72.7 (71.3–74.2)
TiO_2	0.4	0.4 (0.3–0.4)	0.4 (0.4–0.4)	0.3 (0.2–0.4)	0.3 (0.3–0.3)
Al_2O_3	13.2	12.5 (12.5–12.5)	12.9 (12.6–13.2)	13.0 (12.3–14.0)	14.0 (13.3–14.5)
FeO_{total}	4.2	3.8 (3.1–4.5)	3.2 (3.0–3.4)	2.2 (1.4–3.0)	2.1 (0.8–2.9)
MnO	0.1	0.1 (0.1–0.1)	–	0.1 (0–0.2)	–
ZnO	0.5	0.3 (0.3–0.4)	0.3 (0.3–0.4)	0.2 (0.2–0.3)	0.3 (0.1–0.4)
PbO	0.5	0.2 (0.2–0.2)	0.9 (0.8–1.0)	0.2 (0.1–0.3)	0.2 (0.2–0.4)
CaO	0.9	0.3 (0.1–0.5)	1.2 (0.5–1.8)	0.1 (0–0.2)	0.7 (0.2–0.9)
Na_2O	2.8	2.2 (1.8–2.7)	3.1 (3.0–3.3)	1.4 (1.0–1.9)	5.7 (5.4–6.1)
K_2O	5.9	7.3 (6.2–8.5)	5.2 (4.3–6.0)	9.8 (8.0–10.7)	3.6 (3.1–4.1)
n^c		2	2	7	4

[a]The average content of the component
[b]The range of contents of this component
[c]The number of analyses

(0.5 wt% PbO and ZnO each) before crystallization of the melt. The only exception is "$melt_2$", in which an increase in the PbO content correlates with an increase in CaO.

In natural conditions, plagioclases rich in lead in a number of oligoclase-labrador containing up to 2.0 wt% Pb are described in the skarns of Sweden (Christy and Gatedal 2005). They are in association with hyalophane (up to 5.3 wt% Pb and 6.6–14.8 wt% BaO) and scapolite (up to 4.9 wt% Pb). This mineral association was formed at conditions of increased pressure and temperature ($P = 300$ MPa, $T >$ 600 °C). Amazonite (a type of *Kfs*) in a granodiorite containing up to 1.6 wt% Pb is described in paper (Stevenson and Martin 1988). Amazonite-containing association was formed at $T = 560$–620 °C, $P = 500$–800 MPa, at conditions of low f_{O2} and f_{S2} and rapid rise of the massif from the depth.

The problem of formation of amazonites having a bright specific bluish-greenish color (the color of the sea wave) was considered by Zaraysky (2004) on the example of the Orlovka and Etyka granites of the Eastern Transbaikalia. The nature of amazonite coloring is associated with the introduction of a structural admixture of lead into the

Kfs structure (Marfunin and Bershov 1970; Mineeva et al. 1992; Platonov et al. 1984; Petrov et al. 1993). It has be established by the EPR method that the color of amazonite is determined by the presence of dimeric centers of green coloring of $(Pb-Pb)^{3+}$ replacing K^+ ions in two neighboring M positions, in contrast to unpainted *Kfs*. The charge is compensated in result of the ordered substitution of $Al^{3+} \rightarrow Si^{4+}$. A direct correlation between the intensity of signals of EPR centers of $(Pb-Pb)^{3+}$ and the color density of amazonite has been established. Thus, each of the lead ions in amazonite has a kind of valence of $Pb^{1.5+}$: $(Pb^{2+} + Pb^{2+} + e^-)^{3+}$. It can be assumed that these centers are formed in *Kfs* as a result of isomorphic substitution of $2 K^+ \rightarrow 2Pb^+$ in reducing conditions, and then the partial oxidation of lead to the 1.5-valent state with the emergence of color center of $(Pb^{2+} + Pb^{2+} + e^-)^{3+}$ occurs at more oxidizing conditions (Zaraisky 2004). The content of Pb in green, densely colored amazonites of Etyka reaches 300–1200 g/t. The amazonites of the orlovka and etyka granites apparently have the late magmatic origin. These amazonites crystallized at the transition magmatic-hydrothermal stage from the intergranular melt, in which the water fluid was present as an independent phase (Reyf et al. 2000). The enrichment of melt in the lead, the reducing conditions and low chemical potential of sulphide sulfur associated with low concentration of sulfur in the granite melt and high temperature have the decisive importance for the formation of amazonite. The activity of sulphide sulfur increased at the postmagmatic stage with the decrease in temperature, and lead concentrated into galena.

Lead-containing feldspar of $PbAl_2Si_2O_8$ composition (anorthite analogue), in which lead completely replaces calcium, has been synthesized in several ways (Scheel 1971; Bruno and Facchinelli 1972; Bambauer et al. 1974). For example, from a mixture of oxides in platinum crucibles at $T = 1150$ °C, $P = 0.1$ MPa and duration of experiments 12 h, and also by a method of hydrothermal synthesis from gel at $T = 520$ °C, $P_{H_2O} = 120$ MPa and duration of experiments 120 h. Taroev et al. (1990; 1997) in their experimental papers has shown that the occurrence of Pb into potash feldspar is very limited and does not exceed 0.33 (0.5) wt% of PbO in hydrothermal oxidative conditions (buffer $Cu-Cu_2O$) at T up to 500 °C and $P = 100$ MPa, and in reducing conditions (buffer Ni-NiO) the content of PbO < 0.05 wt%. Although it is known that at natural conditions the highest concentrations of Pb (up to 1–2 wt%) are characteristic namely for amazonites. The partition coefficient Pb ($^{fluid/Kfs}D_{Pb}$), which is about 1.25 ± 0.3 at the conditions of $Cu-Cu_2O$ buffer, was also estimated (Taroev et al. 1997).

Filatovite mineral, $K[(Al,Zn)_2(As,Si)_2O_8]$, (aluminoarsenosilicate from the feldspar group, in which pentavalent As^{5+} atoms preferably fill tetrahedral positions; structural type of celsian feldspar), first discovered in the products of fumarole activity of Tolbachik volcano in Kamchatka, contains about 3.1 wt% Zn (Vergasova et al. 2004; Filatov et al. 2004). Besides we do not know, and in the summary (Bambauer et al. 1974) there are no experimental or natural data on the noticeable occurrence of zinc in the feldspars.

Our experimental data presented below make it possible to estimate the content of lead and zinc in feldspars of various compositions crystallizing from granite melt.

It has been established that at high pressure of 500 MPa, the feldspars may contain up to 1.8–1.9 wt% Pb (*Olg* and *Kfs₂*) and up to 0.3–0.4 wt% Zn (*Olg*) (Table 13.10).

Table 13.12 shows the contents of Pb and Zn in the fluid, feldspars, as well as the partition coefficients of polymetals between these phases. Perhaps it is more correct to call them "effective" partition coefficients, as they are calculated from data obtained in non-equilibrium experiments.

The significant influence of pressure on the partitioning of polymetals between the fluid separating from the granite melt and crystallizing feldspars is established. Thus at $P \approx 270$ MPa zinc is preferably concentrated in the fluid, than in the crystal phases, except for *Kfs*; and lead equally enters in different phases, i.e. its partition coefficients are close to one. At $P \approx 510$ MPa the polymetals contents in the fluid decrease by more than an order of magnitude, while the lead content in *Olg* and *Kfs* increases significantly (by 6–7 times) (Table 13.10). At this pressure, polymetals begin to concentrate mainly in feldspars, and a greater extent in *Olg* and *Kfs*, except for Zn in *Ab*. A negative correlation between PbO (ZnO) and SiO₂ concentrations in feldspars can be noted, which probably reflects the change in the content of anorthite component.

Apparently in our experiments lead can enter in the structure of feldspars in the form of a "lead-anorthite" complex end-member ($PbAl_2Si_2O_8$), or replacing two potassium atoms with the formation of one cationic vacancy in *Kfs* ($PbAl_2Si_6O_{16}$). Zinc in these conditions is more likely to remain in the melt or in the *Olg*, minimizing the amount of zinc in the *Ab*. The lowest Zn content into the feldspars is experimentally obtained in Ab, and it is in good agreement with natural geochemical data (Antipin et al. 1984). It can be assumed that the lack of Zn on mass balance is associated with the formation of its own minerals such as franklinite $(Zn,Mn)Fe_2O_4$ or zincite ZnO. Natural geochemical data indicate that the feldspars concentrate

Table 13.12 The contents of Pb and Zn in the crystallizing feldspars and the fluid separated from the melt, as well as the "effective" partition coefficients of Pb and Zn

	$T = 630 \to 580$ °C, $P = 285 \to 255$ MPa		$T = 600 \to 550$ °C, $P = 520 \to 500$ MPa	
	Pb	Zn	Pb	Zn
$^{fluid}C$, wt%	0.27 ± 0.07	0.23 ± 0.05	0.02 ± 0.01	0.016 ± 0.005
^{Ab}C, wt%	0.25	0.04	0.4	<0.015
^{Olg}C, wt%	–	–	1.76	0.24
^{Ancl}C, wt%	0.28	0.03	–	–
^{Kfs}C, wt%	0.3	0.24	1.76	0.08
$^{fluid/Ab}D$	1.1	6	0.05	> 1.1
$^{fluid/Olg}D$	–	–	0.01	0.07
$^{fluid/Ancl}D$	1	8	–	–
$^{fluid/Kfs}D$	0.9	1	0.01	0.2

The initial 0.1N HCl + 1N NaCl solution and the experiments durations of 12–13 days

Pb better than other minerals at crystallization acid magmatic rocks. While Zn is mainly accumulated in micas and amphiboles, replacing divalent magnesium and iron (Antipin et al. 1984).

Figure 13.12 shows the effect of pressure on the partitioning of polymetals. Noticeable difference is seen in the behavior of Pb and Zn in studied system. At the pressure decrease the partition coefficients of Pb and Zn increase, but in varying degrees: for Pb by 1.5–2 orders of magnitude, and for Zn by about 5 times. If concentrations of these elements are close to each other in the fluid, then the feldspars may contain much more Pb than Zn (Tables 13.10 and 13.12). The increase in the anorthite component and the increase in pressure facilitate such predominance of Pb over Zn in the feldspar composition. Thus at $\approx$510 MPa the Pb content in feldspars is 1–2 orders of magnitude higher than in the fluid, and the Zn content in *Ab* is close to its content in the fluid and increases markedly only in *Olg* and *Kfs*, although to a lesser extent than for Pb.

Comparison of Pb and Zn contents in the fluid in these experiments (at decrease of *T* and *P*) and in earlier described experiments on partitioning of Pb and Zn at constants of *T* and *P* between the same fluid and granite melt shows a good similarity.

It can be assumed that the crystallization of granite massifs at different depths, all other conditions being equal, will produce a magmatic fluid with a different degree of potential ore-bearing. The massifs located at less depth (about 250 MPa) will have greater potential productivity with respect to polymetals (Pb, Zn). At conditions of depth crystallization of magma in the zone of its formation (about 500 MPa), Pb practically will not go into the fluid, concentrating in *Pl* and *Kfs*, which as a result of further interaction with fluids and recrystallization can also serve as a source of ore matter at the postmagmatic stage. Zinc, unlike lead, is much less common in feldspars composition. And since in ore deposits these metals occur together, that the formation of ore-bearing fluids associated with leucocratic granites, by a mechanism of solid → fluid at the postmagmatic stage is less preferable as against the formation

Fig. 13.12 Influence of pressure on the "effective" partition coefficients of Pb and Zn between the fluid separated from the granite melt and the crystallizing feldspars

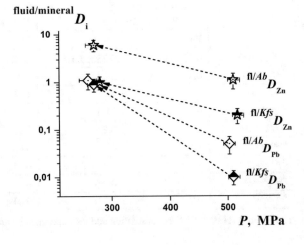

of high-temperature ore-bearing magmatic fluid as a result of the interaction melt $\rightarrow$ fluid, at which the behavior of Pb and Zn is more similar with each other.

13.4.5   Influence of Magmatic Melt Composition on Pb and Zn Partitioning Between Fluid and Melt

Table 13.13 and Fig. 13.13 represent the results of the partitioning of Pb and Zn between acidic chloride fluid (0.1N HCl + 1N NaCl) and leucogranite, granite and granodiorite melts at $T = 800–1000$ °C and $P = 100$ and 500 MPa (Chevychelov 1987, 1998; Chevychelov and Chevychelova 1997). At the studied conditions relatively small changes in partition coefficients were obtained: their average values are greater than one and vary from 1 to 5. With the transition from granodiorite to granite and leucogranite melts these coefficients increase by about two or three times. At $P = 100$ MPa they are slightly higher in the case of granite melt, and at $P = 500$ MPa—in the case of leucogranite melt. The higher concentration of calcium in the granodiorite melt compared to granite and leucogranite melts probably contributed to the retention of Pb and Zn in the melt, and as a result the $^{fluid/melt}D$ coefficients decreased. If the value of A/CNK ratio is less than one, that in addition to minals of feldspars, it can assume the presence of Ca, Na and K in the melt in the form of additional silicate minals (for example, $(Na,K)_4Zn_2(SiO_4)_2$ and others), which could hold Pb and Zn in the melt.

The obtained Zn partition coefficients are slightly higher than for Pb. As the pressure increases, the polymetals partition coefficients decrease, but rather weak (as opposed to a significant change in these coefficients for the melt of the akchatau granite), which is probably due to the oppositely directed influence of increasing pressure and decreasing temperature.

In experiments at $T = 800–850$ °C and $P = 100$ MPa it is shown (Table 13.14) that temperature decrease from 950–1000 to 800–850 °C at pressure 100 MPa causes significant (about 2–5 times) increase of "gross" partition coefficients of Pb and Zn. It should be noted that the lower temperature data in Table 13.14 were obtained at sub-liquidus conditions, at which in the melt there were up to 20–30% of plagioclase (Pl) crystals, so the calculated coefficients are "gross". At $P = 500$ MPa, there is no noticeable influence of temperature decrease in the range of 950–800 °C on the partitioning of Pb and Zn (Table 13.13).

Very strong influence of granite melt composition on the partitioning of Pb and Zn was established (Urabe 1985). At slight change of melt composition from slightly alkaline (melt-1, in wt%: 75.7 SiO_2, 14.7 Al_2O_3, 2.0 CaO, 7.7 Na_2O) to weak aluminous (melt-2, in wt%: 74.7 SiO_2, 16.7 Al_2O_3, 1.8 CaO, 6.7 Na_2O) the $^{fluid/melt}D$ coefficients increase from $\approx 0.0n$ to $\approx n$, and that the pH of the solutions after the experiment decreases from near-neutral (about 7–8) to acidic (about 1–1.5). We have conducted our experiments with both model granite melts, corresponding to the used in the paper (Urabe 1985). Our experiments were carried out at a lower pressure of

Table 13.13 Partitioning of Pb and Zn between acidified chloride fluid and granitoid melts of various compositions ($T = 800$–$1000\ °C$, $P = 100$ and $500\ MPa$[a])

No. of experiment	P, MPa	T, °C	Initial solution, wt%[b]	Initial glass composition[c]	Duration, days	fluid C_{Pb}, wt%	melt C_{Pb}, wt%	fluid/melt D_{Pb}	fluid C_{Zn}, wt%	melt C_{Zn}, wt%	fluid/melt D_{Zn}
165	100	1000	A[b] + 0.65%Pb + 0.56%Zn	grdr[c]	4.0	0.30	0.46	0.65	0.21	0.17	1.2
189	100	990	A + 0.56%Zn	grdr	3.8	0.06	0.03	2.1	0.37	0.18	2.1
190	100	990	A + 0.33%Pb + 0.88%Zn	gm	3.8	0.30	0.09	3.3	0.57	0.19	3.0
167	100	950	A + 0.55%Pb + 0.56%Zn	gm	3.2	0.74	0.25	2.9	0.71	0.10	7.1
164	100	950	A + 0.87%Pb + 0.56%Zn	lcgr	4.1	0.73	0.46	1.6	0.54	0.17	3.1
191	100	950	A + 0.56%Zn	lcgr	3.9	0.06	≤0.02	≥3.2	0.43	0.14	3.1
171	500	900	A + 0.58%Pb + 0.56%Zn	grdr	3.0	0.31	0.31	1.0	0.35	0.15	2.3

(continued)

Table 13.13 (continued)

No. of experiment	P, MPa	T, °C	Initial solution, wt%[b]	Initial glass composition[c]	Duration, days	fluid C_{Pb}, wt%	melt C_{Pb}, wt%	fluid/melt D_{Pb}	fluid C_{Zn}, wt%	melt C_{Zn}, wt%	fluid/melt D_{Zn}
193	500	950	A + 0.56%Zn	grdr	2.8	0.04	0.03	1.1	0.33	0.28	1.2
184	500	800	A + 0.43%Pb + 0.56%Zn	grm	3.8	0.37	0.23	1.6	0.54	0.11	4.8
194	500	950	A + 0.37%Pb + 0.56%Zn	grm	2.8	0.32	0.18	1.8	0.48	0.12	3.9
185	500	800	A + 0.52%Pb + 0.56%Zn	lcgr	3.8	0.38	0.24	1.6	0.51	0.09	5.6
195	500	950	A	lcgr	2.8	0.08	0.02	3.6	0.07	≤0.02	≥3.4

[a] Analyzes were performed using wave dispersion spectrometers. Detection limit is 0.02–0.03 wt%. The composition of the solutions after the experiment was additionally analyzed by the atomic absorption method. The average contents are shown in the table

[b] A—0.1N HCl + 1N NaCl

[c] grdr—granodiorite, grm—granite and lcgr—leucogranite melts. The chemical compositions of these melts are presented in the text. Each of the three initial melts, in addition to the main rock-forming components, contained 0.19 wt% Pb + 0.16 wt% Zn

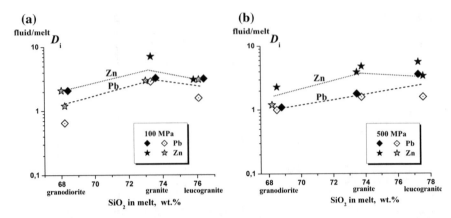

Fig. 13.13 Dependence of the $^{fluid/melt}D_{Pb,Zn}$ partition coefficients on the silica content in the melt. The 0.1N HCl + 1N NaCl initial solution and granodiorite (grdr), granite (grn) and leucogranite (lcgr) melts. $P = 100$ MPa, $T = 950–1000$ °C (**a**) and $P = 500$ MPa, $T = 800–950$ °C (**b**) (Table 13.13). The approach to equilibrium from two sides: filled symbols are experiments with the extraction of metals from the melt into the fluid; open symbols are transitions of metals from fluid to melt. Dotted lines show trends in $^{fluid/melt}D$ coefficients

100 MPa instead of 350 MPa; we also used the method of "approach to equilibrium, both from melt and fluid" (T. Urabe injected Pb and Zn only through the fluid); the temperature and initial composition of the solution coincided with the data of T. Urabe. The exact composition of the initial substances and the results obtained are presented in Table 13.15. In this series similar and rather high $^{fluid/melt}D$ coefficients (4.5–11 for Pb and 3–7 for Zn) were obtained for both melt compositions, which can be compared with the data of T. Urabe for composition 2, and we failed to obtain low ($\approx 0.0n$) $^{fluid/melt}D$ coefficients, presented by T. Urabe for slightly alkaline composition 1.

The dependences of $^{fluid/melt}D$ partition coefficients for Pb and Zn at change of composition of aluminosilicate melts in a wide range from basalt to leucogranite at $P = 100(200)$ MPa, $T = 950–1100$ °C are shown in Fig. 13.14. Data on basalt and andesite melts are taken from (Khodorevskaya and Gorbachev 1993a, b; Gorbachev and Khodorevskaya 1995). The polymetals partition coefficients strongly increase (on 1–2 order) at such change of composition of melt. Positive rectilinear correlation dependences between $^{fluid/melt}D_{Pb,Zn}$ coefficients and silica content in the melt have been established. The calculated correlation coefficients are quite high: $R_{Pb} = 0.95$, and $R_{Zn} = 0.93$.

The obtained results provide convincing experimental evidence of the essential influence of the magmatic melt composition on the partitioning of polymetals in the process of fluid-magmatic interaction. Probably, this effect is connected both with the content and properties of the main rock-forming and possibly volatile components in the silicate melt, with the formation of various components capable of holding metals in the melt at certain conditions, and with changes in the composition and properties of

Table 13.14 Partitioning of Pb and Zn between acidified chloride fluid and granitoid melts of various compositions (sub-liquidus conditions: $T = 800$–850 °C, $P = 100$ MPa[a])

No. of experiment	T, °C	Initial solution, wt%[b]	Initial glass composition[c]	fluidC_{Pb}, wt%	meltC_{Pb}, wt%	fluid/(melt+Pl) D_{Pb}[d]	fluidC_{Zn}, wt%	meltC_{Zn}, wt%	fluid/(melt+Pl) D_{Zn}[d]
132	850	A[b] + 0.92%Pb + 0.56%Zn	grdr[c]	0.54	0.09	6	0.31	0.07	4
134	800	A + 0.88%Pb + 0.56%Zn	grn	0.84	0.06	14	0.87	0.04	20
146	800	A + 0.58%Pb + 0.56%Zn	lcgr	0.45	0.07	6	0.65	0.08	8

[a] Analyzes were performed using wave dispersion spectrometers. Detection limit is 0.02–0.03 wt%. The composition of the solutions after the experiment was additionally analyzed by the atomic absorption method. The average contents are shown in the table

[b] A—0.1N HCl + 1N NaCl

[c] grdr—granodiorite, grn—granite and lcgr—leucogranite melts. The chemical compositions of these melts are presented in the text. Each of the three initial melts, in addition to the main rock-forming components, contained 0.19 wt% Pb + 0.16 wt% Zn

[d] fluid/(melt+Pl) $D_{Pb(Zn)}$—the "gross" partition coefficient of Pb(Zn) between the fluid and the melt containing plagioclase crystals

Table 13.15 Partitioning of Pb and Zn between acidified chloride fluid and two model granite melts coinciding with (Urabe 1985) ($T = 830\,°C$, $P = 100$ MPa[a])

No. of experiment	Initial solution, wt%[b]	Initial glass composition[c]	fluidC_{Pb}, wt%	meltC_{Pb}, wt%	fluid/meltD_{Pb}	fluidC_{Zn}, wt%	meltC_{Zn}, wt%	fluid/meltD_{Zn}
117/1	A[b] + 0.58%Pb + 0.56%Zn	1a[c]	0.6	0.05	11	0.4	0.08	5
117/2	The same	2a	0.6	0.07	9	0.4	0.05	7
117/4	The same	2b	0.6	0.09	7	0.4	0.08	5
118/1	A + 0.09%Pb + 0.08%Zn	1a	0.2	0.05	4.5	0.1	≤0.03	≥3.5
118/2	The same	2a	0.2	≤0.03	≥7	0.1	0.04	3

[a] Analyzes were performed using wave dispersion spectrometers. Detection limit is 0.02–0.03 wt%. The composition of the solutions after the experiment was additionally analyzed by the atomic absorption method. The average contents are shown in the table

[b] A—0.1N HCl + 1N NaCl

[c] 1a—melt-1 + 0.09 wt% Pb + 0.08 wt% Zn; 2a—melt-2 + 0.09 wt% Pb + 0.08 wt% Zn; 2b—melt-2 + 0.19 wt% Pb + 0.16 wt% Zn. The chemical compositions of melts (melt-1 and melt-2) are presented in the text

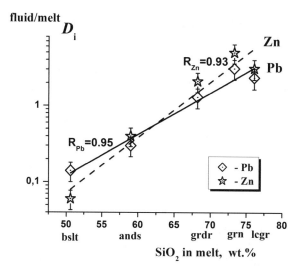

Fig. 13.14 Dependence of the $^{fluid/melt}D_{Pb,Zn}$ partition coefficients on the silica content in the melt in a wide range of magmatic compositions. The about 1N chloride (acidified) solution and granodiorite (grdr), granite (grn), leucogranite (lcgr), basalt (bslt), and andesite (ands) melts. P = 100(200) MPa, T = 950–1100 °C. Data on bslt and ands melts were taken from the studies (Khodorevskaya and Gorbachev 1993b; Gorbachev and Khodorevskaya 1995). The lines show trends in partition coefficients. R—correlation coefficients

the high-temperature magmatic fluid coexisting with the melt and reflecting changes in its composition.

13.5   Conclusion

Detailed experimental studies of the partitioning of Pb and Zn between the fluid and the granite melt showed that relatively low pressure (about 100 MPa) and temperature (about 750 °C) and acidic chloride composition of the (0.1N HCl + 1N NaCl) fluid favor the best extraction of these metals from the melt. In this case the $^{fluid/melt}D_{Pb,Zn}$ partition coefficients are ≥ 10. In systems with acidic chloride fluids these coefficients increase proportionally to the square of the fluid chloride concentration: $^{fluid/melt}D_{Pb} \approx (7 \pm 1.6) \times {}^{fluid}C_{Cl}^2$ and $^{fluid/melt}D_{Zn} \approx (5.7 \pm 2.0) \times {}^{fluid}C_{Cl}^2$, which allows us to assume pure chloride complexes in these systems for Pb^{+2} and Zn^{+2}. Increases in temperature and especially pressure significantly reduce the partition coefficients of these metals. At the conditions of two-phase H_2O–NaCl composition of fluid there is a significant (about 5–7 times) accumulation of Pb and Zn in the dense brine fluid phase with respect low-density vapor phase. The addition of CO_2 to the fluid increases the concentration of these metals in the brine phase of fluid, moreover for Pb it is much higher than for Zn. In the process of crystallization of granite melt at pressure of about 500 MPa polymetals do not pass into the fluid separating from the melt. In this case Pb is concentrated in crystallizing feldspars (*Olg* and *Kfs*), and Zn remains in the uncrystalline melt. At decreasing pressure (up to about 250 MPa) the fluid is visibly enriched with polymetals, and the content of Pb and Zn become almost the same in the fluid and in feldspars, except for the sharply lower content of Zn in *Ab* and *Ancl*. At change of composition of melt from granite to andesite and basalt the $^{fluid/melt}D_{Pb,Zn}$ partition coefficients significantly decrease (on 1.5–2 orders of magnitude); positive linear correlation dependences between $^{fluid/melt}D_{Pb,Zn}$ coefficients and silica content in melt are established ($R_{Pb} = 0.95$, and $R_{Zn} = 0.93$).

The absolute values of the obtained partition coefficients Pb and Zn, their negative baric dependence, as well as the fact that these elements are best extracted from the granite melt by chloride fluids and the statement about the direct proportionality of these coefficients to the square of the fluid concentration are consistent with the experimental data of other authors (Holland 1972; Khitarov et al. 1982; Urabe 1985, 1987a; Kravchuk et al. 1991). The most important of the new results obtained for the first time are apparently quantitative data on the fractionation of Pb and Zn at the fluid-magmatic interaction (1) in the system of heterogeneous $H_2O–NaCl–CO_2$ fluid—granite melt, (2) in the process of depth crystallization of the granite melt (*P* about 500 MPa).

According to the results obtained, the crystallization of granitoid massifs at different depths will produce magmatogenic fluids with different levels of potential ore-bearing in respect of polymetals. Hypabyssal (pressure about 100 MPa) granitoid massifs will have higher potential productivity as compared to depth (*P* about 500 MPa) massifs. It is natural, the ore-bearing of the intrusions due to the influence of many other factors is not directly correlated with the differences in the depth of their formation. But the content of the studied components in granitoids can correlate with the established regularity. Thus, depth intrusions, apparently, should contain large amounts of Pb and Zn in its composition in comparison with the hypabyssal ones. If in general the evolution of the magmatic system is associated with a decrease in temperature and pressure, it can be assumed that the portions of subsequent fluids will be increasingly enriched in Pb and Zn.

At conditions of depth crystallization of granitoid magma in the zone of its formation (about 500 MPa), Pb will practically not go into the fluid, concentrating in *Pl* and *Kfs*, which as a result of further interaction with fluids and recrystallization can also serve as a source of ore matter at the postmagmatic stage. Zinc, unlike lead, is much less common in feldspars composition. And since in ore deposits these metals occur together, that the formation of ore-bearing fluids associated with leucocratic granites, by a mechanism of solid → fluid at the postmagmatic stage is less preferable as against the formation of high-temperature ore-bearing magmatic fluid as a result of the interaction melt → fluid, at which the behavior of Pb and Zn is more similar with each other.

The obtained results confirm the hypothesis about the important role of salt fluid phases in the ore genesis process (Marakushev 1979; Marakushev et al. 1983; Marakushev and Bezmen 1992; Marakushev and Shapovalov 1993, 1994). According to these investigations the ore generating capacity of granite systems arises in the evolution processes of heterogeneous fluid–granite systems, expressed in the separation of dense salt fluid phases from the silicate melt. These fluid phases are effective concentrators of ore metals. Metallogenic specialization arises in relatively closed granite systems, in which the crystallization accumulation of metals is accompanied by a concentration of salt components. On the other hand, the reality of the existence of salt fluid phases containing up to 14–40m Cl and coexisting with granitoid melts at magmatic P-T parameters has been confirmed in a number of studies (e.g., Reynolds and Beane 1985; Chevychelov 1987; Reyf 1990; Solovova et al. 1991).

The potential possibility of formation of a high-temperature ore-bearing magmatic fluid containing about 220 ppm Pb and about 530 ppm Zn is shown due to the mechanism of extraction of these metals by acidic chloride fluid from the granite melt with Clarke concentrations of Pb and Zn. The degree of fluid extraction of these metals is sufficient for the formation of an ore body comparable in scale with a small industrial lead-zinc deposit and, consequently, the presence of metallogenic specialization of magmas for the formation of such deposits is not always necessary. At the same time, for rare-metals such as W, Mo and Sn, in general, a similar mechanism of their extraction from granitoid melts by high-temperature fluids is less typical, due to lower partition coefficients between the fluid and the melt. The apparent contradiction with geological data, according to which the connection with granitoid complexes is more typical for rare-metal deposits, is explained by lower temperatures of polymetallic ore depositions and long-term evolution of ore-bearing fluid, during which the fluid can be significantly removed from the primary source.

The experimental results obtained by us are quite comparable with the partition coefficients between the fluid and the melt determined according to the data of LA-ICP MS analysis of coexisting melt and fluid inclusions in the quartz from different, saturated by volatile components, granite intrusions (from ultra-alkaline to ultra-aluminous by composition, $\log f_{O2}$ from NNO − 1.7 to NNO + 4.5, the fluid separating from the melt contained from 1 to 14 mol/kg $\sum$Cl) (Zajacz et al. 2008). These authors have shown that Pb and Zn are soluble in the fluid mainly in the form of chlorine-containing complexes, since a close to linear positive correlation is in existence between the partition coefficients of these elements and the concentration of Cl in the fluid ($^{\text{fluid/melt}}D_{Pb} \approx 6 \times {}^{\text{fluid}}C_{Cl}$, $^{\text{fluid/melt}}D_{Zn} \approx 8 \times {}^{\text{fluid}}C_{Cl}$).

Studies of melt and fluid inclusions in quartz phenocrysts from rhyodacites and rhyolites of the Uzelga ore field in the Southern Urals by the LA-ICP-MS method allowed us to be established high concentrations of Cu, Zn and Pb in primary magmatogenic fluids and magmatic melts (Vikentjev et al. 2012). Direct data on the potential ore-bearing of acid volcanic complexes, with which sulphide ores deposits of the Urals are spatially associated, have been obtained. Similar high-temperature high-baric ore-bearing aqueous fluids, appeared during degassing of magmatic melts of acid composition, have been established by a number of researchers (e.g., Zaw et al. 2003; Karpukhina et al. 2009; Prokofjev et al. 2011), which suggests the real participation of such magmatic fluids in ore formation.

Acknowledgements The work was carried out with the financial support of RFBR grant № 18-05-01001-a and project of IEM RAS № AAAA-A18-118020590151-3.

References

Anderko A, Pitzer KS (1993) Equation-of-state representation of phase equilibria and volumetric properties of the system NaCl–H_2O above 573 K. Geochim Cosmochim Acta 57:1657–1680

Antipin VS, Kovalenko VI, Ryabchikov ID (1984) Distribution coefficients of rare elements in magmatic rocks. Nauka, Moscow, 254 p (in Russian)

Arutyunyan LA, Malinin SD, Petrenko GV (1984) Solubility of pyrrhotine in chloride solutions at elevated temperatures and pressures. Geokhimiya 7:1029–1039 (in Russian)

Audetat A, Gunther D, Heinrich CA (2000) Causes for large-scale metal zonation around mineralized plutons: fluid inclusion LA-ICP-MS evidence from the Mole Granite, Australia. Econ Geol 95(8):1563–1581

Audetat A, Pettke T (2003) The magmatic-hydrothermal evolution of two barren granites: a melt and fluid inclusion study of the Rito del Medio and Canada Pinabete plutons in northern New Mexico (USA). Geochim Cosmochim Acta 67(1):97–121

Bambauer HU, Kroll H, Nager HE, Pentinghaus H (1974) Feldspat-Mischkristalle - eine Ubersicht. Bull Soc fr Mineral Cristallogr 97:313–345

Barnes HL (1982) Solubility of ore minerals. In: Barnes HL (ed) Geochemistry of hydrothermal ore deposits. Earth Sciences. Mir, Moscow, pp 328–369 (in Russian)

Berndt J, Koepke J, Holtz F (2005) An experimental investigation of the influence of water and oxygen fugacity on differentiation of MORB at 200 MPa. J Petrol 46:135–167

Bodnar RJ, Burnham CW, Sterner SM (1985) Synthetic fluid inclusions in natural quartz. III. Determination of phase equilibrium properties in the system H_2O–NaCl to 1000 °C and 1500 bars. Geochim Cosmochim Acta 49(9):1861–1873

Bogatikov OA, Dmitriev YI, Tsvetkov AA (eds) (1987) Petrology and geochemistry of island arcs and marginal seas. Nauka, Moscow, 336 p (in Russian)

Bowers TS, Helgeson HC (1983) Calculation of the thermodynamic and geochemical consequences of nonideal mixing in the system H_2O–CO_2–NaCl on phase relations in geologic systems: Equation of state for H_2O–CO_2–NaCl fluids at high pressures and temperatures. Geochim Cosmochim Acta 47(7):1247–1275

Bruno E, Facchinelli A (1972) Al, Si configurations in lead feldspar. Ztschr Kristallogr 136:296–304

Chekhmir AS (1988) Express method for determining the partitioning coefficients of elements in the melt—fluid system. Geokhimiya 8:1221–1222 (in Russian)

Chekhmir AS, Simakin AG, Epelbaum MB (1991) Dynamic phenomena in fluid—magmatic systems. Nauka, Moscow, 142 p (in Russian)

Chevychelov VY (1987) Possible mechanism of formation of ore-bearing (lead- and zinc-containing) magmatogenic fluid In: Construction models of ore-forming systems. Nauka, Novosibirsk, pp 71–84 (in Russian)

Chevychelov VYu (1992) The partitioning of polymetals between granitoid melt, fluid-salt and fluid phases. Dokl Akad Nauk SSSR 325(2):378–381 (in Russian)

Chevychelov VYu (1993) Differentiation of Pb and Zn between the phases of fluid-granitic melt system. Exp GeoSci 2(2):18–25

Chevychelov VY (1998) The influence of the composition of granitoid melts on the behavior of ore metals (Pb, Zn, W, Mo) and petrogenic components in the melt-aqueous fluid system In: Experimental and theoretical modeling of mineral formation processes. Nauka, Moscow, pp 118–130 (in Russian)

Chevychelov VY, Chevychelova TK (1997) Partitioning of Pb, Zn, W, Mo, Cl and major elements between aqueous fluid and melt in the systems granodiorite (granite, leucogranite)—H_2O–NaCl–HCl. N Jb Miner Abh 172(1):101–115

Chevychelov VY, Epelbaum MB (1985) The partitioning of Pb, Zn and petrogenic components in the granitic melt–fluid system. In: Essays of physical and chemical petrology (Experimental study of the problems of magmatism), Nauka, Moscow, vol 13, pp 120–136 (in Russian)

Chevychelov VY, Salova TP, Epelbaum MB (1994) Differentiation of ore components (Pb, Zn and W, Mo) in the fluid-magmatic (granitoid) system. In: Experimental problems of geology, Nauka, Moscow, pp 104–121 (in Russian)

Chou IM (1987) Phase relations in the system NaCl–KCl–H_2O. III: Solubilities of halite in vapor-saturated liquids above 445 °C and determination of phase equilibrium properties in the system NaCl–H_2O to 1000 °C and 1500 bars. Geochim Cosmochim Acta 51:1965–1975

Christy AG, Gatedal K (2005) Extremely Pb-rich rock-forming silicates including a beryllian scapolite and associated minerals in a skarn from Langban, Varmland. Sweden Mineral Mag 69(6):995–1018

Epelbaum MB (1986) Fluid-magmatic interaction as process of formation and factor of evolution of granitoid magmas and ore-bearing fluids. In: Experiment in solving actual problems of geology, Nauka, Moscow, pp 29–47 (in Russian)

Filatov SK, Krivovichev SV, Burns PC, Vergasova LP (2004) Crystal structure of filatovite, K[(Al, Zn)$_2$(As, Si)$_2O_8$], the first arsenate of the feldspar group. Eur J Mineral 16:537–543

Gehrig M (1980) Phasengleitgewichte und PVT – daten ternarer mischungen aus wasser, kohlendioxid und natriumchlorid bis 3 kbar und 550 °C. Chemie Band, Freiburg, 108 s

Gorbachev NS, Khodorevskaya LI (1995) Chlorine partitioning between aqueous fluid and basaltic melts at high pressures: the behavior of chlorine and water in magmatic degassing processes. Dokl Akad Nauk SSSR 340(5):672–675 (in Russian)

Hemley JJ, Cygan GL, d'Angelo WM (1986) Effect of pressure on ore mineral solubilities under hydrothermal conditions. Geology 14:377–379

Holland HD (1972) Granites, solutions, and base metal deposits. Econ Geol 67(3):281–301

Ishkov YuM, Reyf FG (1990) Laser spectral analysis of inclusions of ore-bearing fluids in minerals. Nauka, Novosibirsk, 93 p (in Russian)

Karpukhina VS, Naumov VB, Vikentjev IV, Salazkin AN (2009) Melt and high-density fluid inclusions of magmatic water in quartz phenocrysts of acid volcanites of the Verkhneuralsky ore region (Southern Urals). Dokl Akad Nauk SSSR 426(1):90–93 [in Russian]

Khitarov NI, Malinin SD, YeB Lebedev, Shibaeva NP (1982) The partitioning of Zn, Cu, Pb and Mo between the fluid phase and silicate melt of granitic composition at high temperatures and pressures. Geokhimiya 8:1094–1107 (in Russian)

Khodorevskaya LI, Gorbachev NS (1993a) Experimental study of the partitioning of Cu, Zn, Pb, Fe, and Cl between andesite and aqueous-chloride fluid at high parameters. Dokl Akad Nauk SSSR 332(5):631–634 (in Russian)

Khodorevskaya LI, Gorbachev NS (1993b) Pressure dependence of the partitioning of Cu, Zn, Pb, Fe and Cl between fluid and silicate melts In: Proceedings of the 4th international symposium on hydrothermal reactions (4th ISHR). Nancy, France, pp 107–108

Kigai IN (2011) Redox problems of "metallogenic specialization" of magmatites and hydrothermal ore formation. Petrologiya 19(3):316–334 (in Russian)

Kigai IN, Tagirov BR (2010) The evolution of the acidity of ore-forming fluids due to the hydrolysis of chlorides. Petrologiya 18(3):270–281 (in Russian)

Korotaev MY, Kravchuk KG (1985) The heterophasicity of hydrothermal solutions at conditions of endogenous mineral formation. Preprint. Chernogolovka, IEM, 64 p (in Russian)

Kotelnikov AR, Musaev AA, Averina AS, Ganeev IG (1990) Study of the partitioning of tungsten between heterogeneous fluid phases by the method of cation-exchange equilibria. Dokl Akad Nauk SSSR 313(1):164–166 (in Russian)

Kravchuk IF, Malinin SD, Bin Zhao (1991) Zinc fractionation depending on the melt composition and pressure in the fluid—granodioritic melt system at 1000 °C. Geokhimiya 8:1159–1165 (in Russian)

Kravchuk IF, Reyf FG, Malinin SD, Ishkov YuM, Naumov VB (1992) Study of the partitioning of Cu and Zn between the phases of heterogeneous fluid NaCl–H_2O by the method of synthetic fluid inclusions. Geokhimiya 5:735–738 (in Russian)

Kukoev VA, Smirnov MP, Guselnikova NYu, Stulov GV (1982) About the speciation forms of lead in slags of shaft smelting. Non-ferrous metals 11:29–31 (in Russian)

Makrygina VA (2011) Geochemistry of individual elements. Study guide. Academic publishing house "Geo", Novosibirsk, 195 p (in Russian)

Malinin SD, Khitarov NI (1983) On the correlation ratios in the system fluid–melt in connection with the partitioning of elements between the phases. Dokl Akad Nauk SSSR 270(2):427–428 [in Russian]

Malinin SD, Khitarov NI (1984) Ore and petrogenic elements in the system magmatic melt—fluid. Geokhimiya 2:183–196 (in Russian)

Malinin SD, Petrenko GV, Arutyunyan LA (1989) Solubility of pyrrhotine, galena and sphalerite in NH_4Cl–NaCl solutions at high temperatures and pressures. Geokhimiya 10:1479–1488 (in Russian)

Marakushev AA (1979) Petrogenesis and ore formation (geochemical aspects). Nauka, Moscow, 262 p (in Russian)

Marakushev AA, Bezmen NI (1992) Mineralogical and petrological criteria for ore-bearing of igneous rocks. Nedra, Moscow, 317 p (in Russian)

Marakushev AA, Gramenitsky EN, Korotaev MYu (1983) Petrological model of endogenous ore formation. Geologiya rudnykh mestorozhdenii 1:3–20 (in Russian)

Marakushev AA, Shapovalov YuB (1993) Experimental study of the ore concentration process in granitic systems. Dokl Akad Nauk SSSR 330(4):497–501 (in Russian)

Marakushev AA, Shapovalov YuB (1994) Experimental study of the ore concentration in fluorid granitic systems. Petrologiya 2(1):4–23 [in Russian]

Marfunin AS, Bershov LV (1970) Paramagnetic centers in feldspars and their possible crystal chemical and petrographic significance. Dokl Akad Nauk SSSR 193(2):412–414 (in Russian)

Mineeva RM, Bershov LV, Marfunin AS, Petrov I, Hafner S (1992) Dimers $[Pb-Pb]^3$ in the structure of amazonite. Geokhimiya 8:1149–1159 [in Russian]

Nagaseki H, Hayashi K-i (2008) Experimental study of the behavior of copper and zinc in a boiling hydrothermal system. Geology 36(1):27–30

Peretyazhko IS (2010) Conditions for the formation of mineralized cavities (miarol) in granitic pegmatites and granites. Petrologiya 18(2):195–222 (in Russian)

Petrov I, Mineeva RM, Bershov LV, Agel A (1993) EPR of $[Pb–Pb]^3$ mixed valence pairs in amazonite type microcline. Am. Mineral. 78:500–510

Platonov AN, Tarashchan FN, Taran MN (1984) On the color centers in amazonites. Miner J 6(4):3–16 (in Russian)

Popp RK, Frantz JD (1990) Fluid immiscibility in the system H_2O–NaCl–CO_2 as determined from synthetic fluid inclusions. Annual report of the director geophysical laboratory 1989–1990, vol IV. Carnegie inst, Washington DC, pp 43–48

Prokofjev VYu, Kovalenker VA, Elen S, Borisenko AS, Borovikov AA (2011) Metal concentrations in fluid inclusions of the magmatic and hydrothermal stages of the generation of the epitermal Au-Ag-polymetallic deposit of Banská Štiavnica (Western Carpathians) using the LA-ICP-MS method. Dokl Akad Nauk SSSR 440(3):388–391 (in Russian)

Raguin E (1979) Geology of granite. Nedra, Moscow, pp 187–205 (in Russian)

Roedder E (1987) Fluid inclusions in minerals. Mir, Moscow, vol 1, 558 p, vol 2, 632 p (in Russian)

Reyf FG (1990) Ore-forming potential of granites and conditions for its implementation. Nauka, Moscow, 182 p (in Russian)

Reyf FG, Seltmann R, Zaraisky GP (2000) The role of magmatic processes in the formation of banded Li, F-enriched granites from the Orlovka tantalum deposit, Transbaikalia, Russia: microthermometric evidence. Can Mineral 38:915–936

Reynolds TJ, Beane RE (1985) Evolution of hydrothermal fluid characteristics at the Santa Rita, New Mexico, porphyry copper deposit. Econ Geol 80(5):1328–1347

Robie RA, Hemingway BS (1995) Thermodynamic properties of minerals and related substances at 298.15 K and 1 bar (10^5 pascals) pressure and at higher temperatures. U.S. Geological Survey Bulletin. (2131) U.S. Government Printing Office, Washington, 461 p

Romanenko IM (1986) Calculation of the detection limit in X-ray spectral analysis and its relations with the ratio of the analytical signal to the background signal. Zh Anal Khim 41(7):1177–1182 (in Russian)

Salova TP, Epelbaum MB, Tikhomirova VI, Romanenko IM, Akhmedzhanova GM (1983) Analysis methods of trace amounts of petrogenic substance from fluid solution In: Essays of physical and chemical petrology. Nauka, Moscow, vol 11, pp 161–165 (in Russian)

Sato T (1977) Kuroko deposits: their geology, geochemistry and origin. In: Volcanic processes in ore genesis. Institute of Mining Metallurgy, L, pp 153–161

Scheel HJ (1971) Lead feldspar. Ztschr Kristallogr 133:264–272

Simonov VA, Gaskov IV, Kovyazin SV (2010) Physico-chemical parameters from melt inclusions for the formation of the massive sulfide deposits in the Altai-Sayan Region, Central Asia. Austral J Earth Sci 57(6):737–754

Solovova IP, Girnis AV, Naumov VB, Kovalenko VI, Guzhova AV (1991) The mechanism of acid magmas degassing is the formation of two fluid phases during the crystallization of pantellerites of the island of Pantelleria. Dokl Akad Nauk SSSR 320(4):982–985 (in Russian)

Sterner SM, Bodnar RJ (1984) Synthetic fluid inclusions in natural quartz I. Compositional types synthesized and applications to experimental geochemistry. Geochim Cosmochim Acta 48(12):2659–2668

Stevenson RK, Martin RF (1988) Amazonitic K-feldspar in granodiorite at Portman lake, northwest territories: indications of low $f(O)_2$, low $f(S)_2$, and rapid uplift. Can Mineral 26:1037–1048

Takenouchi S, Kennedy JK (1968) The solubility of carbon dioxide in solutions of NaCl at high temperatures and pressures In: Thermodynamics of postmagmatic processes. Mir, Moscow, pp 137–149 (in Russian)

Taroev VK, Tauson VL, Piskunova LF, Abramovich MG (1990) Obtaining crystals of potassium feldspar and the study of the entry into they of lead in hydrothermal conditions. Geologiya i geophizika 2:66–75 (in Russian)

Taroev VK, Tauson VL, Suvorova LF, Pastushkova TM, Proydakova OA, Barankevich VG (1997) Lead partitioning between potassium feldspar and fluid of alkaline composition in the system SiO_2-Al_2O_3-PbO_2-KOH-H_2O at temperature of 500°C and pressure of 100 MPa. Dokl Akad Nauk SSSR 357(6):815–817 (in Russian)

Tugarinov AI (1973) General geochemistry. Short course. Textbook for universities. Atomizdat, Moscow, 288 p (in Russian)

Ulrich T, Gunther D, Heinrich CA (2001) The evolution of a porphyry Cu-Au deposit, based on LA-ICP-MS analysis of fluid inclusions: Bajo de la Alumbrera, Argentina. Econ Geol 96(8):1743–1774

Urabe T (1984) Magmatic hydrothermal fluid and generation of base metal deposits. Mining Geol 34(5):323–334

Urabe T (1985) Aluminous granite as a source magma of hydrothermal ore deposits: an experimental study. Econ Geol 80(1):148–157

Urabe T (1987a) The effect of pressure on the partitioning ratios of lead and zinc between vapor and rhyolite melts. Econ Geol 82(4):1049–1052

Urabe T (1987b) Kuroko deposit modeling based on magmatic hydrothermal theory. Mining Geol 37(3):159–176

Vergasova LP, Krivovichev SV, Britvin SN, Burns PC, Ananiev VV (2004) Filatovite, $K[(Al, Zn)_2(As, Si)_2O_8]$, a new mineral species from the Tolbachik volcano, Kamchatka peninsula, Russia. Eur J Mineral 16:533–536

Vikentjev IV, Borisova AYu, Karpukhina VS, Naumov VB, Ryabchikov ID (2012) Direct data on the ore-bearing of acid magmas of the Uzelginsky ore field (Southern Urals, Russia). Dokl Akad Nauk SSSR 443(3):347–351 (in Russian)

Voitkevich GV, Miroshnikov AE, Povarennykh AS, Prokhorov VG (1977) Brief Guide to Geochemistry. Nedra, Moscow, pp 47–48 (in Russian)

Watson EB (1979) Zircon saturation in felsic liquids: experimental results and applications to trace element geochemistry. Contrib Mineral Petrol 70(4):407–419

Zaitsev VY, Vanyukov AV, Genevska TN, Bagaev IS, Yarygin VI, Yakubov MM (1982) About lead losses with slags of traditional and autogenous processes. Non-ferrous Metals 11:25–29 (in Russian)

Zajacz Z, Halter WE, Pettke T, Guillong M (2008) Determination of fluid/melt partition coefficients by LA-ICPMS analysis of co-existing fluid and silicate melt inclusions: controls on element partitioning. Geochim Cosmochim Acta 72(8):2169–2197

Zaraisky GP (2004) Conditions for the formation of rare-metal deposits associated with granitic magmatism. In: Starostin VI (ed) Smirnovsky collection—2004 (scientific and literary miscellany). VI Smirnov Fund, Moscow, pp 105–192 (in Russian)

Zaraisky GP, Shapovalov YuB, Soboleva YuB, Stoyanovskaya FM, Zonov SV, Ryadchikova EV, Dubinina EO (1994) Physical and chemical conditions of greysenisation at the Akchatau deposit according to geological and experimental data. In: Experimental problems of geology. Nauka, Moscow, pp 371–419 (in Russian)

Zaw K, Hunns SR, Large RR, Gemmell JB, Ryan CG, Mernagh TP (2003) Microthermometry and chemical composition of fluid inclusions from the Mt Chalmers volcanic-hosted massive sulfide deposits, central Queensland, Australia: implications for ore genesis. Chem Geol 194(1–3):225–244

Chapter 14
Experimental Study of Amphibolization of the Basic Rocks of the Tiksheozersky Massif

Northern Karelia, Russia

T. N. Kovalskaya, D. A. Varlamov, Y. B. Shapovalov, G. M. Kalinin, and A. R. Kotelnikov

Abstract The work is devoted to experimental study of postmagmatic transformation of basic rocks (gabbroids) of the Tiksheozersky massif (North Karelia Russia). During the petrological study of gabbroids the formation of edges of alkaline amphiboles around the clinopyroxene grains (diopside-hedenbergite series), and then—low-temperature minerals containing Na, K, Cl. In order to recreate the conditions for the formation of amphibole rims a number of experiments were carried out at a temperature of 850 °C and a pressure of 3 kbar with solutions of KCl and KF concentrations of 0.5 M, 1 M and 2 M, respectively. In the course of experiments, amphiboles corresponding to natural amphiboles from rims around clinopyroxenes, Tiksheozersky massif were obtained and also phlogopites present in the basic rocks of the Tiksheozersky massif were obtained.

Keywords Alkaline magmatism · Gabbro · Carbonatite · Amphibole · Amphibolization · Postmagmatic processes · Experiment

14.1 Introduction

Amphibolite development on the basic rocks within the Fenno-Scandinavian shield is quite common (Khodorevskaya and Varlamov 2018; Safonov et al. 2014), however, the alkaline nature of postmagmatic changes in the basic rocks of the Tickshozersky massif is the most interesting.

The Tickshozersky massif, located in the circumpolar part of Northern Karelia, is a part of the Eletizersko-Tiksheozersky intrusive complex (Fig. 14.1a, b) and lies among the highly dislocated granitoids, migmatites and gneisses of the Archean

T. N. Kovalskaya (✉) · D. A. Varlamov · Y. B. Shapovalov · G. M. Kalinin · A. R. Kotelnikov
D.S. Korzhinskii Institute of Experimental Mineralogy, Russian Academy of Sciences, Academician Osipyan Str. 4, Chernogolovka, Moscow District, Russia 142432
e-mail: tatiana76@iem.ac.ru

© The Editor(s) (if applicable) and The Author(s), under exclusive license to Springer Nature Switzerland AG 2020
Y. Litvin and O. Safonov (eds.), *Advances in Experimental and Genetic Mineralogy*, Springer Mineralogy, https://doi.org/10.1007/978-3-030-42859-4_14

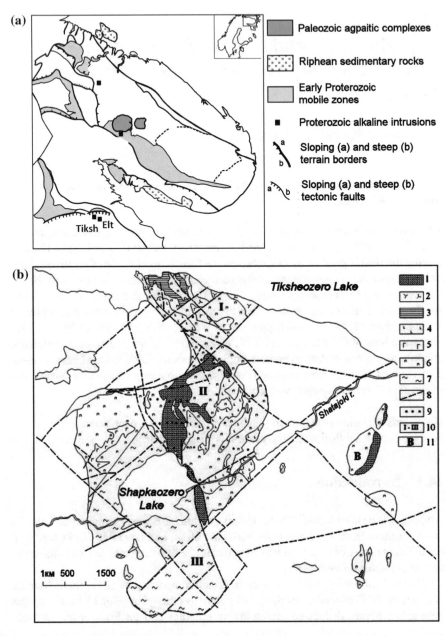

Fig. 14.1 Schematic position of the Tiksheozero massif **a** (**Tiksh**—Tiksheozersky massif, **Elt**—Elet'ozero massif) and **b** geological structure of Tiksheozersky massif (according to Kholodilov 1988): 1—carbonatites, 2—nepheline syenites, 3—teralites, 4—melteigites, ijolite-urtites, 5—gabbros, 6—pyroxenites, 7—olivinites, serpentinized olivinites, 8—fractures, 9—drill holes. Geological blocks: **I**—Tiksheozero, **II**—Central, **III**—Shapkozersky, **B**—Eastern

and Proterozoic (Sharkov et al. 2015). It is represented by a body of a rounded-elliptic shape with a diameter of about 20 km and is composed of olivinites, gabbro, pyroxenites (sometimes with nepheline), ijolites, carbonatites, and amphibole-calcite-cancrinite rocks. The age of the Tikshoozero massif is estimated at about 1.8–1.9 Ga (Shchiptsov et al. 2007). Petrological study of rocks of the massif was carried out by the authors during the field seasons of 2010–2016, in which the rocks composing the massif were described in detail (Kovalskaya et al. 2018a, b), after which the study of the mineral relations and composition in the thin rock sections and polished sections using optic polarization microscope and scanning electron microscope Tescan Vega II XMU was carried out. The change of mineral composition and mineral associations in different types of rocks was traced.

The central part of the massif is composed of olivinites, in some places strongly serpentinized and chloritized, with inclusions of aluminochromites, relics of olivine and clinopyroxene. The X_{Mg} of olivines is 0.83–0.88. The basic rocks are pyroxenites and gabbroids, which are also quite altered. Pyroxenites are composed of clinopyroxene (diopside-hedenbergite and avgite), phlogopite, titanomagnetite. In the gabbroids there is also pyroxene composition of diopside-hedenbergite, plagioclase with a proportion of anorthite component of 70–75%, amphibole of two generations, corresponding to the composition of pargasite and richterite-cataphorite. The first amphibole forms independent grains, the second one is found in the rims around the grains of clinopyroxene diopside-hedenbergite composition, which is probably a result of postmagmatic high-temperature changes in the rocks of the massif. An example of the relationship between clinopyroxene and amphibole is shown in Fig. 14.2. In addition, there are strong secondary changes in the gabbroids—carbonatization of rocks, formation of low-temperature zeolites. Amphibole of different composition can be

Fig. 14.2 Natural amphibole rims around clinopyroxene. The abbreviations of minerals are similar to Table 14.1

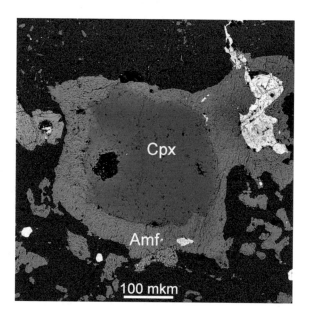

found within the same rock, which is a consequence of the change in physical and chemical conditions of rock formation and the potential of alkaline components.

The carbonatites of the massif are composed intrusive stocks of tens of meters in size, eruptive formations of the breccian texture, veins and veinlets (Safronova 1990). The largest body of carbonatites of the Tiksheozersky massif is traced by drilling to a depth of 450 m. Apatite-calcite ores of complex type were found in carbonatite of the massif (Metallogeny…2001). The mineral composition of the studied rocks is shown in Table 14.1. Paragenetic analysis carried out earlier and geothermometers applied on its basis (Perchuk 1970) allowed to estimate the formation temperatures of amphibole paragenesis. The formation temperatures of amphibole-pyroxene paragenesis of the Tickshozersky massif using clinopyroxene-amphibole, biotite-amphibole and pyroxene-biotite geothermometers (Perchuk and Ryabchikov 1976) were estimated in the range of 710–980 °C. An assessment of the pressure of the formation of amphibole rims using an amphibole geothermometer (Simakin and Shaposhnikova 2017) was not possible due to the alkaline nature of the amphibole. Values of pressure at formation of the Tiksheozersky massif, proceeding from the literature data (Metallogeny… 2001), are estimated as 3–4 kbar. Therefore, for the reconstruction of the mechanism and conditions of the process of amphibolization of the Ticksheozersky massif, an attempt was made to model it experimentally with parameters similar to the calculated ones. This method is described in detail in the paper (Suk et al. 2007). The study of the sodalite-containing paragenesis in the rocks of later phases of formation of the Tiksheozersky massif and carried out on the basis of the obtained thermometry data showed that these associations were formed at a temperature of about 450 °C (Ustinov et al. 2006). The anionic group in sodalite is represented only by Cl^- ion. These data make it possible to use KCl solution of various concentrations as a fluid for modeling the amphibolization process.

14.2   Experimental Technique

Source materials For the experiments, the crushed gabbro of the Lukkulaisvaara massif (Karelia) was used as a starting material, since the composition of the basic rocks of the Lukkulaisvaara massifs is similar in chemical composition to that of the Tiksheozersky massif (Table 14.2), and they were subjected much less to secondary changes than that of the Tiksheozersky massif. Small differences in composition are observed in the content of titanium, aluminum, sodium and potassium. The increased titanium content (relative to the gabbroids of the Lukkulaisvaara massif) is associated with the presence of ilmenite and titanomagnetite in the rocks of the Tiksheozersky massif, and sodium and potassium—with intensive postmagmatic changes observed in the Tiksheozersky massif. KCl and KF solutions with concentrations of 0.5, 1 M and 2 M, respectively, were used as fluids. The ratio "rock/fluid" was 10/1 by weight (Kovalskaya et al. 2018a).

Table 14.1 Mineral composition studied samples from Tiksheozero massif

Rock	Mineral composition	Alteration degree	Structure
Carbonatite	Cal + Bi + Pyr	Weak	Fine-grained
Carbonatite	Cal + Dol + Bi + Ab + Ilm	Weak	Fine-grained
Pyroxenite	Cal + Cpx 1 + CPx 2 + Bi + Am + Ilm + Ap + Ort + Sph	Strong	Medium-grained, traces of cumulative
Gabbro	Pl + Cpx + Amf1 + Amf2 + Ilm +Cal	Average	Medium-grained
Pyroxenite	Cpx 1 + Cpx 2 + Amf + Sph + Ilm + Pyr + Cal	Strong	Medium- and compact-grained (altered areas)
Gabbro	Amf1 + Amf2 + Cpx + Pl + Ilm + Sph +Pyr	Average	Medium- and compact-grained
Altered pyroxenite	Cpx + Amf + Bi +Cal + Ilm + Mt + сульфиды	Very strong	Compact-grained
Syenite	Ab + Ksp + Natr + Cal + Str (мало) + Bi + Mt + Mu	Very strong	Compact-grained
Pyroxenite	Bi + Cpx + Amf1 + Amf2 + Sod + Natr + Cal + Ilm + Mt + Sph	Very strong	Medium- and compact-grained
Ijolite	Sod + Amf + Cpx + Cal +Ap + Sph + Ilm	Average	Compact-grained
Carbonate-amphibolic rock	Cal + Amf1 + Amf2 + Bi + Sod + сульфиды	Not clear	Not defined
Pyroxenite	Cpx + Bi + Cal + Sod + Mt	Average	Medium-grained
Olivinite	Ol + Cpx + Serp + Chrt + Chlt	Serpentinization, average	Medium- and coarse-grained, cumulative

Note Mineral abbreviations: *Ab* albite; *Amf* amphibole, амфибол; *Bi* biotite; *Cal* calcite; *Can* cancrinite; *Chlt* chlorite; *Chrt* chromite; *Dol* dolomite; *Ilm* ilmenite; *Ksp* potassic feldspar; *Mt* magnetite; *Natr* natrolite; *Ort* ortite (allanite); *Cpx* clinopyroxene; *Pyr* pyrrhotite, *Sod* sodalite; *Str* strontianite; *Sph* sphene (titanite); *Zeol* zeolite. The numbers next to the mineral abbreviation indicate the generation of its formation

Table 14.2 The average
chemical composition (wt%)
of gabbroids from the
Lukkulaisvaara and
Tiksheozersky massifs

Oxides	Lukkulaisvaara	Tiksheozersky
SiO_2	49.27	47.13
TiO_2	1.03	3.16
Al_2O_3	13.43	11.46
Cr_2O_3	0.13	0.18
FeO*	14.94	13.98
MnO	0.14	0.67
MgO	5.21	5.98
CaO	6.24	7.02
Na_2O	4.35	6.16
K_2O	1.81	3.86
Total	99.56	99.60

Note FeO*—all iron measured in form FeO

The equipment All experiments were conducted in high gas pressure units with internal heating of UVGD-10000 design of IEM RAS. Temperature control accuracy was ±2 °C; pressure ±50 bar.

Methodology of experiments Source materials were loaded into Ø5 × 0.2 × 50 mm or Ø4 × 0.1 × 50 mm platinum ampoules, the necessary amount of fluid solution was added, weighed and welded. The equipped ampoules were loaded into the reactors of the plants, put into operation and maintained at the parameters of experiments for 10 days. First, the reaction mixture was heated to 1100 °C and set the pressure of 3 kbar, the ampoules were maintained at these parameters for 3 h, then there was isobaric cooling to 850 °C, and the subsequent holding at these parameters for 10 days. Quenching was done in isobaric conditions. After the experience, the ampoules were also weighed to ensure that they were not depressurized.

Methods of analysis To determine the compositions of experimental and natural samples by the method of electron-probe X-ray analysis, the scanning electron microscope Tescan Vega II XMU (Tescan, Czech Republic) was used, equipped with INCA Energy 450 X-ray microanalysis system with energy dispersive (INCA X-sight) and crystal diffraction (INCA Wave 700) X-ray spectrometers (Oxford Instruments, England) and INCA Energy + software platform. The analysis conditions when using only the energy dispersive spectrometer were as follows: accelerating the voltage of 20 kV, the current of absorbed electrons from 150 to 400 pico-amperes, the analysis time at the point of 100 s, beam size 157–180 nm. When using the crystal diffraction spectrometer together with the energy dispersive conditions of the analysis were different: accelerating the voltage of 20 kV, the current of absorbed electrons at Co 20 nA, the total time of analysis at 170 s. Accuracy of analyses (by main elements, using EDS) was 2–3 relative %.

14.3 Experimental Results

In the course of two series of experiments (with KF and KCl solutions with concentration of 0.5, 1, 2 M) the following results were obtained. Products of experiments represented a fine crystalline mass (Figs. 14.3 and 14.4) of greenish-grey color. Separate crystallites were clearly identified during microscopic observation.

Experiments with 0.5 M KCl solution The analysis of products of experiments with 0.5 M KCl fluid concentration did not reveal formation of alkaline amphiboles. In interstitium between newly formed clinopyroxenes there are small grains of potassium feldspar.

Experiments with 1 M KCl solution In experiments with such salt concentration in the fluid clinopyroxenes of diopside-hedenbergite series and amphibole with compositions corresponding to the richterite-cataphorite were obtained, analogues to observed in the Tiksheozersky massif's gabbroids (Tables 14.3 and 14.4, Figs. 14.3 and 14.4). The size of the individual grains reaches 100 μm. Potassium feldspar and titanomagnetite have been diagnosed in some experiments.

Experiments with 2 M KCl solution Among the products of this series of experiments there are alkaline amphibole, which differ greatly from each other in composition, and single grains of clinopyroxene; titanomagnetite was found as an accessory mineral.

Experiments with 0.5 M KF solution As in the experiment with 0.5 M KCl as a fluid, no alkaline amphibole was observed in this experiment. Clinopyroxene and

Fig. 14.3 Results of runs with 1 M KCl solution at a temperature of 850 °C and a pressure of 3 kbar

Fig. 14.4 Results of runs
with 1 M KF solution at a
temperature of 850 °C and a
pressure of 3 kbar

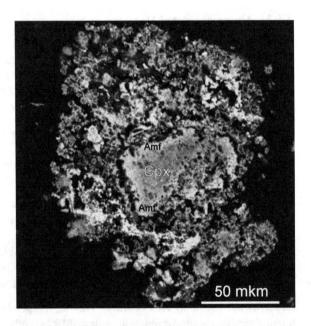

phlogopite individuals were diagnosed in the experimental products. The size of crystallites in experimental products does not exceed 50 μm.

Experiments with 1 M KF solution As well as in the experiments with 1 M concentration of KCl, clinopyroxenes of diopside-hedenbergite series (Fig. 14.5, Table 14.3) and alkaline amphiboles of riherite-cataphorites series were diagnosed in the products of these experiments (Fig. 14.6, Table 14.4). However, we recorded amphiboles of this composition in the ijolite-urtites of the Tiksheozersky massif. Possibly, the differences in the composition of amphiboles were influenced by the subsequent postmagmatic carbonatization of the massif.

Experiments with 2 M solution of KF In the products of these experiments, the formation of a significant amount of fluorite was observed, which is apparently associated with a high concentration of F^- ions in the fluid, as well as needles of fluoride-containing phlogopite.

14.4 Discussion of Results

The conducted series of experiments on modeling the process of amphibolization in the Tickshozersky massif gabbrides and their comparison with the results of the study of natural samples showed the following results:

1. The process took place at the high-temperature postmagmatic stage at a temperature of about 850 °C.

Table 14.3 Composition (wt%) and formula units of clinopyroxenes from Tiksheozero massif and synthesized clinopyroxenes

Oxide	Clinopyroxenes from natural samples of Tiksheozero massif										Synthesized clinopyroxenes		
	Olivinite	оливинит	Pyroxenite	Pyroxenite	Pyroxenite	Gabbro	Gabbro	Syenite	Syenite	Amphibole-cancrinite rock	1 M KF	1M KCl	2 M KCl
SiO_2	48.09	48.00	53.65	52.68	54.03	52.09	52.69	53.31	53.70	52.96	51.83	50.72	51.94
TiO_2	1.95	2.13	–	0.61	1.98	0.42	0.38	–	–	–	–	–	–
Al_2O_3	4.99	6.04	0.11	2.29	–	0.81	1.59	3.07	2.09	0.20	2.43	3.02	2.12
FeO	9.68	7.22	7.97	3.97	10.06	8.07	6.40	8.82	8.67	7.84	8.01	7.45	8.15
MnO	0.16	–	0.16	0.12	–	0.10	0.03	–	–	0.11	–	–	–
MgO	10.98	12.34	14.16	18.66	17.80	14.68	15.04	13.56	14.89	13.22	14.72	17.03	16.01
CaO	22.36	23.67	22.93	21.03	16.13	23.83	24.10	20.79	19.72	25.40	22.12	21.17	21.78
Na_2O	1.55	0.60	1.02	–	–	–	0.08	0.46	0.94	0.23	0.89	0.61	–
K_2O	0.26												
Total	100.00	100.00	100.00	99.94	100.00	100.00	100.02	100.00	100.00	99.99	100.00	100.00	100.00
Formula units (to 6 oxygens)													
Si	1.79	1.78	1.98	1.91	2.08	1.94	1.94	2.02	2.02	1.98	1.94	1.882	1.928
Ti	0.05	0.06	0.00	0.02	0.06	0.01	0.01	0.00	0.00	0.00	–	–	–
Al_{tot}	0.22	0.26	0.00	0.10	0.00	0.04	0.07	0.13	0.09	0.01	0.107	0.132	0.093
Fe_{tot}	0.30	0.22	0.25	0.12	0.31	0.25	0.20	0.27	0.26	0.24	0.251	0.231	0.253
Mn	0.00	0.00	0.01	0.00	0.00	0.00	0.00	0.00	0.00	0.00			
Mg	0.61	0.68	0.78	1.01	0.99	0.81	0.83	1.01	1.02	0.74	0.821	0.942	0.88
Ca	0.89	0.94	0.91	0.82	0.56	0.95	0.95	0.54	0.53	1.02	0.887	0.842	0.86
Na	0.11	0.04	0.07	0.00	0.00	0.00	0.01	0.03	0.07	0.02	0.064	0.044	
K	0.01	0.00	0.00	0.00	0.00	0.00	0.00	0.00	0.00	0.00	–		

Table 14.4 Composition (wt%) of natural amphiboles from Tiksheozero massif and synthesized amphiboles

Oxide	Natural samples						Synthesized samples (fluid)	
	Pyroxenite	Gabbro	Gabbro	Ijoliteurtite	Syenite	Carbonatite	Synthesized amphibole (1 M KF)	Synthesized amphibole (1 M KCl)
SiO$_2$	44.08	41.33	43.27	49.10	40.21	51.91	47.18	46.12
TiO$_2$	0.48	0.06	1.41	0.91	1.07	0.94	1.32	0.86
Al$_2$O$_3$	11.20	17.28	10.91	6.53	14.30	4.06	12.78	10.42
Cr$_2$O$_3$	–	–	–	–	–	–	–	–
FeO*	16.43	20.44	18.57	13.23	19.38	13.75	14.94	15.91
MnO	0.00	0.13	0.05	0.00	0.19	0.09	0.25	0.03
MgO	11.81	6.36	10.81	15.60	9.06	14.70	12.32	13.81
CaO	13.47	11.96	9.00	8.68	10.93	7.30	6.14	8.01
Na$_2$O	1.29	1.85	5.16	5.43	3.57	6.66	2.24	3.71
K$_2$O	1.24	0.58	0.81	0.52	1.30	0.59	1.83	1.13
Total	100.00	100.00	100.00	100.00	100.00	100.00	100.00	100.00

Note FeO*—all iron in microprobe analysis is calculated as FeO

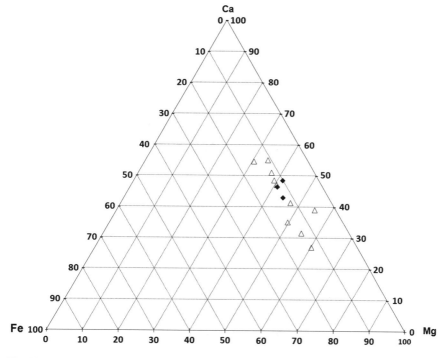

Fig. 14.5 The 'Aeg-Di-Hed' diagram for clinopyroxenes from various rocks of the Tiksheozero massif and experimental products: empty triangles—natural clinopyroxenes, filled rhombus—synthesized clinopyroxenes

2. The concentration of salt (KCl or KF) in the fluid at high-temperature post-magmatic changes of the Tickshozersky massif gabbroids fluctuated within 1 M. Amphiboles were not formed at lower concentrations, and fluorite and phlogopite were formed at higher concentrations.

3. One of the possible mechanisms for the amphibolization of gabbro in the Tick-shozersky massif was the separation of volatiles during the subsequent intro-duction of alkaline rocks, i.e., ijolite-urtites, which is indicated by the similarity of later-generation amphibole compositions in the gabbroids, ijolite-urtites and samples obtained in the course of behavioural experiments.

Thus, the data obtained by us characterize the temperature and fluid regime of formation of amphibole rims around clinopyroxenes in the gabbroids of the Tiksheoz-ersky massif, and also shows that the complex of differentiated rocks of the massif could be formed as a result of the complex evolution of heterogeneous fluid-magmatic system.

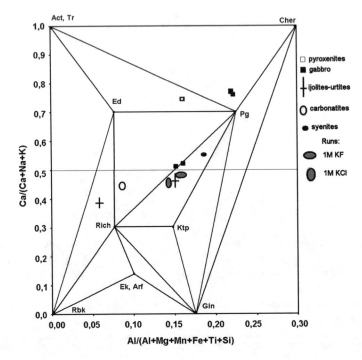

Fig. 14.6 Classification of amphiboles from the Tikshozero massif and synthetic amphiboles (Safonov et al. 2014)

Acknowledgements This research was supported from the project No AAAA-A18-118020590151-3 of the IEM RAS.

References

Khodorevskaya L, Varlamov DA (2018) High-temperature metasomatism of the layered mafic–ultramafic massif in Kiy island, Belomorian mobile belt. Geochem Int 56(6):535–553

Kholodilov NN, Karpatenkov VN (1988) Report on the results of prospecting for apatite and other minerals within the Tikshozero group of arrays of North Karelia for 1985–1988. In: Funds of Central Kola Geological Expedition (in Russian)

Kovalskaya TN, Varlamov DA, Shapovalov YuB, Kalinin GM, Kotelnikov AR (2018a) Experimental study of postmagmatic processes in Tiksheozerskiy massif. Exp Geosci 24(1):117–120

Kovalskaya TN, Varlamov DA, Shapovalov YuB, Kalinin GM, Kotelnikov AR (2018b) Experimental study of post-magmatic processes in the Tikshozero massif. In: Proceedings of the All-Russian annual seminar on experimental mineralogy, petrology and geochemistry, Moscow, GEOKHI, pp 207–210 (in Russian)

Metallogeny of igneous complexes of intraplate geodynamic environments (2001) Moscow, GEOS, p 640 (in Russian)

Perchuk LL (1970) Equilibrium of rock-forming minerals Moscow, Nauka, p 320 (in Russian)

Perchuk LL, Ryabchikov ID (1976) Phase matching in mineral systems. Moscow, Nedra, p 287 (in Russian)

Safonov OG, Kosova SA, Van Reenen DD (2014) Interaction of biotite–amphibole gneiss with H_2O–CO_2–(K, Na)Cl fluids at 550 MPa and 750 and 800 °C: experimental study and applications to dehydration and partial melting in the middle crust. J Petrol 55(12):2419–2456

Safronova GP (1990) Rock-forming carbonates and apatite of the Tiksheozersky massif. In: New in mineralogy of the Karelo-Kola region Petrozavodsk, pp 25–39 (in Russian)

Sharkov EV, Bogina MM, Chistyakov AV, Belyatsky BV, Antonov AV, Lepekhina EN, Shchiptsov VV (2015) Genesis and age of zircon from alkali and mafic rocks of the Elet'ozero complex, North Karelia. Petrology 23(3):259–280

Shchiptsov VV, Bubnova TP, Garanzha AV, Skamnitskaya LS, Shchiptsova NI (2007) Geological-technological and economic assessment of the resource potential of carbonatites of the Tikshozero massif (ultrabasic—alkaline rock formation and carbonatites). In: Proceedings of geology and ore deposit of Karelia, issue 10. Petrozavodsk: Karelian Scientific Center RAS, pp 159–170 (in Russian)

Simakin AG, Shaposhnikova OY (2017) Novel amphibole geobarometer for high-magnesium andesite and basalt magmas. Petrology 25(2):226–240

Suk NI, Kotel'nikov AR, Koval'skii AM (2007) Mineral thermometry and the composition of fluids of the sodalite syenites of the Lovozero alkaline massif. Petrology 15(5):441–458

Ustinov VI, Grinenko VA, Kotelnikov AR, Suk NI, Kovalskaya TN, Smirnova EP (2006) Thermometry of sodalite-containing associations of rocks of the Lovozersky and Tikshozero alkaline massifs. In: Materials of the All-Russian meeting "Geochemistry, Petrology, Mineralogy and Genesis of Alkaline Rocks" Miass, pp 267–272 (in Russian)

Printed in the United States
by Baker & Taylor Publisher Services